Grundlehren der mathematischen Wissenschaften 191

A Series of Comprehensive Studies in Mathematics

Carl Faith

Algebra II
Ring Theory

Springer-Verlag
Berlin Heidelberg New York 1976

Carl Faith
Rutgers, The State University, New Brunswick, N.J. 08903 and
The Institute for Advanced Study, Princeton, N.J. 08540, USA

AMS Subject Classifications (1970): 12-01, 13-01, 15-01, 16-01, 18-01, 20-01

ISBN 3-540-05705-6 Springer-Verlag Berlin Heidelberg New York
ISBN 0-387-05705-6 Springer-Verlag New York Heidelberg Berlin

Library of Congress Cataloging in Publication Data. Faith, Carl Clifton, 1927. –
Algebra. Bibliography: v. 1, p. Contents: 1. Rings, modules and categories. – 2. Ring
theory. 1. Rings (Algebra) 2. Modules (Algebra) 3. Categories (Mathematics). I. Title.
II. Series: Die Grundlehren der mathematischen Wissenschaften in Einzeldarstellungen,
Bd. 191. QA 247.F 34. 512′.4. 72-96724. ISBN 0-387-05551-7 (New York) (v. 1).

Typesetting, printing, and binding: Universitätsdruckerei H. Stürtz AG, Würzburg

This volume is for my family

Mickey, Heidi, and Cindy
Eldridge, Louise, and Frederick
Virginia Nell Caudill Compton
Harold Compton   1895–1964

and the memory of my parents

Herbert Spencer Faith 1895–1952
Vila Belle Foster Faith 1897–1965

Preface to Volume II

I. Ring Theory

The term *The Theory of Rings* seems first used as title of a book by Jacobson [43], and in his preface Jacobson asserts that the theory that forms the subject of the book had its beginning with Artin's extension in 1927 of Wedderburn's structure theory of algebras to rings satisfying the chain conditions.[1]

As the predecessor to his book, Jacobson cites Deuring's *Algebren* (Deuring [35, 68]), and Deuring cites Dickson's *Algebren und ihre Zahlentheorie* (Zürich, 1927). As in his earlier book, *Algebra and its Arithmetic*, Dickson ([23]) extends arithmetic in algebraic number field, that is, arithmetic of the ring of integers in a finite field extension k of the field $\mathbb{Q}$ of rational numbers, to orders in rational algebras, that is, to orders in an algebra over k.

Jacobson also cites nine papers of his teacher, J. H. M. Wedderburn, in the bibliography; the term "algebra" appears in six titles, and "hypercomplex numbers" in another. (Another influential book, appearing shortly before Jacobson's, by A. A. Albert in 1939, was also on the subject of "algebras", that is, algebras with a finite basis over a field.) The study of these so called "hypercomplex number" systems was motivated originally by the desire to discover and classify algebras over the field of real or complex numbers (thus the terminology: hypercomplex).

Hamilton's discovery of the algebra $\mathbb{H}$ of quaternions, the first noncommutative field, motivated by a problem in physics, took him 15 years to find. (There is a legend that he carved the result on a nearby bridge the moment of discovery—see E. T. Bell [37, p. 360].) Another discovery of importance to physics, the Cayley numbers were a nonassociative field (containing $\mathbb{H}$) of dimension 8 over $\mathbb{R}$. (Unlike Hamilton with his Quaternions, Cayley did not write a treatise on the subject purporting to explain the physical universe. Cf. Dyson [72, p. 301, Note 5].)

Wedderburn's theorems apply to the structure of any algebra A of finite dimension over any field k: if $W(A)$ is the maximal nilpotent ideal, then $A/W(A)$ is a finite product of algebras each isomorphic to a total matrix algebra over a (noncommutative or commutative) field. If A is a separable algebra (the case when the center of $A/W(A)$, a finite product of fields, has the property that each of the fields is a separable extension of k), then there is a subalgebra $S \approx A/W(A)$

---

[1] Our convention was stated in Volume I: Jacobson [43], for example, denotes a work by Jacobson published in 1943. When more than one work appears, small letters are used as in Jacobson [45a], [45b], or [45c].

such that

$$A = S \oplus W(A)$$

as vector spaces over k (Wedderburn factor theorem (I,13.18, p. 471)).[2] This reduces the structure theory of A to that of S and $W(A)$ and the effect of the multiplication of $W(A)$ by the elements of S. (The latter effect is mostly conclusive only in determining the structure of A only in low dimensions, however.)

In 1929 R. Brauer showed that the "classes" of simple central algebras over k formed a group $Br(k)$. For each such algebra A, there is a class $[A]$ consisting of all algebras B for which there exist integers m and n such that A_n and B_m are isomorphic total matrix algebras of degrees n and m over A and B respectively. (By Wedderburn's theorem, $[A]$ contains a (not necessarily commutative) field D over k, that is $[A] = [D]$.) In $Br(k)$ we have $[A]^{-1} = [A^{op}]$, where A^{op} is the algebra opposite to A, and, moreover, $Br(k)$ is a torsion abelian group. (In Chapter 13 Exercises, the Brauer group $Br(k)$ of any commutative ring k is discussed.)

In 1921, E. Noether carried over Dedekind's ideal theory (and representation theory) for integral domains (and rings of algebraic integers) to general commutative rings satisfying the ascending chain condition for ideals. Such rings are now called Noetherian rings. These rings are characterized by the condition that every ideal is finitely generated, and they include polynomial rings in any finite number of variables over any field (Hilbert's basis theorem (I, 7.13, p. 341)) in addition to Dedekind rings, and other rings arising in classical mathematics. Moreover, the study of modules over these rings formed an important part of the arithmetic ideal theory (see Noether [21, p. 55 ff]).

In 1927, Noether proved the "Noether homomorphism theorems" for groups with operators, generalizing to modules many of the group theoretical theorems of W. Krull and O. Schmidt (see Noether [27, p. 643 and 645]).

In 1927, Artin generalized some of the Wedderburn theorems for algebras to noncommutative (= not necessarily commutative) rings satisfying the descending chain for right ideals.[3] These have been called right Artinian rings in his honor. This freed the subject from its earlier dependence on an underlying field k of scalars and finite free basis over k. The ascending chain condition for right ideals was required in Artin's rings a restriction which was shown to be superfluous[4] only much later, independently by C. Hopkins [39] and J. Levitzki [39]. (An account of this appears in Chapter 18.)

From all indications, the writing of "Theory of Rings" stimulated Jacobson researches in ring theory to fruition in a series of historically most important papers (Jacobson [45a, b, c]). In essence, for the first time, the full category RINGS was studied in order to more fully understand fundamental concepts and not just because important subcategories arise in classical mathematics: in particular, no chain conditions, ascending or otherwise, were assumed (although

---

[2] For reference to Volume I: (I, 13.18, p. 471) denotes an item (Theorem, Proposition, Exercise, or Corollary) on p. 471. Also: 13.18 (I, p. 471) is a variant reference for this.

[3] Bourbaki [58, Chapter 8, p. 174] gives the interesting history of the notions of minimal ideal of an algebra (H. Poincarë, 1903), one-sided ideals (Noether and Schmeidler, 1920), the maximal or ascending chain condition (Dedekind, 1894), and the descending chain condition (Wedderburn, 1907).

[4] Bourbaki (l.c., p. 175) remarks that Noether [29] dispensed with the ascending chain condition assuming no nilpotent ideals.

applications were given) and, not even an identity element was required. (And, of course, commutativity was not assumed.)

No doubt the single most important of these ideas is that of the Jacobson radical rad R of a ring R, defined as the intersection of all kernels of all (non-trivial) irreducible right representations of R, and characterized by Jacobson as the unique maximal ideal J with the property that

(q.r.) $\forall_{x \in J} \exists_{x' \in R} \, x + x' + x x' = x + x' + x' x = 0$

and containing any one-sided ideal for which the same q.r. condition holds. In characterizing the radical this way, Jacobson seized on the characterization of Perlis [42] of the radical of a finite dimensional algebra over a field, and extended it to an arbitrary ring. (The term quasi-regular element x ($=$ one satisfying the q.r. condition just defined) was coined by Perlis, and a quasi-regular ideal is one in which every element is quasi-regular.)

In other words, Jacobson defined a functor

$$\text{rad}: \text{RINGS} \rightsquigarrow \text{RINGS}$$

and we now list some of its properties.

1. For a ring R, rad R has been defined via irreducible right modules (or representations) of R, and thus should be called the "right" radical of R, but Jacobson proved that rad R coincides with its left-right symmetry, that is, rad R is also the left radical of R. Moreover:

$$\text{rad} \,(R/\text{rad } R) = 0.$$

2. The characterization of rad R as the intersection of all right ideals I such that the right module $V_I = R/I$ is irreducible ($=$ simple) and nontrivial ($V_I R \neq 0$). If R has an identity element, then any such right ideal is a maximal right ideal and conversely. (Then rad R is the intersection of the maximal left ideals.) The existence of these in any ring R with identity element 1 follows by an application of Zorn's lemma; of course, a nilpotent ring does not possess nontrivial simple modules. (Incidentally, in case $1 \in R$, then $x \in R$ is quasi-regular iff $1 + x$ is a unit.)

Before proceeding, we need a definition. An ideal I is right primitive if R/I has a faithful irreducible right module. A ring R is primitive if 0 is a right primitive ideal.

By definition, rad R is the intersection of the right primitive ideals, and by the result in 1., rad R is the intersection of the left primitive ideals. Thus, a simple ring is both right and left primitive.

3. A ring R is right primitive iff R is isomorphic to a dense ring of linear transformations on left vector space V over a field D (Chevalley-Jacobson Density Theorem, proved in Jacobson [45]).

If R is right primitive, with faithful irreducible right module V, then $D = \text{End } V_R$ is a field, and (by writing endomorphisms on the left) V becomes a left vector space over D. Moreover, R imbeds in $L = \text{End}_D V$ canonically, and is dense in L in the finite topology.

Any ring with 1 has right primitive ideals. Any maximal ideal is primitive, since any simple ring (with 1) is right and left primitive (Jacobson [45 b]). Thus, any ring $R \neq \text{rad } R$ is "interesting" in that it has a "good" ring, namely a primitive ring, as an epic image.[5] (Doubtlessly, it is this theorem which establishes the importance of RINGS as a category.)

4. The factor ring $R/\mathrm{rad}\,R$ is either $=0$, or isomorphic to a subdirect product of right primitive rings. (Also the left-right symmetry holds even though not every right primitive ring is left primitive as G. Bergman [64] showed.) Conversely, any subdirect product A of right primitive rings satisfies $\mathrm{rad}\,A=0$.

5. $\mathrm{rad}\,R$ is "functorial" in the sense that any category equivalence

$$T: \mathrm{mod}\text{-}R \rightsquigarrow \mathrm{mod}\text{-}S$$

for rings R and S induces a category equivalence

$$\mathrm{mod}\text{-}(R/\mathrm{rad}\,R) \rightsquigarrow \mathrm{mod}\text{-}(S/\mathrm{rad}\,S)$$

(Jacobson proved this in different language).

6. $\mathrm{rad}\,R$ contains every nil one-sided ideal ($=$ one in which every element x is nilpotent in the sense that $x^n=0$ for an integer n depending on x), and hence $\mathrm{rad}\,R$ contains every nilpotent one-sided ideal ($=$ one in which there is a fixed integer N such that all products $x_1 \cdot x_2 \cdots x_N$ of N elements $x_1,\ldots,x_N$ vanish: $x_1 x_2 \cdots x_N = 0$).

The Wedderburn radical $W(R)$ of a ring R is the maximal nilpotent ideal (if it exists). In a commutative ring R, the "radical" of an ideal I is the ideal

$$\sqrt{I} = \{x \mid \exists_{n=n(x)} x^n \in I\}.$$

Then $\sqrt{0}$ is called the nil radical of R. (If R is Noetherian, then $\sqrt{I}$ is nilpotent modulo I, and hence in this case $W(R/I)=\sqrt{I}/I$. Thus, $W(R)$ is then the nil radical of R, whence the origin of the generic term radical.) In a noncommutative ring, $W(R)$ exists if R is either right or left Noetherian, and, if R is right Artinian $\mathrm{rad}\,R$ coincides with $W(R)$.

Krull [50] pointed out the relationship between Hilbert's Nullstellensatz and the Jacobson radical. It hinges on the question: when is the Jacobson radical of a finitely generated algebra over a field a nil ideal? This holds true for commutative algebras (Krull [51], Goldman [51]), algebras over nondenumerable fields (Amitsur [56]), and algebras satisfying a polynomial identity (Amitsur [57]). The latter theorem is related to a noncommutative Hilbert Nullstellensatz, and many of the foregoing results on Jacobson and Hilbert rings are generalized by Amitsur and Procesi [66] and Procesi [67].

It is tempting to extend this list of contributions of Jacobson, but, of course, this has been done by him much better in his Colloquium volume (and elsewhere) and much of Volume II involves Jacobson's ring-theoretical ideas in an essential way. In addition, two others who exploited and advanced Jacobson's ringtheoretical techniques, notably in work on Kurosch's problem, rings with polynomial identities, topological rings, Lie and Jordan simplicity of simple associative rings, and the so-called "commutativity theorems" (modelled after the famous Wedderburn Theorem on the commutativity of finite fields), among many, many others,

---

[5] "Bad" rings are also interesting from different points of view: do there exist simple rings equal to their radicals? (Yes. Cohn and Saciada [67]). Is the radical of a finitely generated ring nil? (See 6.) When does nil $\Rightarrow$ nilpotence? (See Chapter 17.)

have written wisely and well on the subject—I am speaking of I. Kaplansky and I. N. Herstein from whose books I have reaped so much pleasure and knowledge.[6]

II. Module Theory

One might define module theory to be the structure theory of modules satisfying specific conditions, for example, Noetherian, or Artinian, or semiperfect, or indecomposable, without making stringent requirements on the ring. As an example, any Noetherian or Artinian module may be decomposed over *any* ring into a direct sum of indecomposable modules. It would include, of course, the structure of modules over specified rings, for example, Noetherian, or Artinian, or semiperfect, or indecomposable. Another aspect of module theory is the relation between a module and certain canonical modules, such as the right ideals $\{I_a\}_{a \in A}$ of a ring R, and the cyclic modules $\{R/I_a\}_{a \in A}$. These modules we might say are "at hand", and they form a set as opposed to being a class. (Curiously, "right ideals" and "cyclic right modules" are dual in the sense that one is the class of subobjects, and the other the class of quotient objects of R.) Classically, it has been possible to relate much of the class mod-R to this set. (Heuristically, one might compare the problem of doing this with the problem of knowing the universe with only a telescope at hand.)

Nevertheless, even for $R = \mathbb{Z}$, the ring of rational integers, these modules describe all finitely generated abelian groups. A similar theorem holds for finitely generated modules over a hereditary Noetherian prime (HNP) ring R: every finitely generated module M is a direct sum of uniserial ($=$ has a unique decomposition series) modules and right ideals. This follows since Kaplansky's theorem (I, p. 387) for modules over hereditary rings implies that the torsion submodule $t(M)$ splits off, and that $M/t(M)$ is isomorphic to a direct sum of right ideals; then the theorem of Eisenbud, Griffith and Robson (25.5.1) applies: modulo any nonzero ideal R is a (generalized uni) serial ring, so by Nakayama's theorem (25.4), $M \approx t(M) \oplus M/t(M)$ is isomorphic to a direct sum of uniserial modules and right ideals.

A ring need not be right Noetherian in order that the finitely generated modules have such a decomposition. Indeed, a theorem of Kaplansky [52] states that over any almost maximal valuation ring, any finitely generated module is a direct sum of cyclic modules. In fact, this property characterizes almost maximal valuation rings among commutative local rings (20.49).

However, a stronger requirement does imply right Noetherian. Assume that there is a set S of right R-modules such that every right R-module can be embedded in a direct sum of modules in the set. Then, R is right Noetherian (Faith-Walker (20.7)). The converse also holds.

Moreover, if every right R-module is isomorphic to a direct sum of modules in the set S, then R must be right Artinian (20.23). Warfield [72a] showed that a commutative ring can have this property iff it is a principal ideal ring (PIR).

---

[6] Kaplansky's *Problems in the Theory of Rings*, is indicative of Kaplansky's felicitous influence on these questions (Kaplansky [70]); and, moreover, the second edition of his *Infinite Abelian Groups* contains a prodigious and broad literature commentary on the literature relating directly and indirectly on the first edition (Kaplansky [69]). In a similar vein are Herstein's *Notes from a Ring Theory Conference* (Herstein [71]).

Similarly, Warfield shows in the same paper that every module is a direct sum of indecomposable modules iff R is a PIR.

The theorem of Matlis-Papp (20.5) states that one may decompose injective right modules over R into a direct sum of indecomposable modules iff R is right Noetherian. (What happens if every module is a direct sum of indecomposable modules will be discussed presently.)

Another theorem illustrating the principle that nice properties for the module structure reflect (and are reflected by) nice properties in the ring is a theorem (24.20) which states that every injective right module is projective iff R is quasi-Frobenius (QF) iff every projective right module is injective. For example, every right module can be isomorphic to a direct sum of right ideals only if R is QF, since the condition implies that every injective module is projective. The QF rings are the Artinian rings with a duality between finitely generated right and left modules induced by $\mathrm{Hom}_A(\ , A)$ for the ring A, and can be characterized as right selfinjective rings with the a.c.c. on left (resp. right) annulets. (See Chapter 24.)

The characterization of when does the category mod-R of *right* R-modules satisfying the property that every module has a projective cover [the dual of the property: every module has an injective hull (Bass [60]; cf. Chapter 22)] is interesting because the characterizing property is the d.c.c. on the principal *left* ideals. This class of rings properly contains the class of left Artinian rings.

Next to the basis theorem for abelian groups, the best known example of the kind of theorems we have been examining is the Wedderburn-Artin theorem (I, 8.9 p. 369) which determines the multiplicative structure of a ring for which every right module is semisimple (=a direct sum of simple right modules): the ring must be similar to (that is, Morita equivalent to) a finite product of fields. (This still holds when semisimple in the statement is replaced by injective (projective).)

Nakayama [39, 40, 41] similarly characterized Artinian rings over which every module is a direct sum of uniserial modules: such a ring is a serial ring in the sense that every principal indecomposable (= prindec) right or left ideal is a uniserial module. Nakayama characterized Artinian serial rings as Artinian rings over which every finitely generated indecomposable module is an epic image of a prindec. (These rings, and also more generally, rings over which every finitely generated right module is a direct sum of cyclic modules (= right σ-cyclic rings), are taken up in Chapter 25.) A serial QF ring has the property that every right module is a direct sum of cyclic right ideals (25.4.17).

A good deal of module theory is aimed at the description of the indecomposable finitely generated modules (at least over right Noetherian rings when every finitely generated module decomposes into a direct sum of indecomposable modules!) Let M be an indecomposable module over a right Noetherian ring R, assume that M is finitely generated, and let $g(M)$ be the least cardinal of any set of generators of M. In general, there exist indecomposable modules M with ever larger $g(M)$. Indeed, by Higman's theorem [54], this happens whenever R is the group algebra in characteristic p with noncyclic p-Sylow subgroup G of finite order n; in particular, finite rings can have this property! (However, in the case of cyclic p-Sylow subgroup, n is a bound on the "number" of indecomposable modules (Kasch-Kneser-Kuppisch [57]).)

Next assume a bound on the $\{g(M)\}$. This is a reasonable finiteness condition which one frequently encounters in classical algebra, for example, as we have seen, it holds over serial rings. Such a ring is said to be right FBG, or bounded module type. A commutative local FBG ring R has linearly ordered ideals (Warfield [70]), illustrating the strength of FBG.

Another kind of finiteness condition that frequently occurs in the theory of finite dimensional algebras and Artinian rings: does right FBG imply finiteness of the isomorphism classes of indecomposable finitely generated right modules? A ring with the latter property is said to be right FFM, or finite module type. (Serial rings are right and left FFM rings.) In this notation the question just stated can be stated as the validity of the implication FBG $\Rightarrow$ FFM. For algebras of finite dimension over a field this was called the Brauer-Thrall conjecture, and was proved by Roiter [68]. For Artinian rings, Auslander [74] proved the conjecture utilizing notably different methods.

Although we have not included these theorems in the text, they are typical of many theorems in the text, in fact, extensions of them, and because of their importance we take this opportunity to acquaint the reader with these results.

Auslander [74, Cor. 4.8] and Ringel and Tachikawa in Tachikawa [73, p. 129, Cor. 9.5] prove: Let R be a right Artinian right FFM ring. Then, every indecomposable right R-module is finitely generated, and every right module is a direct sum of indecomposable modules.

Moreover, Tachikawa [73] also shows that all modules have decompositions which complement direct summands (cds) in the sense if $M = \bigoplus_{i \in I} M_i$ is such a decomposition, then for any direct summand P, there is a subset J of I such that [7] $M = (\bigoplus_{j \in J} M_j) \oplus P$. Fuller-Reiten prove a converse for rings over which right and left modules have decompositions which cds. Auslander [74] showed that Artin algebras are FFM provided only that every indecomposable left module is finitely generated. Moreover, a theorem of Faith-Walker [67] puts on the finishing touch: if every injective left module is a direct sum of finitely generated modules, then R is left Artinian 20.17. (This property characterizes commutative Artinian rings 20.18: as stated earlier, if every left module decomposes into a direct sum of modules of bounded cardinality, then R is left Artinian 20.23.)

To return to cyclic modules: why are they so important to many structure theories? A possible answer: every right FBG ring R is similar to a ring A over which every generated module is a direct sum of cyclic modules. (This is trivial to prove: see 20.39.) Moreover, in this case, R is right FFM iff A has at most finitely many nonisomorphic indecomposable cyclic modules. Since isomorphic modules have the same annihilating ideal, in some cases, for example, when R is right Artinian, then right FFM implies that the lattice of ideals is finite 20.4.4. This notwithstanding, the right ideal structure of a right FFM ring has yet to be determined that would make the theory comparable to that for serial rings, and appears to be a problem to which a solution will have a reasonable expectation of clearing out a jungle of present-day special cases.

---

[7] This concept of Anderson-Fuller [72], and its relationship to ideas of Crawley-Jonsson [64] and Warfield [72b], is discussed in Notes for Chapter 21.

III. Algebra

I will close the Preface with a few generalities and some specifics. Algebra, like other branches of mathematics, systematically exploits quite general geometric properties—I am thinking of simple things like up-down, left-right, co-ordinations, sequences, symmetries-asymmetries, subdivisions, partitions, the "pigeon-hole" principle, equivalence, dualities, and the like. (To continue the list would make omissions appear more ominous than I intend!)

Because of the generality which mathematical statements are capable of, the term "abstract" is often applied to what in reality is quite specific. For example, theorems are published, but "theories" rarely, if at all. (The German use of the word *Satz*, or sentence, for theorem illustrates this point nicely, I think.)

The confusion between what is abstract and what is concrete arises, I believe, from the mathematician's passion for making the concrete as general as possible, by eliminating unnecessary, that is, unused, hypotheses from the statements. But it is, first of all, and above all, the concrete, the real, and indeed the useful, that involves the mathematician. (I do not mean to exclude beauty—the beauty of mathematical statements is a useful organizing principle for the sensitive mind.) The ethic is to eliminate waste, or the wasted, to determine the real, by making the vague, or imprecise, meaningful (if possible!).

Let me illustrate this with an example: G. Köthe proved that an Artinian commutative ring R with the property

(right $\sum$-cyclic) every right module is a direct sum of cyclics

is a uniserial (einreihig) ring. Cohen and Kaplansky [51] countered with the observation that it was redundant to assume that R is Artinian. S. U. Chase [60], then a student of Kaplansky, proved commutativity is not necessary to assert the ring is right Artinian, and moreover, that finitely generated modules can replace the cyclics in the statement. (But, then, the ring is no longer necessarily serial, of course.) Finally, it was noticed that finite cardinality of the modules in the direct summands played no role; if there exists a *set* of modules such that every right module decomposes into a direct sum of modules isomorphic to modules in that set, then the ring is right Artinian. The proof of this, given in Chapter 20, makes heavy use of another theorem of Chase [60] on direct sum decompositions of modules: If there is a cardinal number c not less than the cardinal of R such that the product R^c is a pure submodule (for example, a direct summand) of a direct sum of right R-modules having cardinal not exceeding c, then R satisfies the d.c.c. on principal left ideals. These latter rings are in fact rings which Bass [60] (then another Kaplansky student!) studied in the connection with the requirement that all right modules have projective covers. Bass called these rings right perfect rings, and much of the structure theory of Artinian nonsemisimple rings was extended by Bass to perfect rings. (An account of this is given in Chapter 22.)

The complete structure of $\sum$-cyclic rings, a problem posed by Köthe [35] is still unknown. Nakayama's papers on (generalized uni) serial rings (Nakayama [39, 40, 41]) showed the rings to be more general than serial rings. Kawada gave an exhaustive study, and complete solution in a very special case, but even so there were 19 (or so) formidable conditions deemed necessary and sufficient.

In turn, Kaplansky [69] asked for the structure of right σ-cyclic rings, or those over which every finitely generated module is a direct sum of cyclic modules.

As noted, Köthe's theorem solved the problem for commutative $\sum$-cyclic rings, and while Kaplansky [49, 52] *et al.* solve the problem for commutative local rings, it still remains open for arbitrary commutative rings. (Some of these matters are taken up in Chapters 20 and 25.)

In the meantime, research continues on the problems treated here, and the related problems on the right ideal structure of rings of finite module type (FFM rings) discussed in the Introduction to Volume II.

Having succeeded, at the very least, in connecting the first two parts of the preface, and having already described a number of ideas from Nakayama, let us remember his closing remark to the International Congress of Mathematicians, Amsterdam, 1950 (reprinted in Nakayama [50b]) on a closely related subject:

It seems to the writer that our topics possess a somewhat deeper connection with each other than was said in the beginning.

IV. Principal Contributors

The principal contributors to the contents are: S.A. Amitsur, E. Artin, K. Asano, M. Auslander, G. Azumaya, R. Baer, H. Bass, J.A. Beachy, J.E. Björk, R. Brauer, G. Burnside, S.U. Chase, A.W. Chatters, C. Chevalley, I.S. Cohen, P.M. Cohn, I. Connell, R. Croisot, J.H. Cozzens, R. Dedekind, J.A. Dieudonné, B. Eckmann, S. Eilenberg, D. Eisenbud, G.D. Findlay, H. Fitting, G. Frobenius, E. Feller, K.R. Fuller, C.F. Gauss, D.T. Gill, A.W. Goldie, K.R. Goodearl, P. Griffith, M. Harada, I.N. Herstein, O. Hölder, D. Hilbert, C. Hopkins, M. Ikeda, N. Jacobson, R.E. Johnson, I. Kaplansky, E. Kolchin, G. Köthe, L. Kronecker, W. Krull, J.P. Lafon, C. Lanski, J. Lambek, L. Lesieur, J. Levitzki, L.S. Levy, E. Maschke, E. Matlis, N.H. McCoy, Y. Miyashita, K. Morita, T. Nakayama, E. Noether, O. Ore, B.L. Osofsky, Z. Papp, S. Perlis, R. Remak, J.C. Robson, F.L. Sandomierski, O. Schmidt, A. Schopf, O. Schreier, I. Schur, R. Shock, L. Small, E. Steinitz, R. Swan, E. Swokowski, H. Tachikawa, Y. Utumi, P. Vámos, J. von Neumann, E.A. Walker, R.B. Warfield, Jr., D.B. Webber, E.T. Wong, and J.H.M. Wedderburn.

Readers familiar with my research interests will not be surprised to see to what extent the text is a delineation of the dominant roles that injective and projective modules have played in the simplification, clarification, extension, and deepening of much of classical algebra. It would be pointless to adduce specific examples here, since so much of the text is devoted to such examples, but even the Chevalley-Jacobson density theorem fits into this framework!

It goes without saying that another author would have made other choices for inclusion in the text; but some papers, especially "break-throughs" like Roiter [68], touch on a number of important theorems of classical mathematics, and therefore invite a *rewriting* of mathematics. To relegate papers of such power to the status of an "inclusion" would be a mutilation not only of the potential of such a paper to *revise* mathematics, but also of what I had planned to do. *Of course*, mathematics has not stood still!

V. Acknowledgements

Albert Einstein has been quoted as saying that teachers should set an example for their students—of what to avoid if they cannot be the other kind. In this context

I must confess: I began this book in Summer 1965 at the Institute for Advanced Study, continued it at Berkeley in 1965–1966 (where I finished the prototype of Volume II in Summer 1966), made revisions and additions, notably category theory (described in the Introduction to Volume I), in Princeton in 1969 (both years provided for by Faculty Fellowships awarded by the Rutgers Research Council), and I am writing this in my sabbatical year at the Institute for Advanced Study, where I began. *Sic semper scriptor!*

Mere mention of the splendid faculty and facilities of the Institute for Advanced Study would not sufficiently convey their importance to my work. The dedication to mathematics and rational thought of those who study there provided me with an unending source of inspiration, and an inexpressible joy of being, which sustains me in all my work.

More than that, there is a freedom of inquiry and thought that is powerfully unique even among other truly fine institutions. May the present faculty accept this small tribute as a sign of may admiration for, and involvement with, them in their constant struggle to preserve this fierce intellectual freedom.

Without the understanding interest of the editors of the Springer-Verlag, this book might never had been printed. There are enormous expenses involved in the typesetting required for the fine Springer books, and a comittment to publish is not a lighthearted one to make. Most of all I am indebted to one of the two Chief Editors of the Grundlehren volumes, Professor Beno Eckmann, for making this book possible. I also offer my grateful thanks to Professor Albrecht Dold for his part in making this decision.

I am grateful to the staff of the Institute for Advanced Study for much help in assembling the manuscript (in countless editions), and without the unstinting assistance of Ms. Caroline D. Underwood, the School Secretary, and Ms. Evelyn Laurent at the Electronic Computer Project (E.C.P.), I would have despaired of completing it. I cannot thank them enough.

Ms. Judith Friday Lige has smoothed over many technical problems for me with her friendly advice and executive clout in the Mathematics Department of Rutgers University, and Ms. Ann-Marie McGarry translated as much of what I wrote into plain English as I would permit. They have my great appreciation.

I have the added pleasure of thanking Mss. Mary Anne Jablonski, Annette Roselli, Alice Weiss (of Rutgers), and Mss. Irene Abagnale and Johanna Rodkin (of the Institute), all of whom contributed much time and effort in helping me.

For answering specific queries on topics contained in the text, I am indebted to R. Baer, H. Bass, J. A. Beachy, A. K. Boyle, R. Bauer, J. H. Cozzens, E. Formanek, K. R. Fuller, L. Fuchs, K. R. Goodearl, G. Ivanov, A. V. Jategaonkar, I, Kaplansky, E. Kolchin, T. S. Shores, R. B. Warfield, Jr., R. Wiegand, and W. Vasconcelos. I also owe many favors to D. Gorenstein and C. Neider.

I doubt that anyone has ever found a way to equitably thank everyone who has helped him or her by their encouragement, through friendship or personal example. So let me simply say *thank you* to those whose names ought to be here— names of many who have helped me immeasurably in those ways.

September 1975   Carl Faith
                                                     Institute for Advanced Study
                                                     Princeton, NJ

Contents

Contents of Volume I

---

* Because of the increase in size (in revision) of Part V, Part VI (Commutative Rings, Hereditary Rings, Separable Algebras and the Brauer Group), comprising of Chapters 27–32, will not appear as part of Volume II as announced in Volume I. Thus, Chapters 17–26 comprise all of Volume II.

Special Symbols and Terms

These are listed in Volume I, p. XXIII, but the list does not contain the conventions: mod-R denotes the category or class of all right R-modules, for a ring R, and R-mod is the left-right symmetry. Unless specified otherwise, a ring R will have an identity element 1, and mod-R denotes the category of unital modules in the sense that for every M of mod-R, $x1 = x \: \forall \: x \in M$. Usually, homomorphisms are written on the side opposite scalars, as discussed in Volume I, pp. 119–120.

A non-standard term used throughout is the word similar, applied to two rings A and B, to denote an equivalence mod-$A \approx$ mod-B of categories. (In the literature, the expression A is Morita equivalent to B is used.) Similarity is taken up in Volume I on p. 217, The Morita Theorem, 4.29.[1] The notation $A \sim B$ denotes the similarity relation, and it is reflexive, symmetric, and transitive.

We employ a now standard symbol: $A \hookrightarrow B$ indicates an embedding of a group or module A in B. However, in some places the printers have substituted a symbol $C \rightsquigarrow D$ to indicate a functor, replacing the symbol used in Volume I!

The symbol ring-1 indicates a ring in which an identity element is not assumed. (Thus, any proper ideal is a ring-1.)

As in Volume I, I have found it convenient to quote relevant literature in the form of "exercises", and I wish to emphasize that this is indeed a convenience, as well as a stimulation to the imagination of the intending reader, and in no way is to be interpreted as a relegation to exercises some very important theorems. (Many of results of mine and coauthors are found thus!) It is unlikely that many of the proofs of these will be discovered by the neophyte, yet I do believe this is the way mathematics ought to be learned: *do or die!* [As compensation many papers are published!!]

---

[1] Ordinarily in the text I will abbreviate such a reference by (I, 4.29, p. 217).

Introduction to Volume II

This book is primarily a survey of aspects of ring theory since the publication of Jacobson's Colloquium publication [55, 64] and all of Volume II is devoted to ring theory. A few brief indications of the overlap with Jacobson might be helpful. The revised edition [64] of Jacobson [55] contained three appendices, which overlaps with us in the main Goldie-Lesieur-Croisot theorem (Chapter 9), the Faith-Utumi theorem (Chapter 10), the Wedderburn Factor Theorem and the Amitsur theorems on the Jacobson radical of polynomial and group rings (Chapter 26). However, some extensions and applications of the Goldie theorems and the Faith-Utumi theorem given in Chapters 9, 10, 18 and 19 are new with us.

In the other direction we have included numerous theorems not in Jacobson's revision. We take the opportunity to introduce Part V briefly by listing a few of these:

1. Theorems of Morita [58] on category equivalences mod-$A \approx$ mod-B (Chapters 4 and 12, and applied throughout Volume II).

2. (Chapter 18) Krull-Schmidt theorems for "idemsplit" additive categories, the structure, and characterization of semilocal lift/rad rings (as endomorphism rings of modules having finite Azumaya, or Krull-Schmidt, "diagrams"), the Hopkins-Levitzki-Köthe-Eilenberg-Nagao-Nakayama-Asano-Chase structure theory of semiprimary rings, and rings with semilocal right quotient rings.

3. (Chapter 19) Quasinjective modules and selfinjective rings, the Lambek-Findlay notion of a rational extension, and the Johnson [51]-Utumi [58] maximal right quotient ring of a ring. Also, the regular rings of von Neumann *as quotient rings*, and the prime ideal spectrum of the selfinjective ones of Goodearl [73].

4. (Chapters 20 and 21) Direct decomposition theorems for rings and modules of Asano, Azumaya, Baer, Bass, Cailleau, Cartan, Chatters, Chase, Eilenberg, Faith, Gill, Goldie, Kaplansky, Koehler, Kurshan, Lafon, Levy, Matlis, Papp, Renault, Robson, Swan, and Warfield. Also, the unique decomposition theorem of Azumaya (=infinite Krull Schmidt theorem).

5. (Chapter 22) (Semi)perfect rings [=rings over which (finitely generated) modules have projective covers, that is, the dual of an injective hull]. The transfinite (T) nilpotency of the Jacobson radical (Bass [60]), and the Brauer theory of blocks generalized to perfect rings.

6. (Chapter 23) Morita [58], Azumaya [59], and Tachikawa [58] duality.

7. (Chapter 24) The structure of quasi-Frobenius rings of Nakayama [39–42], Ikeda [51, 52], Eilenberg-Nakayama [55, 57], Dieudonné [58], Morita [58], Faith-Walker [67], and others.

8. (Chapter 25) The Köthe [35]-Asano [39]-Nakayama [39–41] structure of serial rings, in the generalized form of Warfield [75] characterizing when every finitely presented module is a direct sum of uniserial modules.

The two chapters most closely related to the subject of "radical and semi-simplicity" in the sense of Jacobson [45, 55, 64] are Chapters 18 and 26. (Other chapters are distinctly more homological.)

We now outline Volume II. More specific references are found in the introductions to the various chapters. (Also see Chapter Notes.)

Chapter 17: Modules of finite Jordan-Hölder length, the Kurosch-Ore theorem for modular lattices, the Schreier refinement theorem, Fitting's Lemma, the Köthe-Levitzki and Kolchin theorems placing matrices simultaneously in upper (lower) triangular form, and theorems of Levitzki, Herstein, Lanski, Shock and Small on when nil $\Rightarrow$ nilpotence for ideals (subrings).

Chapter 18: The Jacobson radical of a ring, and the Perlis-Jacobson characterization, Nakayama's lemma, superfluous submodules, power series rings, the p-adic numbers, the Chinese remainder theorem, and other topics already outlined above.

Chapter 19: The Johnson-Wong characterization of quasinjective (QI) modules as fully invariant in the injective hull, the quasinjective hull of a module, the Chevalley-Jacobson density theorem for simple of irreducible modules obtained as a consequence of the double annihilator condition for QI modules of Jacobson [64] and Johnson-Wong [61].

We prove Utumi's theorem:

If E is quasinjective, and if $A = \operatorname{End} E_R$, then $A/\operatorname{rad} A$ is von Neumann regular, and rad A is the set of endomorphisms with essential kernels. Moreover $A/\operatorname{rad} A$ is right selfinjective by a theorem of Wong-Johnson, Utumi and Osofsky. Chapter 19 also contains Lambek's theorem: the maximal Johnson-Utumi right quotient ring $\bar{R}$ of a ring R is the biendeomorphism ring of the injective hull of R. When R is right nonsingular, then $\bar{R}$ is right selfinjective and regular (see 19.35).

A number of theorems on modules finite over endomorphism ring, or finendo modules as we call them, are proved including the fact that when quasinjective they are injective modulo annihilator. (A necessary and sufficient condition for this is a theorem of Fuller: each product of copies of the module is quasinjective.) A prime right Goldie ring R, that is, rings with semisimple Artinian right quotient ring, can thereby be characterized by the existence of a single module, namely, a finendo indecomposable injective module E with no fully invariant submodules (except trivial ones).

Chapter 20: A ring is right Noetherian iff one of the following holds:

(a) There exists a cardinal c such that every injective right module is (or contained in) a direct sum of modules generated by at most c elements (Faith-Walker [67]).

(b) Every right module is a direct sum of indecomposable modules (Matlis [58]-Papp [59]).

If every module can be so decomposed (as in (a)), then R is right Artinian (Griffith [70], Vámos [71] and Faith [71 d]). In fact, this follows from the above, and Chase's theorem which implies that a module has d.c.c. on principal left [*sic*] left ideals if every product R^b of copies of R is a pure submodule of a direct sum of modules generated by c elements (for a cardinal c depending on R).

While Chapter 25 is specifically on the subject of right σ-cyclic rings, or rings over which every module is a direct sum of cyclic modules, some specifics do appear in Chapter 20, for example, the theorems of Kaplansky [52] and Matlis [66] (for domains) and Lafon [70], Gill [71], and Warfield [71] stating that a commutative local ring is σ-cyclic iff an almost maximal valuation ring (in the sense of Kaplansky [52]). Some non-commutative generalizations of this theorem by Roux [72] and Warfield [75] appear in Chapter 25. Moreover, in Chapter 24 one proves that a commutative ring R has the property that every injective module is a direct sum of cyclic modules iff R is quasi-Frobenius.

Also included in Chapter 20 is the Decomposition Theorem of Chatters [72] for a hereditary Noetherian ring R into a product of prime or Artinian rings. The same decomposition also holds for principal ideal rings by a theorem of Krull-Asano-Goldie, but neither theorem implies the other. However, we give Robson's criterion for such a decomposition in Noetherian rings, which (with modest reinforcements) does imply both theorems.

Chapter 21: Definition of an Azumaya Diagram (AD) for an object M in Abelian Category: a direct sum decomposition into objects with local endomorphism rings. Azumaya's theorem: An AD is unique up to an automorphism. When the direct sum is finite, this is just the Krull-Schmidt (KS) diagram and theorem.

Over a ring R complete in the J-adic topology ($J = \text{rad } R$), every finitely generated module has a KS diagram (Swan [68]).

Also included in Chapter 20 are theorems on direct summands of Azumaya diagrams for a module M: are they Azumaya diagrams? The answer is trivially true for finite AD's, or when the indecomposable summands are injective (Matlis [58], Faith-Walker [67]), or countable (Faith-Walker [67] and Warfield [69 b]; for the latter, only the case M is countably generated is considered here, however).

Chapter 22: A right ideal I is (eventually) right vanishing if every sequence $\{a_1 \cdot a_2 \cdots a_n\}_{n=1}^{\infty}$ of products of elements of I is eventually zero. (Called left T-nilpotent by Bass [60].) If R satisfies the d.c.c. on principal right ideals, then every radical right ideal I is right vanishing (an application of Nakayama's lemma readily shows). A converse: if $R/\text{rad } R$ and rad R satisfy the d.c.c. on principal right ideals, then so does R. Bass [60] proved the latter rings are precisely the left perfect rings, that is, rings over which every left module has a projective cover. Other characterizations: every left module has "the same weak as projective dimension", or every flat left module is projective, or the direct limit of projective left modules is projective.

The projective cover is dual to the notion of injective hull. Unlike the injective hull, however, the projective cover can be specified concretely in *all* cases, and in fact, it is the cover of the "top" M/JM of the module M, where $J = \text{rad } R$. (Over a left perfect ring, $JM = \text{rad } M$.) Thus, $M/JM = \sum \oplus V$, where V is simple, and $V \approx Re$ for a principal indecomposable left module Re. Then, p.c. $V = Re/Ie$ for some left ideal $I \subseteq$ rad R, that is, p.c. V is a principal cyclic module, and then p.c. M is a direct sum of principal cyclic modules. In particular, every projective is a direct sum of the principal indecomposable modules (Bass [60]). See Notes for Chapter 22 for generalizations.

Björk's theorem is proved: the d.c.c. on principal submodules of a module implies that for the finitely generated submodules.

Chapters 22 ends with the block structure of perfect rings.

Chapter 23: This is on the subject of the possible dualities between subcategories $\mathscr{S}_B$ and $\mathscr{S}_A$ of B-mod and mod-A which contain B and A respectively, and closed under quotients, submodules, and finite products. Morita [58] determined that every such duality $T: \mathscr{S}_A \rightsquigarrow \mathscr{S}_B$ is induced by the canonical (B, A)-bimodule $U = TA$. Thus, $TX \approx \mathrm{Hom}_A(X, U)$ in B-mod, and $TY \approx \mathrm{Hom}_B(Y, U)$ in mod-A, for all $X \in \mathscr{S}_A$ and $Y \in \mathscr{S}_B$. Then T is called a U-duality, and we have the duality context $_B U_A$. For example, the Pontryagin duality on Ab $(A = B = \mathbb{Z})$ is induced by $U = \mathbb{R}/\mathbb{Z}$, where $\mathscr{S}_A = \mathrm{Ab}$ are the discrete abelian, and $\mathscr{S}_B$ are the groups compact in the uniform topology. (See Notes for Chapter 24 [*sic*].)

To continue, Morita determined that there is a duality context $_B U_A$ iff U is an injective cogenerator both in B-mod and A-mod. A necessary condition is that A and B be semiperfect (= semilocal lift/rad) rings, and if A is right or left perfect, then A is right and B is left Artinian (Osofsky [66]).

Chapter 24: Much of this is an application of Chapter 23 to the case of an $_R R_R$ duality context, that is, when R is an injective cogenerator on both sides. Injective cogenerator rings, say, on the right, have been characterized by Azumaya [66], Osofsky [66] and Utumi [67] as right selfinjective rings with finite essential right socle. They coincide with the class of rings over which every faithful right module is a generator in the category of all right modules. These are called right pseudo-Frobenius (PF) rings, and they generalize quasi-Frobenius (QF) rings, namely the left or right Artinian right PF rings.[1] However, the QF rings are right and left PF. The classical examples of QF rings are group algebras kG of a finite group G over an arbitrary field k, and factor rings of principal ideal domains modulo any nonzero ideal. The QF rings have an explicit ideal-theoretical characterization: every one-sided ideal is the annihilator of a finite subset of elements of R. Thus, there is a duality between the left and right ideals as well as the $_R R_R$ duality context

$$\begin{cases} \text{fin. gen. mod-}R \rightsquigarrow \text{fin. gen. } R\text{-mod} \\ \qquad X \mapsto X^* = \mathrm{Hom}_R(X, R) \end{cases}$$

given by taking dual modules.

Another aspect of duality in QF rings are that these are precisely the rings over which every projective module is injective. Dually for injective $\Rightarrow$ projective. (See Notes for Chapter 24 for other aspects of duality in QF rings.)

Chapter 25: We prove Warfield's theorem: Every finitely presented left module over a ring R is a direct sum of uniserial modules iff R has that same structure both in R-mod and mod-R (Warfield [75]). In this case R is said to be a serial

---

[1] The prefixes *pseudo* and *quasi* are unfortunate since they are actually the most useful concepts, that is, the *real* things.

ring. Any Noetherian serial ring is a direct sum of prime and Artinian rings. (The proof we give employs Robson's criterion which, as we have remarked, also applies to the theorems of Chatters and Krull-Asano-Goldie.) The structure of Nakayama's Artinian serial rings is also given: then, every module is a direct sum of principal cyclic uniserial modules. Moreover, every dominant principal indecomposable module eR is injective modulo the annihilator. This is Fuller's characterizing property of Artinian serial rings when the left-right symmetry of this condition is also assumed; and indicates how one gets the decomposition of a module M into a direct sum of uniserial modules: the dominant principal cyclic submodules split off!

For Noetherian serial rings, there remains, then, only to determine the structure of the prime ones. This has been done by Michler [69 b] when rad $R \neq 0$, and Warfield (loc. cit.) gives a short proof, but we omit this. Instead we give the Eisenbud-Griffith-Robson theorem: Any hereditary Noetherian prime ring R is an Artinian serial ring modulo any ideal $I \neq 0$. The proof has two ingredients, the first being the theorem of Webber [70] – Chatters [70], which states that A/J is Artinian for any essential right ideal J of any hereditary Noetherian ring A. The second ingredient is the quite short (but ingenious!) proof of Eisenbud-Griffith that the dominant principal indecomposable modules are injective modulo annihilator.

Primary-decomposable Artinian serial rings, namely finite products of full matrix rings over local Artinian principal ideal rings, are characterized a la Nakayama [40, 41] by the property that every factor ring is PF.

Chapter 26 is concerned with prime and primitive ideals, and their intersections. The former is the prime radical which contains every nilpotent ideal, and is characterizable (after Levitzki) as the set of all "strongly" nilpotent elements of R, whereas the latter intersection is the Jacobson radical. (The prime radical coincides with the Baer lower nil radical. See 26.11.)

The main result of Chapter 26 is Amitsur's theorem on the semiprimitivity (Jacobson radical $= 0$) of arbitrary group algebras over transcendental fields of characteristic zero, in particular, over nondenumerable fields of characteristic zero. (See 26.20–21.)

V. Ring Theory

Chapter 17. Modules of Finite Length and their Endomorphism Rings

Topics in this chapter are: (1) the Kurosh-Ore theorem for modular lattices 17.4; (2) the Schreier refinement theorem 17.6; (3) the Jordan-Hölder theorem 17.7; (4) Fitting's lemma 17.16; (5) theorems of Köthe-Levitzki and Kolchin on putting matrices simultaneously in triangular form 17.19 and 17.30; and (6) nilpotency of nil submonoids of monoids satisfying various chain conditions 17.19–25.

In 17.22 we present a theorem of Shock [71 b] which has two notable consequences: (1) a theorem of Levitzki [64] and Herstein-Small [64, 66]: if R is a semigroup with zero and if R satisfies the a.c.c. on annihilator right and left ideals, then every nil submonoid of R is nilpotent 17.23; (2) Lanski's theorem stating that in a right Goldie ring multiplicative nil submonoids are nil 17.24.

Other nil $\Rightarrow$ nilpotent theorems are discussed in exercises, and in chapter notes.

Modular Lattices

A lattice L is said to be **modular** provided that for any three elements A, B, and C, the equivalence holds:

$$A \cap (B \cup C) = (A \cap B) \cup C \Leftrightarrow A \ge C.$$

The implication $\Rightarrow$ is trivial, and holds in general, whereas the implication $\Leftarrow$ is called the **modular identity.**

17.1 Exercise. A lattice L is modular if and only if the following implication holds for any two elements $y' \ge y$:

$$(\exists x) \qquad y' \cap x = y \cap x \ \& \ y' \cup x = y \cup x \Rightarrow y' = y.$$

17.2 Proposition. *The lattice of (normal) subobjects of any object M of a category C is modular, when C is abelian ($C = $ GROUPS).*

Proof. Let $Y' \supseteq Y$, and X be subobjects of M such that

$$Y \cap X = Y' \cap X \ \& \ Y \cup X = Y' \cup X.$$

If C is abelian, then the 5-lemma 5.34 applied to the row exact commutative diagram

$$
\begin{array}{ccccccccc}
0 & \longrightarrow & X/X \cap Y & \longrightarrow & M/Y & \longrightarrow & M/X+Y & \longrightarrow & 0 \\
 & & \Big\downarrow {\scriptstyle =} & & {\scriptstyle g}\Big\downarrow {\scriptstyle \text{canon}} & & \Big\downarrow {\scriptstyle =} & & \\
0 & \longrightarrow & X/X \cap Y' & \longrightarrow & M/Y' & \longrightarrow & M/X+Y' & \longrightarrow & 0
\end{array}
$$

yields that the canonical morphism $g: M/Y \to M/Y'$ is an equivalence. Thus, $\ker g = Y'/Y = 0$, and $Y' = Y$. (The case $C = \text{GROUPS}$ is an exercise.) $\square$

Projective Intervals

Let L be a lattice. A **quotient** $[x, y]$ of two elements x and y is the closed interval $[x, y] = \{t \in L \mid x \leq t \leq y\}$. The quotient is **simple** if it consists of two elements x and y. Then y **covers** x. Two quotients $[x \cap y, x]$ and $[y, x \cup y]$ are called **transposes**. Two quotients $[x, y]$ and $[x', y']$ are **projective** provided that there is a finite sequence

$$[x, y], [x_1, y_1], \ldots, [x_t, y_t], [x', y']$$

such that consecutive quotients are transposes.

17.3 A Proposition (Dedekind [1900, XI, p. 259]). *If L is a modular lattice, and $u, v \in L$, then*

$$\begin{cases} [u \cap v, v] \to [u, u \cup v] \\ \quad x \mapsto x \cup u \\ y \cap v \leftarrow y \end{cases}$$

is a lattice isomorphism which carries any quotient in this interval into its transpose.

Proof (see Birkhoff [48, p. 73]). If $u \cap v \leq x \leq v$, then

$$u = u \cup (u \cap v) \leq u \cup x \leq u \cup v$$

and if $x \leq x'$, then $u \cup x \leq u \cup x'$. This shows that the mapping is an order injection. Moreover,

$$u \cap v \leq x \leq v \implies v \cap (u \cup x) = (v \cap u) \cup x = x$$

so that mapping is bijective, hence a lattice isomorphism. If $u \cap v \leq x \leq x' \leq v$, then $[x, x']$ and $[u \cup x, u \cup x']$ are transposes, and dually. This follows, since

$$x' \cup (x \cup u) = (x' \cup x) \cup u = x' \cup u,$$

and

$$x' \cap (u \cup x) = (x' \cap u) \cup x = [(x' \cap v) \cap u] \cup x$$
$$= [x' \cap (v \cap u)] \cup x = (v \cap u) \cup x = x. \quad \square$$

17.3 B Corollary. *Projective quotients in a modular lattice are order isomorphic.* $\square$

Kurosch-Ore Theorem

An intersection $\bigcap_{i \in I} x_i$ is **redundant** if there is a subset $J \neq I$ such that $\bigcap_{i \in I} x_i = \bigcap_{j \in J} x_j$. Otherwise, the intersection is **irredundant**. An element x is ("meet") **irreducible** provided that

$$x = y \wedge z \implies x = y \quad \text{or} \quad x = z, \qquad \forall \, y, z \in L.$$

The next theorem was proved by E. Noether [21] for ideals of a commutative ring.

17.4   Proposition (Kurosch [35], Ore [36]). *If L is a modular lattice, then any two irredundant finite intersections of an element $a \in L$ as an intersection (meet) of irreducible elements, say,*

$$a = x_1 \cap x_2 \cap \cdots \cap x_r = y_1 \cap y_2 \cap \cdots \cap y_s$$

have the same number of elements $r = s$. Moreover, for any subset I of $\{1, \ldots, n\}$, there is an injection $p: I \to \{1, \ldots, n\}$ such that

$$a = \bigcap_{i \in I} x_i \cap \bigcap_{i \in I} y_{p(i)} \qquad \text{(replacement lemma)}$$

is an irredundant intersection.

Proof (see Birkhoff [48, p. 93]). The proof is by induction on the number of elements in I, and it suffices to let I consist of a single element, say $I = \{1\}$. Set $y = \bigcap_{i>1} x_i$. By irredundancy, $y > a$, and $x_1 \cap y = a$. Set $z_j = y \cap y_j$. Then $y \geq z_j \geq a$. Also, since $z_j \leq y_j$, then

$$a \leq z_1 \cap \cdots \cap z_s \leq y_1 \cap \cdots \cap y_s = a.$$

Then, by Dedekind's theorem 17.3A, there is a lattice isomorphism

$$[a, y] = [x_1 \cap y, y] \to [x_1, x_1 \cup y].$$

Thus, since x_1 is irreducible in the quotient $[x_1, x_1 \cup y]$, then a is irreducible in $[a, y]$, and so there exists j such that $a = z_j$

$$a = z_j = y_j \cap y = y_j \cap \bigcap_{i>1} x_i$$

as claimed. Since every x_i can be replaced by some y_j, by the irredundancy of the representations, necessarily $r \leq s$. Then, $s \geq r$ by symmetry, so $s = r$. $\square$

A union $a = a_1 \cup \cdots \cup a_n$ is **direct** provided that the elements $\{a_i\}_{i=1}^n$ are independent in the sense that

$$0 = a_i \cap \left(\bigcup_{j \neq i} a_j \right), \qquad i = 1, \ldots, n.$$

Then one writes $a = a_1 \times \cdots \times a_n$.

17.5   *Exercise.* If L is a modular lattice, and if there are two representations

$$a = a_1 \times \cdots \times a_n = b_1 \times \cdots \times b_m$$

of an element a as a direct union of indecomposable elements, then $m = n$, and there is a permutation p of $\{1, \ldots, n\}$ such that a_i and $b_{p(i)}$ are projective, $i = 1, \ldots, n$. (Birkhoff [48, p. 94] associates the following names with this theorem: Wedderburn (for finite groups), Kronecker (for abelian groups), Remak (Remak corrected Wedderburn), Krull, Schmidt, Fitting, Korinek, Ore, and Kurosch. Ore gave the first purely lattice theoretical proof (cf. 18.18).)

Composition Chains

An ordered set L is **Noetherian** if and only if L satisfies the a.c.c., or equivalently, the maximum condition (Foreword Prop. 12). The dual concept is **Artinian.** An object x in an ordered set L is Noetherian (Artinian) if the ordered subset $\{t \in L \mid t \leq x\}$ is such. A Noetherian lattice L has a greatest element P, and $P = \max\{x \in L\}$. The least element of a lattice is denoted 0. Thus, a lattice which is both Artinian and Noetherian has a least element 0 and a greatest element P. In this case

$$(A) \qquad P = M_0 \geq M_1 \geq \cdots \geq M_{n-1} \geq M_n = 0$$

is a **composition chain** for L provided that the interval $[M_i, M_{i+1}]$ is simple, $i = 0, \ldots, n-1$. The integer n is called the length $d[A]$ of the chain (A). Let (A) be a chain of subobjects of an object M of a category C, where C is abelian, or $C = \text{GROUPS}$. In the latter case, assume that (A) is a chain which is subnormal in the sense that M_{i+1} is a normal subgroup of M_i, $i = 0, \ldots, n-1$. Then, in either case, the quotient objects M_i/M_{i+1} are defined, $i = 0, \ldots, n-1$. Moreover, an interval $[M_i, M_{i+1}]$ is simple if and only if the quotient object M_i/M_{i+1} is simple.

If (A) is any finite chain in L, then, the set $\{M_i/M_{i+1}\}_{i=0}^{n-1}$ is called the **cycle of factors** of (A). If (A) is a composition chain, this is called the **cycle of simple factors.** Define:

$$\text{gr}(C) = \sum_{i=0}^{n-1} \oplus M_i/M_{i+1}$$

to be the coproduct in case L is a lattice of subobjects of an abelian category, or the direct sum in GROUPS in case (A) is a subnormal chain of groups. Two chains (A) and

$$(B) \qquad N_0 \geq N_1 \geq \cdots \geq N_{n-1} \geq N_n$$

are said to be **equivalent** provided that their respective cycles of factors determine the same equivalence class of quotient objects, that is, in case there is a permutation p of $\{1, \ldots, n\}$ such that $M_i/M_{i+1} \approx N_{p(i)}/N_{p(i+1)} \ \forall\, i$. In this case, write $(A) \approx (B)$. The relation $(A) \approx (B)$ is an equivalence relation, but the important fact is the Jordan-Hölder theorem which states that any two composition series are equivalent.

The chain (B) is a **refinement** of (A) provided that every element of (A) is an element of (B). Then one says that (A) **can be refined** to (B).

17.6   Refinement Theorem (Schreier). *If M is an object of an abelian category C (or if M is a group) then any two finite (subnormal) chains (A) and (B) of subobjects of M can be each refined to equivalent finite (subnormal) chains (A') and (B').*

Proof. The proof depends on Zassenhaus's (butterfly) lemma 1.12, which in turn depends on the first and second Noether isomorphism theorems, which hold in GROUPS, and, by 5.33 and 5.36, in any abelian category. We give the proof in the case M is an object of an abelian category, and (A) and (B) are finite chains of subobjects

$$(A) \qquad M_0 > M_1 > \cdots > M_{n-1} > M_n = 0;$$

$$(B) \qquad N_0 > N_1 > \cdots > N_{m-1} > N_m = 0.$$

Without loss of generality, we may assume that $M = M_0 = N_0$. For each $i = 0, \ldots, n$, and $j = 0, \ldots, m-1$, define

$$M_{ij} = M_{i+1} + (N_j \cap M_i).$$

Then $M_{im} = M_{i+1}$, and so

$$M_0 = M_{00} \supseteq M_{01} \supseteq \cdots \supseteq M_{0, m-1} \supseteq M_1 = M_{10} \supseteq M_{11} \supseteq \cdots$$
$$\supseteq M_{n-1, 0} \supseteq \cdots \supseteq M_{n-1, m-1} \supseteq M_n = 0$$

is a refinement (A') of (A). Similarly, define

$$N_{ji} = N_{j+1} + (M_i \cap N_j)$$

for $j = 0, \ldots, m-1$, $i = 0, \ldots, n$. This gives a refinement (B') of the chain (B). By the butterfly lemma 1.12,

$$M_{ij}/M_{i, j+1} \approx N_{ji}/N_{j, i+1}$$

$\forall i, j$. Since there are $(n-1)(m-1)+1$ subobjects in (A') and (B'), these isomorphisms show that (A') and (B') are equivalent. $\square$

17.7   Theorem (Jordan-Hölder). *Let M be an object of a category C, where C is abelian (or C = GROUPS). Assume that M has a finite composition chain. Then, any two composition chains of M are equivalent, and any finite (subnormal) chain of subobjects of M can be refined to a composition series.*

Proof. If (A) is a composition chain, and if (B) is any finite chain, then by Schreier's theorem, (A) and (B) have equivalent refinements (A') and (B'). Since (A) is a composition chain, then $(A') = (A)$, and (B') is a composition chain that refines (B). If (B) is a composition chain, then $(B) = (B')$ is equivalent to (A). $\square$

The height $d[x]$ of an element x of an ordered set L is the maximum integer d such that there is a chain

(C) $x = x_0 > x_1 > \cdots > x_{d-1} > x_d.$

Then (C) must be a composition chain for the segment $\{t \in L \mid t \leq x\}$ terminating with x, since otherwise (C) can be refined to a chain (C') with more elements. If no such integer d exists, then x is said to have infinite height. If the integer $d[x]$ is unambiguously defined, for every x, then the ordered set L is said to satisfy the **Jordan-Dedekind chain condition** (J. D. c. c.). This means that any two composition chains for the same ordered subset of L have the same length.

17.8   Exercise

17.8.1   Let L be an ordered set in which all chains are finite. Then L satisfies the J.D.c.c. if and only if there is a function $d: L \to \omega$ such that the condition holds:

$$x \text{ covers } y \iff x > y \quad \text{and} \quad d[x] = d[y] + 1.$$

17.8.2   The J.D.c.c. holds for the lattice of (normal) subobjects of any object M of finite length of a category C, where C is abelian ($C = $ GROUPS).

17.8.3   Any ordered set L which is both Artinian and Noetherian has a composition series. If the J.D.c.c. holds, then conversely.

17.8.4   Let L be any lattice in which all (bounded) chains are finite. Then the conditions (a)–(c) are equivalent:

(a) L is a modular lattice.

(b) The J.D.c.c. holds, and there is a function $d: L \to \omega$ such that

$$d[x]+d[y]=d[x\cup y]+d[x\cap y], \qquad\qquad\qquad \forall\, x, y \in L.$$

(c) (ξ') If x and y both cover a, and if $x \neq y$, then $x \cup y$ covers x and y.

(ξ'') If a covers x and y, and $x \neq y$, then x and y cover $x \cap y$.

(Hint: Condition (ξ') implies the inequality $d[x]+d[y] \geq d[x\cup y]+d[x\cap y]$, whereas the reverse inequality is implied by (ξ''). Consult Birkhoff [48, pp. 66–68].)

17.8.5   Show that there is an ordered set of cardinality 5 which obeys the J.D.c.c. but which fails to have a function $d[x]$ such that $d[x]=d[y]+1$, if x covers y.

17.8.6   If L and P satisfy the J.D.c.c., does $L \times P$?

17.8.7   Show that there is a unique non-modular 5-element lattice.

17.8.8   The modular law is self-dual. State the dual of the Kurosch-Ore theorem 17.4.

17.9   Exercise

17.9.1   As defined in Chapter 15, a subclass S of an abelian category C is a Serre class if for every $0 \to X \to Y \to Z \to 0$ exact in C it is true that $Y \in S$ if and only if $X, Z \in S$. For any abelian category, the class of Artinian (resp. Noetherian) objects is a Serre class.

17.9.2   If L is a modular lattice, then L is Noetherian (Artinian) if and only if for every $x \in L$ it is true that the lattice of elements above x, and the lattice of elements below x, are both Noetherian (Artinian).

17.9.3   An object M of mod-R is Noetherian if and only if every submodule of M is finitely generated.

17.9.4   If R is right Artinian (Noetherian), then every finitely generated object M of mod-R is Artinian (Noetherian).

17.9.5   If M, N are objects of mod-R of finite lengths, then $\mathrm{Hom}_R(M, N) \neq 0$ implies the existence of a common simple factor.

Polynomials over Modules of Finite Length

17.10   Theorem. *If an R-module M has finite length n, then every $R[x]$-submodule of $M[x]$ is generated by n or fewer elements.*

Proof. Let S be an $R[x]$-submodule of $M[x]$. Define the submodule $L_i(S)$ as in the lemma (I, 7.12, p. 341) used in the proof of the Hilbert basis theorem. Since M is Noetherian, there is an index k such that $L_k(S)=L_{k+1}(S)=\cdots$. Hence, there is a composition series $0=M_0\subset M_1\subset\cdots\subset M_n=M$ of M such that each $L_k(S)=M_p$ for some p, and each $L_i(S)$ is one of M_q, $q\leq p$. Since M_i/M_{i-1} is simple, there is an element $a_i\in M_i$ such that $M_i=a_iR+M_{i-1}$, and then $M_i=\sum_{j=1}^{i}a_jR$, $i=1,\ldots,n$.

If $i\leq p$, then $a_i\in L_k(S)$, so there is a polynomial $g_i(x)$ of smallest possible degree t, $t\leq k$, such that a_i is the coefficient of x^t. We assert that S is generated as an $R[x]$-module by $g_1(x),\ldots,g(x)$: $S=\sum_{i=1}^{p}g_i(x)R[x]$.

Let f be a nonzero element of S of degree d, let a be the coefficient of x^d, and let t be such that $a\in M_t$, $a\notin M_{t-1}$. The degree of any polynomial of S with leading coefficient not in M_{t-1} is at least as large as the smallest subscript c such that $M_{t-1}\subset L_c(S)$ properly. The degree of g_t is precisely c, since $a_t\in M_t\subseteq L_c(S)$, and g_t has minimal degree among the polynomials having leading coefficient a_t. Therefore, $d_1=\text{degree }g_t\leq\text{degree }f=d$.

Now $M_t=a_tR+M_{t-1}$, and since $a\in M_t$, there exists $r_t\in R$ such that $b=a-a_tr_t\in M_{t-1}$. Consequently, the polynomial $f_1=f-g_tr_tx^{d-d_1}$ lies in S, and either has degree less than d (provided that $d_1<d$) or else has leading coefficient b in M_{t-1}. Applying this procedure a finite number of times, we eventually reach 0, showing that $f\in\sum_{i=1}^{p}g_i(x)R[x]$. $\square$

17.11   Corollary. *Let R be a ring that has finite length n as a right R-module. Then each right ideal in the polynomial ring $R[x]$ is generated by n or fewer elements.* $\square$

As we stated earlier, a theorem of Hopkins and Levitzki 18.13 asserts that a right Artinian ring is right Noetherian, and hence has finite length as required, so the corollary can be stated for right Artinian rings.

If R is a field, then length $R_R=1$, and we have the following result.

17.12   Corollary. *If R is a field, then the polynomial ring $R[x]$ is a principal right ideal ring, and a principal left ideal ring.* $\square$

17.13   Exercise

17.13.1   For a right ideal I of a ring R, define $n(I)$ to be the minimal number of elements in any generating set of I. Set $n(I)=\infty$, if $n(I)$ is not finite. (a) If $\mathbb{Z}$ denotes the ring of integers, then for each natural number t, let I_t denote the ideal of $\mathbb{Z}[x]$ generated by $\{2^t, 2^{t-1}x,\ldots,2^{t-i}x^i,\ldots,x^t\}$. Then $n(I_t)=t+1$. (b) If R is any ring for which there exists a natural number N such that $n(I)\leq N$ for all right ideals I of the polynomial ring $R[x]$, then $n(J)\leq N$ for all right ideals J of R.

17.13.2   Let R be a ring, and M a Noetherian R-module. (a) Show that the power series module $M\langle x\rangle$ is a noetherian $R\langle x\rangle$-module. (Hint: Define the degree of $f(x)=\sum_{n=0}^{\infty}a_nx^n\in M\langle x\rangle$ to be the smallest k such that $a_k\neq 0$. If S is a $R[x]$-submodule, let $L_n(S)$ be the set of elements of M consisting of 0 and those elements $a\in M$ that appear as a coefficient of a polynomial of degree n.)—(b) Can a theorem be proved for $M\langle x\rangle$ corresponding to 4.5?

17.13.3   If N is a submodule of M, show that $(M/N)[x]\approx M[x]/N[x]$.

17.13.4   If M is a simple module, and if Q is an $R[x]$-submodule of $M[x]$, then Q is generated by a polynomial of least degree in x.

17.13.5   Investigate what properties of rings which have been encountered in the text thus far are preserved by polynomial extensions. (If you run dry, check some of the items listed on pp. 220-1 of Vol. I.) Also, try some properties of commutative rings, such as principal ideal domains, unique factorization domains, Dedekind domains, Prüfer domains, etc.) Similarly for power series rings: If R is a ring with one of the stated properties, is $R\langle x\rangle$?

The Endomorphism Ring

If A is an endomorphism of M, we investigate the effect on A of various chain conditions on M.

17.14   *Exercise.* Let A be an endomorphism of a module M.

17.14.1   If k is a natural number such that $\ker A^k = \ker A^{k+1}$, then $A^k M \cap \ker A^k = 0$, A induces a monomorphism in $A^k M$, and A induces a nilpotent endomorphism in $\ker A^k$.

17.14.2   If t is a natural number such that $A^t M = A^{t+1} M$, then A induces an epimorphism in $A^t M$, A induces a nilpotent endomorphism in $\ker A^t$, and $M = A^t M + \ker A^t$. $\square$

If M is any R-module, then an element $f \in S = \text{End } M_R$ has a (two-sided) inverse in S if and only if f is an automorphism of M, that is, if and only if f is an isomorphism of M.

17.15   *Exercise.* Let M be an R-module, let $S = \text{End } M_R$, and let $A \in S$. Then:

17.15.1   If M is Artinian, A is an automorphism if and only if A is a monomorphism.

17.15.2   If M is Noetherian, A is an automorphism if and only if A is an epimorphism. $\square$

Fitting's Lemma

Applying 17.15 to the situation of 17.14, if M is Artinian, then so is the submodule $A^k M$ of 17.14.1, and A induces an automorphism a of $A^k M$. If M is Noetherian, then so is the submodule $A^t M$ of 17.14.2, and A induces an automorphism b of $A^t M$. Now let M be both Artinian and Noetherian, and assume A is not an automorphism. Then there exist integers k and t such that

$$M \supset AM \supset \cdots \supset A^t M = A^{t+1} M = \cdots$$

$$0 \subset \ker A \subset \cdots \subset \ker A^k = \ker A^{k+1} = \cdots.$$

Since A induces an automorphism of $A^k M$, necessarily $A^{k+1} M = A^k M$, and therefore $k \geq t$. On the other hand, A induces an automorphism b of $A^t M$. Hence, if $x \in \ker A^{t+1}$, then $0 = A^{t+1} x = b A^t x$, and $A^t x = 0$, so $x \in \ker A^t$. Thus, $\ker A^{t+1} =$

ker A^t, showing that $t \geq k$ and $t = k$. If length $M = s$, then clearly $t = k \leq s$. Then $M = A^s M \oplus \ker A^s$ by 17.14.1 and 2. This proves the next result.

17.16   Theorem (Fitting's Lemma). *If M is a module of finite length s, then any endomorphism A induces an automorphism in the submodule $A^s M$, induces a nilpotent endomorphism in $\ker A^s$, and $M = A^s M \oplus \ker A^s$.* $\square$

17.17   Corollary. *Let M be an indecomposable R-module of finite length s, and let $S = \operatorname{End} M_R$.*

17.17.1   *Every element of S is either a unit, or is nilpotent.*

17.17.2   *The set N of nilpotent endomorphisms of M is a nilpotent ideal of S of index $\leq s$ (cf. 17.20).*

Proof of 17.17.1. If $A \in S$, then by Fitting's lemma, $M = A^s M \oplus \ker A^s$. Since M is indecomposable, either $M = A^s M$, or else $M = \ker A^s$. By Fitting's lemma, A is an automorphism in the former case, and A is nilpotent in the latter case.

Proof of 17.17.2. Now let $A, B \in S$ be such that AB is an automorphism, that is, such that AB has an inverse $C \in S$. Then $A(BC) = (CA)B = 1$, and consequently neither A nor B is nilpotent; that is, both A and B are automorphisms. Thus, if $B \in N$, then $AB \in N$ and $BA \in N$. Hence, in order to show that N is an ideal of S, it remains only to show that N is an additive subgroup of S. Let $A, B \in N$, and suppose for the moment that $A - B = C \notin N$. Then C is an automorphism, so $A_1 - B_1 = 1$, where $A_1 = AC^{-1}$, $B_1 = BC^{-1}$. We already know that $A_1, B_1 \in N$. Thus, $A_1^s = 0$, and, by the binomial theorem,

$$0 = A_1^s = 1 + s B_1 + \cdots + B_1^s = 0.$$

Since $B_1 \neq 0$, there exists a natural number k such that $B_1^{k-1} \neq 0$, and $B_1^k = 0$. Then

$$0 = 0 \cdot B_1^{k-1} = B_1^{k-1} + s B_1^k + \cdots + B_1^{s+k} = B_1^{k-1}$$

a contradiction, which proves that $A - B = C \in N$ for all $A, B \in N$. Therefore N is an ideal of S, and the following Proposition 17.20 proves that N is nilpotent of index $\leq s$. $\square$

In volume I, p. 172, we defined (but did not make much use of) a **local ring** R as a ring in which the set of nonunits is an ideal, or equivalently, $R/\operatorname{rad} R$ is a field (see 18.10 A). In this terminology we rephrase 17.17.

17.17′   Corollary (Fitting [33, Satz 3 and 8]). *The endomorphism ring of an indecomposable module of finite length s is a local ring with nilpotent radical of index $\leq s$.* $\square$

Let K be a ring, and let S be a multiplicative submonoid of the $n \times n$ matrix ring K_n. Then, S is said to be in **(strict) upper triangular form**, provided that for some set $\{e_{ij}\}_{i,j=1}^n$ of matrix units of K_n, and unit $x \in K_n$ every element $s \in x^{-1} S x$ has the canonical form

$$s = (s_{ij}) = \sum_{i,j=1}^n s_{ij} e_{ij}$$

with $s_{ij} = 0$ whenever $i \geq j$ (resp. whenever $i > j$). Then, the elements of S are said to be placed **simultaneously** into (strict) upper triangular form.

17.18 *Exercise*

17.18.1 If K is a field, and if $T_n(M)$ is the set of (strict) upper triangular matrices of K_n relative to some set M of $n \times n$ matrix units, then a submonoid S can be placed into (strict) upper triangular form if and only if there is a unit $x \in K_n$, and a set M such that $x S x^{-1} \subseteq T_n(M)$.

17.18.2 Show that S can be put into (strict) upper if and only if S can be placed in (strict) lower triangular form.

17.18.3 Let F be a field, and let R be the ring of all $n \times n$ (lower) triangular matrices over F. Let $\{e_{ij}\}_{i,j=1}^n$ be matrix units in F_n and let $x = \sum_{i,j=1}^n a_{ij} e_{ij} \in F_n$, where $a_{ij} \in F$, $i, j = 1, \ldots, n$. Then $x \in R$ if and only if $a_{ij} = 0$ whenever $j > i$. Show that the set $N = \{x \in R \mid a_{ij} = 0$ whenever $j \geq i\}$ is an ideal of R, which is nilpotent of index n. Show that $N = \operatorname{rad} R$. (Hint: R/N is a direct product of fields, each $\approx F$.)

17.18.4 Let R be a ring, and Q an ideal such that R/Q is a field. Show that for any integer n, R/Q^n is a local ring (for example, $\mathbb{Z}/p^n \mathbb{Z}$, where p is a prime number).

17.18.5 Every commutative Artinian ring is a direct product of finitely many local rings. (Hint: Write the ring A as a direct product of finitely many indecomposable right ideals. Let B be one of these. Then B is a ring, and $B \approx \operatorname{End} B_R$.)

17.19 Proposition (Köthe [30b], Levitzki [31]). *If K is a field, then any multiplicative nil submonoid S of $K_n \approx \operatorname{End}_K K^n$ can be placed simultaneously into strict triangular form, and $S^n = 0$.*

Proof. For $n = 1$, there is nothing to prove. Assume the proposition for vector spaces K^m of dimension $m < n$. Let $V = K^n$. Then, there is a vector subspace U of V of dimension $< n$ such that $US \subseteq U$. Otherwise, $VS = V$, and then there exist elements $s_1, \ldots, s_k \in S$ such that

$$V = \sum_{i=1}^k V s_i.$$

Then, for any integer $j \in \mathbb{Z}^+$,

$$V = \sum_{1 \leq i_1, \ldots, i_j \leq k} V s_{i_1} \cdots s_{i_j}.$$

This implies the existence of a sequence $\{s_{i_t}\}_{t \in \mathbb{Z}^+}$ such that $s_{i_t} \in \{s_1, \ldots, s_k\}$, and

$$s_{i_1} \cdots s_{i_t} \neq 0, \qquad\qquad \forall\, t \in \mathbb{Z}^+.$$

Now one of the elements $s_1, \ldots, s_t$, say $a = s_1$, occurs infinitely often in the sequence $\{s_{i_t}\}_{t \in \mathbb{Z}^+}$. Then, there is a sequence $\{s'_{i_t}\}_{t \in \mathbb{Z}^+}$ of elements of S such that

$$s'_1 a s'_2 a \ldots s'_t a \neq 0, \qquad\qquad \forall\, t \in \mathbb{Z}^+.$$

Put $W = Va$. Since a is nilpotent, $\dim W < \dim V$. Furthermore,

$$W S a \subseteq V a = W$$

so that Sa induces a nil submonoid of $\mathrm{End}_K W$. By the induction hypothesis,

$$s'_1 a s'_2 a \ldots s'_{n-1} a s'_n a = 0$$

which is a contradiction.

Since, therefore, there exists $t \in \mathbb{Z}^+$ such that

$$V \supset VS \supset VS^2 \supset \cdots \supset VS^{t-1} \supset VS^t = 0$$

is strictly decreasing, then there is a basis $x_1, \ldots, x_n$ of $V = K^n$, such that $x_1, \ldots, x_{n_i}$ is a basis for VS^{t-i}, $i = 1, \ldots, t$, and the matrix representations of the elements of S relative to this basis all have the strict lower triangular form. $\square$

17.20   Theorem (Levitzki [31], Fitting [33]). *If M is an R-module of length n, then any nil submonoid of $A = \mathrm{End}\, M_R$ is nilpotent of index $\leq n$.*

Proof. We prove it first under the assumption that M is semisimple. If M is homogeneous, that is, if M is a direct sum of isomorphic simple modules, then $A \approx K_n$, where $K = \mathrm{End}\, V_R$ is the endomorphism ring of a simple submodule V. In this case, the theorem follows from 17.19. Otherwise, $M = H \oplus G$ is a direct sum of two fully invariant submodules of lengths h and $n - h$ respectively. If S is a nil submonoid of A, then S induces a nil submonoid $\bar{S}$ of $\mathrm{End}\, H_R$, and a nil submonoid S' of $\mathrm{End}\, G_R$. Then $\bar{S}^h = 0$, $S'^{n-h} = 0$, and $S^n = 0$.

In the general case, we may assume that the socle H of M has length $m < n$. Since H is fully invariant, an element $s \in S$ induces $\bar{s} \in \mathrm{End}(M/H)_R$, and $s' \in \mathrm{End}\, H_R$, both of which elements are nilpotent. By the induction hypothesis, $\bar{S} = \{\bar{s} \mid s \in S\}$ and $S' = \{s' \mid s \in S\}$ are nilpotent of indices $n - m$, and m respectively. Thus, given a sequence $\{s_i\}_{i=1}^n$ of elements of S, then

$$s_n s_{n-1} \cdots s_{n-m+1} \cdots s_2 s_1 M \subseteq s_n s_{n-1} \cdots s_{n-m+1} H = 0.$$

Thus, $S^n = 0$, completing the induction and proof. $\square$

17.21   Lemma. *Let R be a monoid with zero element, and $x, y \in R$. Then*

(a) $(y x)^k = 0$ & $x y x \neq 0$ $\Rightarrow$ $x^\perp \subset (x y x)^\perp$.

Proof. $A \supset B$ denotes proper inclusion of sets. Let $x y x \neq 0$, and $(y x)^k = 0$, and $(y x)^{k-1} \neq 0$. Then:

(b) $x(y x)^{k-1} \neq 0$ $\Rightarrow$ $(x y x)(y x)^{k-1} = 0$, so that (a) holds.

(c) $x(y x)^{k-1} = 0$ $\Rightarrow$ $(x y x)(y x)^{k-2} = 0$ and $x(y x)^{k-2} \neq 0$, so (a) holds. $\square$

17.22   Proposition (Shock [71b]). *Let R be a semigroup with zero and $(\mathrm{acc})^\perp$. If R has a nil submonoid N which is not nilpotent, then there is a set $\{a_i\}_{i \in \omega}$ of elements of R such that*

$$^\perp A_1 \subset {}^\perp A_2 \subset \cdots \subset {}^\perp A_n \subset \cdots$$

where $A_n = \{a_i\}_{i \geq n}$. If, moreover, R is a ring, then the set $\{a_i\}_{i \in \omega}$ can be chosen so that the sum $\sum_{i \in \omega} a_i R$ is direct, that is, such that the set $\{a_i R\}_{i \in \omega}$ of right ideals is independent.

Proof. For some t,

$$K=(N^t)^\perp=(N^{t+j})^\perp, \qquad\qquad\qquad \forall\, j\geq 0.$$

Choose $n_0\in N$ such that $n_0^\perp$ is maximal in

$$\{x^\perp\,|\,x\in N \ \ \& \ \ xR\nsubseteq K\}.$$

If $Nn_0\,R\subseteq K$, then $N^t N n_0\,R\subseteq N^t K=0$, which implies that $n_0\,R\subseteq K$, a contradiction to the choice of n_0. Hence, choose n_1 such that $n_1^\perp$ is maximal in $\{x^\perp\,|\,x\in N \ \& \ x n_0\, R\nsubseteq K\}$, and inductively choose n_k such that $n_k^\perp$ is maximal in

$$\{x^\perp\,|\,x\in N \ \& \ x n_{k-1}\dots n_1\, n_0\, R\nsubseteq K\}.$$

Then:

(1) $\qquad\qquad (n_{j+m}\dots n_{j+1})^\perp\cap n_j R=0, \qquad\qquad \forall\, m\geq 1,\ j\geq 0.$

Otherwise, $(n_{j+m}\dots u_j)^\perp\supset n_j^\perp$. Define $a_i=n_i\,n_{i-1}\dots n_1\,n_0$, $i\geq 0$, and suppose, for some $j\geq 0$, that $n_k\,a_{k+j}R\neq 0$. Then (a) of 17.21 implies that $n_k^\perp\subset(n_k\,n_{k+j}\dots n_{k+1}\,n_k)^\perp$. Since $n_k\,a_{k+j}R\nsubseteq K$, this contradicts the maximality of $n_k^\perp$. For $j=0$, the fact that $n_k^\perp\subset(n_k^2)^\perp$ is contradicted. Therefore $n_k\,a_{k+j}=0$ $\forall\, j\geq 0$. Since $n_k\,a_{k-1}\neq 0$, then $^\perp A_{k-1}\subset {}^\perp A_k$, for every k, where $A_k=\{a_i\}_{i\geq k}$.

Next, assuming that R is a ring, if $a_{s+1}x_{s+1}=\sum_{i=1}^s a_i\,x_i$, with $x_i\in R$, and $1\leq i\leq s$, then multiply on the left by n_2 and get $n_2\,a_1\,x_1=0$. Now $n_2\,a_1^\perp=a_1^\perp$, and so $a_1\,x_1=0$. Then multiply by n_3 to get $a_2\,x_2=0$. By induction $a_{s+1}x_{s+1}=0$, and so $a_{s+1}R\cap\sum_{i=1}^s a_i R=0$. This proves the last assertion. $\quad\square$

17.23 Corollary (Levitzki [64], Herstein and Small [64, 66]). *If R is a semigroup, with zero element, and if R satisfies the a.c.c. on annihilator right and left ideals, then every nil multiplicative submonoid of R is nilpotent.* $\quad\square$

17.24 Corollary (Lanski [69]). *If R is a right Goldie ring, then every nil multiplicative submonoid of R is nilpotent.*

Proof. As defined in Chapter 9 (I, p. 396), a ring is right Goldie if it satisfies $(\mathrm{acc})^\perp$ and $(\mathrm{acc})\oplus$, so the proposition applies. $\quad\square$

*17.25 *Exercise* (Small). If R is the endomorphism ring of a Noetherian module M over a ring A, then every nil subring-1 of R is nilpotent.

17.26 (Fisher [72]). Same as 17.25 for M Artinian. Dual proof.

17.27 (Fisher and Small, in Fisher [72, p. 77, Theorem 2.1]). The indices of nilpotency of the nil subring-1 in 19.25 are bounded.

Kolchin's Theorem

A matrix A is **unipotent** if $A=1+B$, where B is nilpotent. (An equivalent formulation: the characteristic roots of A are all $=1$.) Kolchin's theorem states that any multiplicative semigroup S of unipotent matrices over a commutative field k

can be placed simultaneously in triangular form. It is tempting to derive Kolchin's theorem from 17.19; however, the B's do not form a multiplicative submonoid in general. (They do, obviously, when S consists of commuting matrices, but this is an unnecessary restriction. Moreover, there is a theorem which permits the diagonalization of commuting matrices over an algebraically closed field. See Jacobson [53, p. 134].)

Kolchin's proof, which we give, requires two theorems of Burnside, and devolves into two cases, k algebraically closed and k arbitrary, and, in fact, bears little resemblance to the proof of the theorem 17.19 on triangularizing nilpotent matrices. Conversely, Kolchin's theorem does not imply the latter, since if N is a multiplicative monoid of matrices, there is no reason to suppose that the corresponding class of unipotent matrices is a semigroup.

17.28   Burnside's Theorem. *Any irreducible semigroup S of linear transformations on a vector space V of dimension n over an algebraically closed field k contains n^2 linearly independent vectors.*

Proof. Let R be the subalgebra of $L = \text{End}_k V$ spanned by S. Then, V is a simple R-module, so (I, 13.10 A, p. 465) applies, so that $R = L$, so S has $n^2 = \dim_k L$ linearly independent vectors. $\square$

17.29   Corollary. *If t is the number of distinct traces of elements of S in 17.19 B, then S has at most t^{n^2} elements.*

Proof. On the algebra of all linear transformations on V, introduce the inner product $(A, B) = \text{Tr}(AB)$. (This is easily seen to be nonsingular.) Let $c_1, \ldots, c_t$ be the distinct traces that occur. Let A_i $(i = 1, \ldots, n^2)$ be n^2 linearly independent elements in S (Theorem 17.28). Each X in S satisfies equations $\text{Tr}(A_i X) = b_i$ where $b_1, \ldots, b_{n^2}$ are chosen from the c's.

These equations determine X uniquely, so there are at most t^{n^2} choices for X. $\square$

17.30   Theorem (Kolchin [48]). *Let S be a multiplicative semigroup of unipotent matrices. Then the elements of S can be put simultaneously into triangular form.*

Proof. Let n be the size of the matrices. We argue by induction on n. The case $n = 1$ is trivial.

Case I. The scalar field is algebraically closed. If S is irreducible, then by 17.29, S has only one element, and one matrix is always reducible (here $n > 1$). Hence S is reducible. Then by choosing a basis for the invariant subspace of S and extending it to a complete basis, all the elements of S will have matrices of the block form:

$$\begin{pmatrix} B & C \\ 0 & D \end{pmatrix}.$$

Now the sets S_L, of the upper left corners B, and S_R, of the lower right corners D, form multiplicative semigroups of unipotent matrices of dimension less than n.

One can then use the induction hypothesis to triangulate simultaneously these matrices, and all elements of S will then have been put in triangular form.

Case II. An arbitrary scalar field k. Form the algebraic closure of k and triangulate the elements of S simultaneously as matrices over the extension field. Then any product of n matrices $(T-1)$, where T is in S and 1 is the identity matrix, must be zero. Let r be the smallest integer such that the product of any r elements $(T-1)$ is zero. Then there exist elements $T_1, \ldots, T_{r-1}$ in S such that

$$(T_1 - 1)(T_2 - 1)\ldots(T_{r-1} - 1) \neq 0.$$

Find a vector x such that

$$x(T_1 - 1)\ldots(T_{r-1} - 1) = y \neq 0.$$

Then for any T in S, $y(T-1) = 0$, or $yT = y$. This shows that S is reducible. The argument can now proceed as in Case I.   □

17.31 *Exercise.* (a) The group T of nonsingular upper triangular matrices of degree n over the field k is a solvable group, and in fact, T is an extension by an abelian group of a nilpotent group. [Hint: Let $U = I + N$, consisting of all unipotent matrices, where I is the identity matrix, and N is the ideal of strictly upper triangular matrices. Then, $I + N^i$, $(1 \leq i \leq n)$ is a normal subgroup of U, and the commutator formula $(I + N^i, I + N^j) \subseteq I + N^{i+j}$ for the case of $i = 1$, and variable j, shows that U is a nilpotent group. Moreover, U is a normal subgroup of T, and T/U is abelian.]

*(b) (Shoda, Köthe [30], Levitzki [31]) In an Artinian ring R, any nil sub-ring-1 is nilpotent, and contained in a maximal nil subring-1. Moreover, any two maximal nil subrings are conjugate.

*(c) (Shoda, Köthe [30], Barnes, and Michler [66]) If R is right Noetherian, then the intersection of all of the maximal nilpotent subrings-1 is the maximal nilpotent ideal of R.

Exercises

1. Uniqueness of the basis number. Let R be a right Noetherian or right Artinian ring, and let M be a free module with a finite basis $x_1, \ldots, x_n$. (a) If $y_1, \ldots, y_m$ is any basis of M, then $m = n$, and there exists an automorphism f of M such that $f(x_i) = y_i$, $i = 1, \ldots, n$. (b) If $y_1, \ldots, y_m$ generates M, then $m \geq n$. Furthermore, $y_1, \ldots, y_m$ is a basis if and only if $m = n$.

2. A right R-module M is **divisible** provided that $Mr = M$ for each regular element $r \in R$, and **torsionfree** provided that $xr = 0$, with $x \in M$, $r \in R$, implies either $x = 0$, or r is a zero divisor of R. Now let R be a commutive ring. (a) In what sense are divisibility and torsion freeness dual concepts? (b) If M is divisible, and Noetherian, and if R is a (commutative) integral domain, then M is torsion free and M is a vector space over the quotient field K of R. (c) State and prove a dual to (b).

3. Let S be a ring, let N denote the set of nilpotent elements of S, and let S^* denote the group of units of S. Prove the equivalence of the following two statements: (a) $S = N \cup S^*$ (set union); (b) N is a (nil)ideal and S/N is a field.

4. Let S be any ring $\neq 0$. An element $x \in S$ is **right** (resp. **left**) **regular** provided that x is not a left (resp. right) divisor of zero of S, that is, when $xy = 0$ (resp. $yx = 0$) implies $y = 0$ $\forall y \in S$; x is **regular** provided that x is both right and left regular. Let S' denote the totality of right regular elements of S. Prove the equivalence of the following two statements, where N denotes the set of nilpotent elements of S. (a) $S = N \cup S'$ (set union); (b) N is a (nil) ideal of S, and S/N is an integral domain.

5. (a) If I_i is a right (or left) ideal of a ring S, which is nilpotent of index k_i, $i = 1, 2$, then $I_1 + I_2$ is nilpotent of index $\leq k_1 + k_2$. (b) If I is a nilpotent right ideal, then $I + SI$ is a nilpotent ideal of S containing I. (c) The sum of all nilpotent right ideals of S is a nil ideal that contains each nilpotent left ideal of S.

6. (Vasconcelos) If M is a finitely generated module over a commutative ring R, then any R-epimorphism of M is an automorphism.

7. (Camillo [75]) A module is **distributive**, if for any submodules A, B, C, we have $A \cap (B + C) = A \cap B + A \cap C$. Show that a module is distributive if and only if every quotient has at most one copy of every simple submodule in its socle.

8. (Jensen [66]) A commutative ring R is distributive as a module over itself iff for any two elements a and b

$$(a : b) + (b : a) = R$$

where $(a : b) = \{r \in R \mid a r \in b R\}$. Then R is said to be arithmetical. A local commutative ring is arithmetical iff the ideals are linearly ordered.

9. (Camillo [73]) If R is commutative, and the ideal generated by elements a and b is projective, then $(a : b) + (b : a) = R$.

10. (Camillo [73]) For a commutative ring R, the following conditions are equivalent: (a) R is semihereditary; (b) R is arithmetical, and for every $a \in R$, the ideal $aR + \mathrm{ann}_R a$ contains a regular element; (c) Every ideal generated by two elements is projective.

11. (Jensen-Camillo) Let R be a commutative ring, and let Q be its (classical) quotient ring. Then every ring between R and Q is semihereditary iff R is semihereditary (see Camillo [73]).

12. (Small [67]) A **right PP ring** R is one in which every principal right ideal is projective. A ring R is right semihereditary iff for every integer $n \geq 1$, the full matrix ring R_n is right PP.

13. (Small [67]) In a right PP ring with no infinite sets of orthogonal idempotents, every right or left annulet is generated by an idempotent, and hence R satisfies the a.c.c. and d.c.c. on right and left annulets. Moreover, the maximal nil ideal is a nil ideal containing every nil right or left ideal. Furthermore R is also left pp.

14. (Small [67]) If R is right semihereditary, and if no full $n \times n$ matrix ring over R contains an infinite set of orthogonal idempotents (e.g. if R is right or left Noetherian), then R is left semihereditary. Thus, a right Noetherian right hereditary ring is left semihereditary, and a left Noetherian right semihereditary ring is left hereditary.

15*. (Chatters [70]) A (right and left) Noetherian hereditary ring is a finite product of rings each of which is either a prime ring, or an Artinian ring.

16. (Levy [63 a]) A right Noetherian hereditary semiprime ring is a finite product of prime rings (see Exercise 15).

17. (Fuller [70]) A module U generates a module V in case V is the trace of U in V (I, p. 143), that is, iff V is an epic image of a direct sum of copies of U. Dually, U cogenerates V if $\bigcap_{f: V \to U} \ker f = 0$, that is, iff V can be embedded in a direct product of copies of U. Consider the commutative diagram:

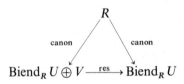

Then, the restriction map (denoted res) is an isomorphism iff U both generates and cogenerates V. As a corollary deduce that any generator U of mod-R is balanced (theorem of Morita; see I, 7.1, p. 326, and also, 4.1.3, p. 190).

18. (Chase [61]) Let S be a commutative von Neumann regular ring which is not semisimple (e.g., let S be a finite product of fields). Then, there is an ideal I which is not a direct summand of S, and the idealization $A = \begin{pmatrix} S' & S' \\ 0 & S \end{pmatrix}$ of the (S', S)-bimodule is a left semihereditary ring which is not right semihereditary. (Consult (I, pp. 334–5) for the idealization of a module.)

Notes for Chapter 17

The purely algebraic theorem of Kolchin [48 b], given in 17.30, was preceded by the theorem of Lie-Kolchin which states that any connected solvable algebraic matrix group over an algebraically closed field can be placed simultaneously into triangular form (Kolchin [48 a]). This has been proved by Borel [69, p. 243] in a way which yields as an immediate corollary Mal'cev's theorem [49, 51] which states that if M is any solvable subgroup of the group $GL(n, k)$ [the general linear group of nonsingular $n \times n$ matrices over a field k] over an algebraically closed field k, then M is an extension by a finite group of a group which can be placed simultaneously into triangular form. Moreover, a theorem of Zassenhaus [38] states that the derived series of any solvable subgroup of $GL(n, F)$, for any field F, is bounded by a number $z(n)$ independent of F.

A theorem of Tits [72] on a conjecture of Bass and Serre states that any finitely generated subgroup G of $GL(n, F)$, for any field F, and $n > 1$, contains either a free group of rank >1, or else is an extension by a finite group of a solvable group. (Over characteristic 0, the subgroup G is not required to be finitely generated.) A number of theorems on the structure of solvable subgroups of $GL(n, F)$ of Zassenhaus, e.g. a maximal irreducible subgroup has a unique maximal Abelian normal subgroup when F is infinite, Mal'cev and others, is presented in Suprunenko [63].

Theorem 17.19, the placing nil submonoids of K_n simultaneously into triangular form is attributed to Levitzki by Jacobson [43], but the theorem can be deduced from a theorem of Köthe [30], the discussion of which is postponed to Notes for Chapter 18. (However, see Exercise 17.31 (b).)

A theorem generalizing both Kolchin's theorem and the commutative field case of 17.19 is proved by Kaplansky [69a, p. 137].

There is by now a very long list of theorems of the

$$\text{nil} \ \Rightarrow \ \text{nilpotent}$$

variety. Besides the ones already mentioned, and other ones contained in this chapter (e.g. Köthe-Levitzki 17.19, Levitzki-Fitting 17.20, Shock 17.22, Levitzki-Herstein-Small 17.23, Lanski 17.24, Small 17.25, Fisher 17.26), there are the theorems of Kaplansky [46] (and Levitzki [46]) solving Kurosch's problem for algebraic algebras of bounded degree (nil algebras of bounded index of nilpotency): thus every finite subset generates a finite dimensional (resp. nilpotent) subalgebra. In particular, if a finitely generated ring satisfies the identity $x^n = 0$, for a fixed integer $n \geq 1$, then the ring is nilpotent. A related theorem of Nagata-Higman states that in any associative algebra A over a field k of characteristic p that any subalgebra B satisfying the identity $x^n = 0$ for an integer $n < p$ when $p \neq 0$ is nilpotent of index $\leq 2^n - 1$. A short proof of this by P.J. Higgins appeared in Jacobson [64, p. 274]. (The theorem of Nagata [52] is for nilpotency of B over characteristic 0.) Jacobson [64, p. 260] also cites the example of Golod of a finitely generated nonnilpotent nil algebra B over a field F. Thus, B is not finite dimensional (the main point of the example). By a theorem of Goldman and Krull cited by Jacobson, p. 23, any such example would have to be noncommutative: that is, nil $\Rightarrow$ nilpotency for finitely generated commutative nil algebras. (A theorem of Golod-Shafarevitch enables one to find over any countable field an infinite dimensional nil algebra generated by three elements. (For an exposition, see Herstein [68, p. 192].))

Lenagen [73] proved nil $\Rightarrow$ nilpotency in rings with Krull dimension. This answered a question raised by Robson and Gordon, and also answered by them independently.

In Chapter Exercises, we have cited a number of theorems of Camillo [73, 75], Chatters [70], Levy [66a], Jensen [66c], and Small [67] on semihereditary and hereditary rings.

A theorem of Fuller [70] on double centralizers (= biendomorphism rings) of modules, which we also have put down as an exercise (# 17), ought to have been included in Volume I not later than about 7.1, p. 326.

Notes on local rings, and other aspects of Fitting's lemma may be found in Notes for Chapter 18.

General References

Birkhoff, G.: Lattice Theory, Colloquium Publication, Vol. 25 (revised). Amer. Math. Soc., Providence 1948, 1967.

Dedekind, R.: Über Zerlegungen von Zahlen durch ihre größten gemeinsamen Teiler. Festschrift TH Braunschweig 1897. *Also:* Ges. Werke, Vol. 2 (1900) 103–148.

— Über die von drei Moduln erzeugte Dualgruppe. Math. Ann. **53**, 371–403 (1900). *Also:* Ges. Werke, Vol. 2 (1900) 236–271.

Fitting, H.: Die Theorie der Automorphismenringe Abelscher Gruppen und ihr Analogon bei nicht kommutativen Gruppen. Math. Ann. **107**, 514–542 (1933).

— Über die direkten Produktzerlegungen einer Gruppe in direkt unzerlegbaren Faktoren. Math. Z. **39**, 19–41 (1935).

Hölder, O.: Zurückführung einer beliebigen algebraischen Gleichung auf eine Kette von Gleichungen. Math. Ann. **34**, 26–56 (1889).

Jordan, C.: Recherches sur les substitutions. J. Math. Pures Appl. (2) **17**, 351–363 (1872).

Köthe, G.: Über maximale nilpotente Unterringe und Nilringe. Math. Ann. **103**, 359–63 (1938).

Kronecker, L.: Auseinandersetzung einiger Eigenschaften der Klassenanzahl idealer complexer Zahlen. Monatsber. Berl. Akad. **1870**, 881–889. *Also:* Werke, Leipzig: Teubner 1895, pp. 273–282.

Krull, W.: Über verallgemeinerte endliche Abelsche Gruppen. Math. Z. **23**, 161–196 (1925).

Kurosch, A.: Durchschnittsdarstellungen mit irreduziblen Komponenten in Ringen und sogenannten Dualgruppen. Mat. Sbornik **42**, 613–616 (1935).

— Isomorphisms of direct decompositions. Izv. Akad. Nauk S.S.S.R. **7**, 185–199 (1943). (English Summary pp. 199–202).

Levitzki, J.: Über nilpotente Unterringe. Math. Ann. **105**, 620–27 (1931).

Noether, E.: Idealtheorie in Ringbereichen. Math. Ann. **83**, 24–66 (1921).

Ore, O.: Direct decompositions. Duke Math. J. **2**, 581–596 (1936).

— On the foundations of abstract algebra, I, II. Ann. Math. **36**, 406–437 (1935); **37**, 265–292 (1936).

Remak, R.: Über die Zerlegung der endlichen Gruppen in direkte unzerlegbare Faktoren. J. reine angew. Math. **139**, 293–308 (1911).

Schmidt, O.: Über unendliche Gruppen mit endlicher Kette. Math. Z. **29**, 34–41 (1928).

Schreier, O.: Über den Jordan-Hölderschen Satz. Abh. Math. Sem. Univ. Hamburg **6**, 300–302 (1928).

Wedderburn, J. H. M.: On the direct product in the theory of finite groups. Ann. Math. **10**, 173–176 (1909).

Borel [68], Camillo [73, 75], Chase [61], Chatters [70], Fisher [72], Fuller [70], Goldman [51], Goldie and Small [73], Herstein-Small [64, 66], Higman [56], Jacobson [43, 64], Jensen [66c], Kaplansky [46, 69a], Kolchin [48], Köthe [30], Krull [51], Lanski [69], Lenagen [73], Levitzki [46, 64], Levy [63a], Mal'cev [49, 51], Michler [66], Nagata [52], Procesi and Small [68], Shock [71b], Small [67], Suprunenko [63], Tits [72], Zassenhaus [38].

Chapter 18. Semilocal Rings and the Jacobson Radical

The main topics of this chapter are (1) the Jacobson radical of a ring and of a module 18.0; (2) the Perlis-Jacobson characterization 18.6 of the radical of a ring; (3) local rings 18.10; (4) semiprimary rings 18.12; (5) the theorem of Hopkins and Levitzki 18.13; (6) the Krull-Schmidt or Unique Decomposition Theorem 18.18; (7) the basic module and ring 18.21–23; (8) the Chinese remainder theorem 18.30–32 and primary decomposable rings 18.36–37; and (9) the characterization 18.47 of rings with semilocal right quorings. (As defined by (I, p. 4.10), R is semilocal if $R/\operatorname{rad} R$ is semisimple. Also see 18.10 A.)

The Jacobson Radical of a Module

One of the basic techniques in determining the structure of a module M is to examine certain pieces of it. For example, the socle of a module M, defined in Chapter 8, is the sum of the simple submodules of M. The structure of socle M is transparent since socle M is a semisimple module. Dual to notion of a simple submodule is that of an epimorphism $M \to V$, where V is simple (supposing that such exists). This dual notion of looking at the simple homomorphic images of M is a fruitful one in the structure theory of rings and modules. With this sketchy motivation, we begin to systemize the study of the simple homomorphic images of M by making the following definition.

18.0 Definition and Proposition

18.0.1 *The **radical** of a module M, written $\operatorname{rad} M$, is defined to be the submodule described by the following equivalent statements:*

(a) *The intersection of the kernels of all epimorphisms $f: M \to V$ such that $V = \operatorname{im} f$ is a simple module.*

(b) *The intersection of all of the maximal submodules of M.*

If M has no maximal submodules then we define $\operatorname{rad} M = M$. For any module M, $\operatorname{rad}(M/\operatorname{rad} M) = 0$.

18.0.2 *The following conditions on a module $M \neq 0$ are equivalent:*

(a) $\operatorname{rad} M = 0$.

(b) *If $x \in M$, and $x \neq 0$, there exists a simple module V and a homomorphism $f: M \to V$ such that $f(x) \neq 0$.*

(c) *Let* $\{M_\alpha\}_{\alpha \in I}$ *denote the set of maximal submodules of* M. *Then the (direct) product mapping*

$$M \to \prod_{\alpha \in I} (M/M_\alpha)$$

is a monomorphism.

(d) *M can be embedded in a product of simple modules.*

Proof. (a) $\Leftrightarrow$ (b) follows from the correspondence theorem for modules (I, p. 70), since a submodule M' is maximal if and only if M' is the kernel of a simple epic $M \to M/M'$.

(b) $\Rightarrow$ (c). Let $h: M \to \prod_\alpha M/M_\alpha$ be the direct product map. If $x \in M$, and $x \neq 0$, then (b) implies that $x \notin M_\alpha$ for some α, whence $h(x)(\alpha) = x + M_\alpha \neq 0$, and $h(x) \neq 0$. Thus h is monic.

(c) $\Rightarrow$ (d) is trivial.

(d) $\Rightarrow$ (b). If we say $M \subseteq \prod_{\beta \in J} W_\beta$, where W_β is simple $\forall \beta \in J$, and if $x \in M$ is nonzero, then there exists $\beta \in J$ such that $p_\beta(x) \neq 0$, where $p_\beta: \prod_{\beta \in J} W_\beta \to W_\beta$ is the projection. If $f = p_\beta | M$, then $f: M \to W_\beta$ and $f(x) \neq 0$. $\square$

The **radical** of a ring R is defined to be $\mathrm{rad}(R_R)$, and is denoted rad R. Since R always has maximal right ideals (by Zorn's lemma), rad $R \neq R$. The **left radical** of R is defined to be $\mathrm{rad}(_R R)$. One of the main results to be established in this section is that rad R coincides with the left radical. Since the (left) radical is a right (left) ideal, being the intersection of right (left) ideals, this result will imply that rad R is an ideal of R. Actually, we begin by proving this latter result first 18.1.

Another important concept for the structure theory of rings is that of a primitive ideal. An ideal I is said to be (right) **primitive** provided that I is the annihilator of a simple right R-module, or equivalently, the factor ring R/I has a faithful simple right R-module. Jacobson's theorem 18.2 states that rad R is the intersection of all right primitive ideals. Together with the theorem of Jacobson (18.5) stating that the left radical is equal to the right radical, this implies that the radical is also the intersection of the left primitive ideals.

A remark on terminology: a ring R is (right) primitive if 0 is a (right) primitive ideal, that is, iff R has a faithful simple right module. Thus there is a duality between primitive rings and primitive ideals: an ideal I is (right) primitive iff R/I is a (right) primitive ring. This leads to some confusion, especially when rings without identity elements are introduced, for in this case the ideal I can be considered as a ring, and, as a ring-1 may be primitive. (This is the case for any nonzero ideal of a primitive ring R.)

Examples

Any simple ring is primitive since any nonzero module is faithful. Thus, every maximal ideal is a primitive ideal. Moreover, any full right linear ring $L = \mathrm{End}\, V_D$ is (right and) left primitive since, as is shown in 19.23, the right vector space V over D is a canonical simple left L-module, and *a fortiori* faithful. It can be shown that V embeds in L (in general in many ways), that there is an idempotent e in L such that $V \approx Le$, and that $S = LeL$ is the least nonzero ideal. (In fact, S is the set of all l.t.'s of finite rank. See p. 93.) Well, V is a simple faithful left S-module, so S is a primitive ring-1. If dim V_D is countably infinite, then S is the only nontrivial ideal of L, hence is a maximal ideal, so that S is a primitive ideal. (Actually, S is always a primitive ideal, regardless of the dimension of V, as is every ideal of L.)

It may be verified that $W = eL$ is a faithful simple right L module, so that L is a right primitive ring as well.

Once the appropriate theorems are proved the discussion above becomes much clearer, namely: (1) L is a von Neumann regular ring in the sense of (I, p. 434) and 19.24, that is, every finitely generated onesided ideal is generated by an idempotent (this is proved by Theorem 19.28); (2) The radical contains no nonzero idempotents 18.9.3, so rad $L/I = 0$ for any ideal I of L; (3) The ideals of L are well ordered by inclusion (19.64).

Thus, using (3), and the fact that I is the intersection of the primitive ideals containing it, one concludes that every ideal of L is left and right primitive.

This way of looking at this fact is not a complete waste of time since the same proof shows that any ideal I in any regular ring with a well ordered set of ideals is right and left primitive. We return to these ideas again in Chapter 19, especially in Goodearl's theorems (19.62–65) on the linear ordering of the set of ideals containing any prime ideal in a right selfinjective regular ring.

18.1 Proposition. *If R is a ring, and M a module, then* rad M *is a fully invariant submodule of M, and* rad R *is an ideal of R.*

Proof. Let $T = \operatorname{End}_R(M)$. Now rad $M = \bigcap \ker f$, where f ranges over all maps $f: M \to C$, and C is a simple module. If $g \in T$, then $f \circ g: M \to C$, where C is simple, so $fg(\operatorname{rad} M) = 0$; that is, $g(\operatorname{rad} M) \subseteq \bigcap \ker f = \operatorname{rad} M$, proving that rad M is fully invariant. If $M = R$, if $a \in R$, and if a_s is the left homothetic, then

$$\operatorname{rad} R \supseteq a_s(\operatorname{rad} R) = a \cdot \operatorname{rad} R$$

so rad $R \supseteq R \cdot \operatorname{rad} R$, proving that rad R is an ideal. $\square$

We now come to a mathematically and historically important problem. If x is an element of a ring R, give a criterion for x to lie in the radical of R. One criterion to be given is that $Vx = 0$ for every simple R-module V. A theorem of Perlis and Jacobson (18.6) is another criterion.

18.2 Theorem (Jacobson). *Let R be a ring. Then* rad R *is the intersection of the (right) primitive ideals, that is,*

(1) $$\operatorname{rad} R = \bigcap_{\text{simple } V \in \operatorname{mod-}R} \operatorname{ann}_R V.$$

Proof. Let Q denote the intersection on the right. Since $\operatorname{ann}_R V$ is an ideal of R for every V, Q is an ideal of R, and $Q = RQ$. If M is a maximal right ideal, then $W = R/M$ is a simple R-module. Thus, $WQ = 0$, whence $Q = R \cdot Q \subseteq M$. This proves $Q \subseteq \operatorname{rad} R$.

Conversely, let $x \in R$, and $x \notin Q$, that is, $Vx \neq 0$ for some simple R-module V. Since V is simple, $VxR = V$, so $VxRx = Vx \neq 0$. Pick $v \in V$, $a \in R$ so that $vxax \neq 0$.

Since $vxa \neq 0$, simplicity of V implies $vxaR = V$. Thus we have an epimorphism

$$\begin{cases} f \colon R \to V \\ f \colon r \mapsto vxar \end{cases} \qquad \forall\, r \in R.$$

Since $f(x) = vxax \neq 0$, this proves that $x \in \operatorname{rad} R$. Thus, $\operatorname{rad} R \subseteq Q$, so $\operatorname{rad} R = Q$ as asserted. $\quad\square$

This theorem is only a partial characterization of $\operatorname{rad} R$. We want to classify the elements of $\operatorname{rad} R$ purely ring theoretically, that is, in terms of the ring operations. This is done in 18.4.

Superfluous Submodules

A submodule S of a module M_R is **superfluous** provided that $M = S + K$ for some submodule K only if $K = M$. The symbol $M \bigcirc S$ denotes the statement that S is a superfluous submodule. An element $x \in M$ is a **nongenerator** of M provided that it generates a superfluous submodule. Obviously: (1) Any submodule of a superfluous submodule of M is a superfluous submodule of M. (2) A sum of finitely many superfluous submodules is superfluous.

The sum of all the superfluous submodules of M is a submodule denoted by **superfl** M. By (2), superfl M consists of the elements of M that are nongenerators. In general, superfl M is not a superfluous submodule; for example, every element of the $\mathbb{Z}$-module $\mathbb{Q}$ is a nongenerator (proof?), so superfl $\mathbb{Q} = \mathbb{Q}$.

Examples. 1. If P is a submodule (ideal) of $\mathbb{Z}$, then $P = x\mathbb{Z}$, with $x \in \mathbb{Z}$. If $P \neq 0$, and if $y \in \mathbb{Z}$, $y \neq 1$, is such that g.c.d. $(x, y) = 1$, then $\mathbb{Z} = x\mathbb{Z} + y\mathbb{Z}$, and $y\mathbb{Z} \neq \mathbb{Z}$. Thus, superfl $\mathbb{Z} = 0$.

2. If $M = \mathbb{Z}_{p^\infty}$, then every proper subgroup is a finite subgroup (see Vol. I, p. 55). Hence the sum of two proper subgroups is proper, so

$$\text{superfl } \mathbb{Z}_{p^\infty} = \mathbb{Z}_{p^\infty}.$$

Similarly, superfl $\mathbb{Q}/\mathbb{Z} = \mathbb{Q}/\mathbb{Z}$, and superfl $\mathbb{Q} = \mathbb{Q}$.

3. If M is a semisimple right R-module, then every submodule is a summand, so superfl $M = 0$.

4. A product of simple modules need not be semisimple. For example, if F is any field an infinite product $R = F^I$ is a nonsemisimple ring which is a product of (nonisomorphic) simple right ideals $\{F_i\}_{i \in I}$, where F_i is the image of the canonical injection $u_i \colon F \to F^I$ $\forall\, i \in I$. (Hint: isomorphic modules have the same annihilating ideal.)

18.3 Proposition. *Let M be a right R-module. Then*

(a) $\operatorname{rad} M = \text{superfl } M \supseteq M(\operatorname{rad} R)$.

(b) *An object M of* mod-R *is called a* **B-object** *(after Bass [60]) provided that the equivalent conditions hold:*

(1) *Every submodule $N \neq M$ is contained in a maximal submodule.*

(2) *For every submodule N of M,*

$$\mathrm{rad}\,(M/N) \neq M/N.$$

When this is true, then rad M *is a superfluous submodule of* M, *and* $M(\mathrm{rad}\,R) = M$ *only if* $M = 0$. *If every object of* mod-R *is a B-object, then* R *is called a right* **B-ring**. *A semilocal ring with nilpotent radical is a B-ring (cf. 22.8 and 18.12).*

(c) *Any finitely generated module over any ring is a B-object.*

(d) *If* $R/\mathrm{rad}\,R$ *is a right V-ring 7.32 (e.g. if* R *is semilocal), or if* $M/M(\mathrm{rad}\,R)$ *is semisimple, then* rad $M = M(\mathrm{rad}\,R)$.

(e) $R/\mathrm{rad}\,R$ *is a right V-ring if and only if* rad $M = M(\mathrm{rad}\,R)$ *for every object* M *of* mod-R. *In this case, if* M *is any B-object of* mod-R, *then* rad $M = M(\mathrm{rad}\,R)$ *is a superfluous submodule.*

This holds, in particular, for any finitely generated module M *over any ring* R *such that* $R/\mathrm{rad}\,R$ *is a right V-ring (e.g. over any semilocal ring* R*).*

Proof. We first show that superfl $M \supseteq$ rad M. For if $y \in M$ and $y \notin$ superfl M, then $M = B + yR$, for some submodule B. Then $yR/yR \cap B \approx M/B$ is cyclic with generator $[y + yR \cap B]$. Now any cyclic module $C \neq 0$ has radical $\neq C$, so there is an epic $f: M \rightarrow V$, where V is simple, such that $f(y) \neq 0$. Thus, $y \notin$ rad M. The reverse inclusion is trivial, for if $M \supset S$, and if M' is a maximal submodule of M, then $S + M' \neq M$, consequently, $S \subseteq M'$, and $S \subseteq$ rad M.

Let $J = $ rad R. Since J is an ideal of R, we have $R \cdot J = J$. To complete the proof of (a), we must show that rad $M \supseteq MJ$. This is trivial if rad $M = M$. If M' is then a maximal submodule, the factor module $\overline{M} = M/M'$ is simple, hence cyclic, so there is an epimorphism $\varphi: R \rightarrow \overline{N}$. Since $Q = \ker \varphi$ is a maximal right ideal of R, $J = $ rad $R = R \cdot J \subseteq Q$, and therefore $(R/Q) J = 0$. Applying φ, we see that $\overline{M}J = 0$, which can happen only if $M' \supseteq MJ$. This proves that rad $M \supseteq MJ$.

(b) If K is a proper submodule of M, then (1) implies that K is contained in a maximal submodule M'. But $M' \supseteq$ rad M, hence $M' \supseteq K +$ rad M. This proves that rad M is a superfluous submodule of M. The equivalence of (1) and (2) is trivial. Moreover, (2) implies that rad $M = M$ if and only if $M = 0$. Thus, $M = M(\mathrm{rad}\,R)$ only if $M = 0$.

(c) The set of proper submodules of a nonzero finitely generated module M is inductive, and hence every factor module M/N contains a maximal submodule.

(d) In either case, $M(\mathrm{rad}\,R)$ is an intersection of maximal submodules, and hence contains rad M, reversing the inclusion (a).

(e) By 7.32, R is a V-ring if and only if rad $M = 0$ for every object M of mod-R. Thus, R/J is a right V-ring, if and only if rad $M/MJ = 0$, that is, rad $M \subseteq MJ$, for every M. Clearly, this is equivalent to asserting that rad $M = MJ$, for every M. If M is a B-object, then rad $M = MJ$ is superfluous, so that (e) holds.

To complete the proof of (b), if R is semilocal with nilpotent radical J, then rad $M = MJ = M$ only if $M = MJ^n$ for all n, whence $M = 0$. Thus, R is a B-ring. □

A multiplicative monoid M is said to be **left vanishing** provided that every sequence of products of elements of M of the form

(S) $\{a_n a_{n-1} \cdots a_1\}_{n \in \omega}$

terminates in zeros, that is, $a_n a_{n-1} \ldots a_2 a_1 = 0$ for some $n = n(S)$ depending on the sequence S. Thus, every nilpotent ideal of a ring is left and right vanishing, and $n(S)$ is bounded by the index of nilpotency[1]. The radical of any right B-ring is left vanishing 22.7.

Any right V-ring is a B-ring, since rad $M = 0$ for every right R-module. The ring $R = k[x, D]$ of differential polynomials over a universal field k is an example which is a principal right and left ideal domain 7.42.

18.4   Theorem. *Let R be a ring, and let $J = $ rad R. For any right ideal I of R, the following conditions are equivalent:*

(a) $I \subseteq $ rad $R = J$.

(b) $1 + I = \{1 + x \mid x \in I\}$ *consists of units of R.*

(c)[2] *If M is any nonzero finitely generated module, then $MI \neq M$.*

(d) *If M is finitely generated, then MI is a superfluous submodule of M.*

Proof. (a) $\Rightarrow$ (b). Superfl $R = $ rad R by (b) of 18.3. Thus, if $x \in I \subseteq J = $ rad R, then x lies in some superfluous right ideal, so xR is superfluous. Since $R = xR + (1 + x)R$, this implies that $R = (1 + x)R$; that is, $1 = (1 + x)y$ for some $y \in R$. Therefore, if $x \in I$, then $1 + x$ has a right inverse in R. Since $-xy \in I$, then $1 - xy = y$ has a right inverse w in R. Since $1 + x$ is a left inverse of y, then y is a unit, and $y^{-1} = 1 + x = w$. Hence $1 + x$ is a unit.

(b) $\Rightarrow$ (a). Let $x \in I$, and assume that $R = xR + K$ for some right ideal K. Then $1 = xa + k$, with $a \in R$, $k \in K$, whence $k = 1 - xa \in 1 + I$. Then k is a unit in R, proving that $K = R$. This shows that x is a nongenerator of R_R, and $x \in $ superfl $R = $ rad R, by (c) of 18.3. Hence $I \subseteq $ rad R.

(a) $\Rightarrow$ (c). If M is finitely generated, and if $MI = M$, then $MJ = M$, so $M = 0$ by (c) of 18.3.

(c) $\Rightarrow$ (d). Let K be a submodule such that $M = K + MI$, and set $N = M/K$. Then N is finitely generated, and $NI = N$, so (c) implies that $N = 0$, that is, that $M = K$. Hence MI is a superfluous submodule.

(d) $\Rightarrow$ (a). Take $M = R$. Then $RI \supseteq I$, and RI is a superfluous right ideal of R, so $I \subseteq $ rad R by (c) of 18.3. $\square$

The **left radical** of R is defined to be the intersection of the maximal left ideals.

18.5   Proposition (Jacobson [45]). *The left radical of a ring R is the radical of R. Moreover, for any ring R,*

$$\mathrm{rad}(R/\mathrm{rad}\, R) = 0.$$

Proof. By 18.4, $1 + $ rad $R \subseteq $ units R. Since rad R is an ideal, hence a left ideal, by the right-left symmetry of the theorem, rad R is contained in and contains the left radical of R. Also, 18.1 yields rad$(R/\mathrm{rad}\, R) = 0$. $\square$

---

[1] Compare right essentially nilpotent ideals 19.13 C.

[2] (c) is sometimes referred to as Nakayama's lemma. Scholarly work by Nagata [62, p. 213] shows that this is a misnomer. Like most misnomers, this is as popular as the *Songs of David*! Nakayama could not remember whether Azumaya or himself first discovered it, and then paradoxically suggested the names Azumaya-Krull for commutative R, and Azumaya-Jacobson, for non-commutative R.

18.6 **Proposition and Definition** (Perlis [42] and Jacobson [45]). *An element* x *of a ring* R *is* **quasiregular** *provided the equivalent requirements* (a) *and* (b) *hold:*

(a) $1 + x$ *is a unit.*

(b) *There exist* $x' \in R$ *such that* $x + x' + xx' = 0$ *and* $x + x' + x'x = 0$. *Every nilpotent element* x *of index* n *is quasiregular and* $(1 + x)^{-1} = \sum_{i=0}^{n-1} (-1)^i x^i$. *A onesided ideal is* **quasiregular** *provided that it consists of quasiregular elements. The radical of a ring* R *is a quasiregular ideal containing every quasiregular, and every nil, right or left ideal. Thus,* $R/\mathrm{rad}\, R$ *is a semiprime ring containing no nonzero nil or quasiregular onesided ideals.*

Proof. The equivalence of (a) and (b) follows with $x' = (1 + x)^{-1} - 1$ (called the **quasinverse** of x). If $x^n = 0$, then $x' = -x + x^2 + \cdots + (-1)^{n-1} x^{n-1}$ satisfies (b). The rest follows from 18.4. The last part follows from $\mathrm{rad}\,(R/\mathrm{rad}\, R) = 0$. ☐

18.7 **Corollary.** *Let* R *be a ring with radical* J, *and let* $\bar{x}$ *denote the image of any* $x \in R$ *under the canonical map* $R \to \bar{R} = R/J$. *Then* x *is a unit in* R *if and only if* $\bar{x}$ *is a unit in* $\bar{R}$.

Proof. One way is trivial. Conversely, suppose $\bar{x}$ is a unit in $\bar{R}$. Then $\bar{x}\,\bar{y} = \bar{y}\,\bar{x} = 1$ for some $y \in R$, and therefore $xy = 1 + t_1$, and $yx = 1 + t_2$, with $t_1, t_2 \in J$. Then $u_i = 1 + t_i$, $i = 1, 2$, is a unit in R, so $x y u_1^{-1} = u_2^{-1} y x = 1$. Thus, x has a right and left inverse that (in associative systems) coincide, so x is a unit in R.

18.8 **Corollary.** *If* R *is right Artinian, then* $\mathrm{rad}\, R$ *is a nilpotent ideal that contains each nil right or left ideal of* R.

Proof. $J = \mathrm{rad}\, R$ contains each nil right or left ideal by a preceding corollary. Since R is right Artinian, $J^{n+1} = J^n$ for some n. If $J^n \neq 0$, then the set P of those nonzero right ideals K for which $KJ = K$ is nonempty. Hence, there exists a right ideal Q that is minimal in P. Since $QJ^n = Q \neq 0$, then $q J^n \neq 0$ for some $q \in Q$. Since $(q J^n)\, J = q J^n$, the minimality of Q implies $q J^n = Q$. Then $-q = q u$ for some $u \in J^n$. Since $u \in \mathrm{rad}\, R$, $1 + u$ is a unit, so $q(1 + u) = 0$ implies $q = 0$. This contradiction forces us to the conclusion that $J^n = 0$, so J is nilpotent as stated. ☐

18.9 *Exercises and Examples*

18.9.1 Let $\{e_i\}_{i=1}^n$ be finitely many orthogonal sum-1 idempotents in a ring R. Define $J(A) = \mathrm{rad}\, A$, for any ring A, and put

$$H(e_i\, R\, e_j) = \{a \in e_i\, R\, e_j \,|\, a e_j\, R\, e_i \subseteq J(e_i\, R\, e_i)\}.$$

Then:

$$\mathrm{rad}\, R = J(R) = \sum_{i,j=1}^n H(e_i\, R\, e_j)$$

and

$$\mathrm{rad}\,(e_i\, R\, e_i) = J(e_i\, R\, e_i) = H(e_i\, R\, e_i) = J(R) \cap e_i\, R\, e_i = e_i\, J(R)\, e_i,$$

$$i, j = 1, \ldots, n.$$

18.9.2 In the notation of Exercise 18.9.1, (a) $\mathrm{rad}\, R$ is left vanishing (resp. nilpotent of index $\leq k$) if and only if $\mathrm{rad}\,(e_i\, R\, e_i)$ is left vanishing (resp. nilpotent of index $\leq k$) for $i = 1, \ldots, n$.

Moreover, (b) R is semisimple if and only if $\operatorname{rad} R = 0$, and $\exists \{e_i\}_{i=1}^n$ such that $e_i R e_i$ is a field, $i = 1, \ldots, n$. Conclude that if R is any ring such that $e_i R e_i / e_i J(R) e_i$ is a field, $i = 1, \ldots, n$, then R is semilocal (in the sense of 18.10 A).

18.9.3   The radical of a ring contains no nonzero idempotent elements. Thus, any full linear ring, any boolean ring, and any algebraic algebra without nilpotent elements over a field have zero Jacobson radicals.

18.9.4   Let F be a field, and let $R = F\langle x \rangle$ be the ring of all formal power series over F in the indeterminate x. If $a = x b \in x R$, $b \in R$, then $u = a + a^2 + \cdots + a^n + \cdots$ is an element of R (prove this!) and $(1-a)(1+u) = (1+u)(1-a) = 1$. Since $1 + xR$ consists of units, by the theorem the (right) ideal $I = xR$ is contained in $\operatorname{rad} R$. But I is a maximal right ideal, since the factor ring R/I is a field isomorphic to F. Thus, $I = \operatorname{rad} R$.

18.9.5   We may define the ring $F\langle x, y \rangle$ of formal power series inductively as $A\langle y \rangle$, where $A = F\langle x \rangle$. The elements of $F\langle x, y \rangle$ are power series

$$t = \sum_{i,j=1}^{\infty} a_{ij} x^i y^j$$

and t is a unit in $F\langle x, y \rangle$ if and only if $a_{00} \neq 0$. Thus, the radical of $R = F\langle x, y \rangle$ contains the ideal $J = xR + yR$. Since R/J is a field $\approx F$, we conclude that $\operatorname{rad} R = xR + yR$.

18.9.6   If M is an indecomposable quasiinjective module then the radical of $\operatorname{End} M_R$

$$\operatorname{rad} \operatorname{End} M_R = \{f \in \operatorname{End} M_R \mid \ker f \neq 0\}$$

(see the proof of 18.11 following).

18.9.7   The ring p-adic$(\mathbb{Z})$ of p-**adic numbers** may be defined as the set S of all sequences

$$a = a_0 + a_1 p + a_2 p^2 + \cdots + a_n p^n + \cdots$$

with integers

$$0 \leq a_n < p, \qquad\qquad\qquad \forall n \in \omega.$$

If also

$$b = b_0 + b_1 p + \cdots + b_m p^m + \cdots$$

then we define addition and multiplication by the formulas:

$$a + b = \sum_{i \in \omega} q_i p^i \quad \text{and} \quad ab = \sum_{i \in \omega} r_i p^i$$

where the coefficients q_i and r_i $\forall i \in \omega$ are defined inductively: by the division algorithm in $\mathbb{Z}$, there exist: unique integers $q_0 \leq p$ and $k_0 \leq p$ such that $a_0 + b_0 = q_0 + k_0 p$; and unique integers q_1, k_1 such that $a_1 + b_1 + k_0 = q_1 + k_1 p$; and inductively, integers q_i and k_i $\forall i \in \omega$, such that $q_n = a_n + b_n + k_{n-1} - k_n p$ is the desired coefficient of p^n.

Similarly, there exist integers r_i and m_i $\forall$ $i \in \omega$, each $\leq p$, such that $a_0 b_0 = r_0 + m_0 p$, and inductive defined so that

$$r_n = a_0 b_n + a_1 b_{n-1} + \cdots + a_n b_0 + m_{n-1} - m_n p, \qquad \forall n \in \omega.$$

is the desired coefficient of p^n in the definition of $a\,b$.

We now prove that

$$J = \mathrm{rad}\,(p\text{-adic}(\mathbb{Z})) = \{a \mid a_0 = 0\}.$$

Thus, $a \in J \Leftrightarrow a = a_1 p + a_2 p + \cdots$. The proof may be carried out computationally, as in the proof of 18.9.4, or using the characterization of the radical stated in Example 18.9.6. We do the latter:

There is a canonical isomorphism

$$\begin{cases} p\text{-adic}(\mathbb{Z}) \rightarrow \mathrm{End}\,\mathbb{Z}_{p^\infty} \\ \quad a \mapsto \bar{a} \end{cases}$$

where $\bar{a}\,x = \sum_{i \in \omega} a_i p^i x$ is defined $\forall$ $x \in \mathbb{Z}_{p^\infty}$ since there exists $n = n(x)$ such that $p^n x = 0$, and then $\bar{a}\,x = \sum_{i=1}^{n} a_i p^i x$. If $a_n \neq 0$, then since there exists an element y of order $> p^n$, $\bar{a}\,y \neq 0$. This shows that the indicated mapping is a monic. Finally, $\mathbb{Z}_{p^\infty}$ is generated by elements $\{y_n\}_{n \in \omega}$ such that $p\,y_1 = 0, \ldots, p\,y_n = y_{n-1}$, and y_n generates a cyclic group of order n. Now any endomorphism f maps each y_n into a subgroup of (y_n). Thus, $f = \bar{a}$, where $a = a_0 + a_1 p + \cdots$ is inductively defined as the sum $a = \sum_{n \in \omega} b_n$ where b_n is that unique p-adic integer such that $f(y_n) = b_n y_n$. Finally, using 18.9.6, we see that $a \in J$ iff $a_0 = 0$.   □

The examples 18.9.4 to 7 are all rings which are local rings in the sense of (a)–(d) of the next definition.

18.10 A   Definition and Proposition. *A ring R is said to be semilocal provided that $R/\mathrm{rad}\,R$ is semisimple* (I, p. 401).

*A ring R is said to be **local** (sometimes called scalar local) provided that R satisfies the equivalent properties (a)–(d):*

(a) *The set of nonunits of R is an ideal.*

(b) *R has a unique maximal right ideal.*

(c) *$R/\mathrm{rad}\,R$ is a field.*

(d) *The sum of two nonunits of R is a nonunit.*

Proof. (a) $\Rightarrow$ (b). Suppose I is the ideal consisting of nonunits of R. Then I contains every (maximal) right ideal. Since I is contained in a maximal right ideal, I is the unique maximal right ideal of R.

(b) $\Rightarrow$ (c). If R has a unique maximal right ideal I, then $I = \mathrm{rad}\,R$. Since the factor ring $Q = R/(\mathrm{rad}\,R)$ has but two right ideals, namely 0 and Q, Q is a field.

(c) $\Rightarrow$ (d). Recall that $J = \mathrm{rad}\,R$ has the property (18.7) that $x \in R$ is a unit in R if and only if $\bar{x}$ is a unit in $\bar{R} = R/J$, where $x \rightarrow \bar{x}$ is the natural map $R \rightarrow R/J$. Since R/J is a field, this implies that any element of R not in J is a unit of R. Thus, the set of nonunits of R coincides with J. Since $J \neq R$, J contains no units, and the sum of nonunits lies in J.

(d) $\Rightarrow$ (a). We must show that the set I of nonunits of R is an ideal. Since we are given that $a-b\in I$, $\forall a, b\in I$, we must show that $ar\in I$ and $ra\in I$ for $\forall a\in I$, $r\in R$. Assume momentarily that $a\in I$, $r\in R$ is such that $ar\notin I$. Then ar is a unit, so $arv=var=1$ for some $v\in R$. Let $u=rv$. Then $au=1$. If $uc=1$ for some $c\in R$, then $a=auc=1c=c$, and $au=ua=1$, which contradicts the fact that a is a nonunit. Hence $uR\neq R$, so $uR\subseteq I$; in particular, $ua\in I$. But $a(1-ua)=0$, so $1-ua\in I$. Then $1=ua+(1-ua)\in I$, which is a contradiction. Therefore, $IR\subseteq I$, and by symmetry $RI\subseteq R$, showing that I is an ideal. $\square$

18.10B   *Exercise.* Let R be a local ring. (a) A right R-module G is a generator of mod-R if and only if R is isomorphic to a summand of G.

(b) If $R\to S$ is a surjective ring homomorphism, then S is a local ring.

18.11   Corollary. *Let M be an indecomposable object of mod-R which is either quasiinjective, or has finite J-H length. Then, $\operatorname{End} M_R$ is a local ring.*

Proof. The second case follows from 17.16 and 18.10. Now assume that M is injective, and S the endomorphism ring, and let J denote the set of elements of S with nonzero kernel. If $f\in S$, and $\ker f=0$, then by injectivity of M there exists $g\in S$ sending $x\mapsto fx$ $\forall x\in M$. Then $gf=1$, and $e=gf$ is therefore an idempotent which by indecomposability of M must be 1. Since every nonzero submodule N of M has a nonzero injective hull $\hat{N}$ contained in M as a summand, then indecomposability of M implies that $M=N$. Thus, every nonzero submodule of M is an essential submodule. Therefore, given endomorphisms f and g, then $\ker(f+g)\supseteq \ker f\cap\ker g$ and so the sum $f+g$ of two nonunits of S is a nonunit. By 18.10, this implies that J is the radical.

We leave the case M is quasiinjective for an exercise. $\square$

Artinian Rings are Noetherian

The theorem stated in the heading was proved by C. Hopkins [39] and independently by J. Levitzki [39]. The proof is so simple that its age scarcely shows. The main ingredient in his proof is the Wedderburn-Artin theorem, which does not appear to say anything about nonsemisimple rings. As we shall see many times, appearances are deceiving. This theorem is *the* fundamental tool in the study of nonsemisimple Artinian rings. Why is this so? A study of the proof of the following proposition will reveal the answer.

18.12   Proposition and Definition. *A ring R is **semiprimary** if R is a semilocal ring with nilpotent radical. Any right Artinian ring is semiprimary. Moreover, a semiprimary ring is right Artinian if and only if it is right Noetherian.*

Proof. If R is right Artinian, then so is $R/\operatorname{rad} R$. Since 18.6, $R/\operatorname{rad} R$ is semiprime, then by the W-A Theorem 8.8, $R/\operatorname{rad} R$ is semisimple. Now $\operatorname{rad} R$ is nilpotent by 18.8, so R is semiprimary.

In order to prove the equivalence of Noetherian and Artinian for semiprimary rings, first note that any semisimple module M (over any ring) is Artinian if and only if it is Noetherian. This follows because M is a direct sum of simple modules (which are both Artinian and Noetherian), and the number of these simple summands is finite if and only if M is either Artinian or Noetherian. Now let t be the index of nilpotency of $N = \operatorname{rad} R$. Defining $N^0 = R$, for any integer i between 0 and $t-1$, N annihilates the right R-module $M = N^i/N^{i+1}$, hence M is a right R/N-module. Since R/N is a semisimple ring, M is a semi-simple R/N module 8.9. Since the R-submodules of M coincide with its R/N-submodules, M is a semisimple R-module. Consider the sequence

$$R \supset N \supset N^2 \supset \cdots \supset N^t = 0$$

of right R-modules. If R is right Noetherian (resp. Artinian), then each factor module M_i, $i = 0, \ldots, t-1$, being semisimple, has finite length. Therefore R has a composition series, so R is right Artinian (resp. Noetherian). $\square$

18.13 **Theorem** (Hopkins [39], Levitzki [39]). *A right Artinian ring is (semi-primary and) right Noetherian.* $\square$

The following also can be used to deduce that any Noetherian semiprimary ring is Artinian.

18.14 *Exercise.* (a) (Chase). If R is semiprimary, then each left R-module, and each right R-module, satisfies the d.c.c. on finitely generated submodules (compare 22.9).

(b) If M is an object of mod-R of finite length, then $A = \operatorname{End} M_R$ is a semi-primary ring satisfying the a.c.c. on right and left annulets (compare Exercise 17.25).

Krull-Schmidt Theorems and Azumaya Diagrams

As was remarked upon in the introduction to Volume II, there is symbiotic relationship between the categories RINGS and the categories mod-R, for any object R of RINGS. Purists are inclined to neglect the one aspect for the glory of the other. Group theorists are understandably reluctant to resort to the machinery of ring theory, applied to the group ring RG of a finite group over a ring R usually taken to be $\mathbb{C}$; but, alas, some theorems are comprehensibly proved only in that way. Similarly, ring theory occasionally has to take back-seats, and let group theory (formerly abelian groups, but now more often abelian categories) drive.

Two problems of module theory are:

(1) When does a module M decompose into a direct sum of indecomposable modules? In this case M is said to be **completely decomposable.** (2) When are two such decompositions unique in the sense that they differ only by an auto-morphism of M. (This will be described precisely in 18.8 and 21.6.) A **Krull-Schmidt theorem** is one which states that a module (object) has a unique direct decomposition. (Krull [25], Schmidt [28], referred to in Noether [29, p. 645 and p. 651]: every group G with minimal condition on subgroups is a direct product of indecomposable groups; and if the maximal condition holds, then the decomposition is unique up to an automorphism of G.) The Krull-Schmidt

theorem addresses itself mainly to finite direct sums (products) of indecomposable objects; however, Azumaya [50] obtained a unique decomposition theorem for infinite direct sums of modules each of which had local endomorphism ring (something which attained in the Krull-Schmidt theorems). (See 21.6.) Such a module is said to have an **Azumaya Diagram (AD)**, and we have found this a convenient language to apply even when the direct sum is finite (when the Krull-Schmidt theorem applies). The next section contains, then, the Krull-Schmidt theorem 18.18 for "idemsplit" categories (defined in 18.15), and applications for the structure of rings with AD (see 18.23–18.28).

The afore-mentioned symbiosis between rings and modules is complete with respect to characterizing when a right A-module M has a finite AD, since Proposition 18.23.2 (cf. 18.26) states that this happens iff the endomorphism ring R of M has a finite AD, and, moreover, this happens iff R is a lift/rad semi-local ring. (Curiously, the latter are precisely those rings over which finitely generated modules have projective covers 22.23.)

18.15 Proposition and Definition[3]. *An **idemsplit** category is an additive category C in which, for any object A, every idempotent $e \in \text{End}_C A$ is split in the sense dual to 3.35, that is, there exist morphisms $B \xrightarrow{p} A \xrightarrow{q} B$ such that $pq=1$ and $e=qp$. (Any abelian category is idemsplit.)*

If $A \xrightarrow{q} B \xrightarrow{p'} A$ are morphisms in an idemsplit category such that $u=p'q$ is an automorphism of A, then there is a monic $q_1: A_1 \to B$ such that 1_B is the co-product of q and q_1, that is, $B \approx A \oplus A_1$.

Proof. If $p=(p'q)^{-1}p'$, then $pq=1_A$, and $e=qp$ is idempotent, so by assumption there is a diagram $A_1 \xrightarrow{q_1} B \xrightarrow{p_1} A$ such that $p_1 q_1 = 1$, and $q_1 p_1 = 1-e$. By (2) of 3.34, q is the kernel of $1_A - e$, and q_1 is the kernel of e. Then, (3) of 3.34 implies that

$$B = \ker e \oplus \ker(1-e) \approx A \oplus A_1$$

and the same result implies that any abelian category is idemsplit. $\square$

18.16 Lemma. *Consider the morphism*

$$A \oplus B \xrightarrow{\binom{a\ b}{c\ d}=\alpha} C \oplus D$$

in an additive category X. If both α and a are equivalences, then $B \approx D$.

Proof. If $c=0$, then the lower right coefficient of α^{-1} is an inverse for d. In general, replace α by

$$\begin{pmatrix} 1_C & 0 \\ -ca^{-1} & 1_D \end{pmatrix} \alpha = \begin{pmatrix} a & b \\ 0 & d' \end{pmatrix}$$

and reduce to the first case. $\square$

---

[3] Here, and in 18.16–17, and the first part of 18.18, I am following Bass [68]. In 18.18(c) I follow Swan [68].

An object A is **indecomposable** if

$$A \approx B \oplus C \implies B = 0 \quad \text{or} \quad C = 0.$$

In an additive category X, this is equivalent to the assertion that $R = \text{End}_X A$ has no idempotents other than 0 or 1_A, that is, that R is indecomposable in mod-R.

An **Azumaya diagram** is a coproduct $\coprod_{i \in I} A_i$ such that $\text{End}_X A_i$ is a local ring $\forall i \in I$. Thus, in an Azumaya diagram (AD), each A_i is indecomposable.

18.17 Exchange Lemma. *Let C be an idemsplit category. If*

$$A \oplus B = \coprod_{i=1}^n C_i$$

and if $R = \text{End}_C A$ is a local ring, then there is an $i \le n$ such that $C_i \approx A \oplus C_i'$, and $B \approx C_i' \oplus \coprod_{j \ne i}^n C_j$. In particular, if each C_i is indecomposable, e.g. if $\coprod_{i=1}^n C_i$ is an AD, then $C_i \approx A$, and $B \approx \coprod_{j \ne i} C_j$.

Proof. Let q_A and q_B (p_A and p_B) be the respective injections (projections) of the first coproduct $A \oplus B$, and $\{q_i \colon C_i \to X\}_{i=1}^n$ ($\{p_i \colon X \to C_i\}_{i=1}^n$) be those of the second. Then,

$$1_A = p_A q_A = p_A \left(\sum_{i=1}^n q_i p_i \right) q_A = \sum_{i=1}^n p_A q_i p_i q_A.$$

Since R is local one of the $p_A q_i p_i q_A$ must be a unit. Relabeling, if necessary, we can assume $(p_A q_1)(p_1 q_A)$ is an automorphism of A. According to 18.15 there is a $q_1' \colon C_1' \to C_1$ such that C_1 is the coproduct of $p_1 q_A$ and q_1', $C_1 = A \oplus C_1'$. Using this to refine the decomposition $C_1 \oplus \cdots \oplus C_n$ to $A \oplus C_1' \oplus \cdots \oplus C_n$, we obtain an isomorphism of the latter with $A \oplus B$ such that the composite

$$A \xrightarrow{\ q_A\ } A \oplus B = A \oplus C_1' \oplus C_2 \oplus \cdots \oplus C_n \xrightarrow{\ \text{canon}\ } A$$

is an equivalence. It follows therefore from 18.15 that $B \approx C_1' \oplus C_2 \oplus \cdots \oplus C_n$. $\square$

Let C be an abelian category. An object A of C satisfies the **bichain condition** *(b.c.c.) if every pair of sequences of morphisms*

$$A \underset{f_1}{\overset{p_1}{\rightleftarrows}} A_2 \underset{f_2}{\overset{p_2}{\rightleftarrows}} A_3 \rightleftarrows \cdots A_n \underset{f_n}{\overset{p_n}{\rightleftarrows}} A_{n+1} \rightleftarrows \cdots$$

with every p_n epic, and f_n monic, implies that there is an n_0 such that p_n and f_n are equivalences $\forall n \ge n_0$.

A Noetherian and Artinian object A in an abelian category need not have the b.c.c. (see Swan [68, p. 76]).

Unique Decomposition Theorem

The following names are associated with the next theorem: Kronecker, Wedderburn, Remak, Krull, Schmidt, Fitting, Korinek, Ore, and Kurosch (the latter for lattices; cf. 17.5).

18.18 **Finite Unique Decomposition Theorem**[4]. *Let C be idemsplit.*

(a) *If $A = \coprod_{i=1}^{n} A_i$ is a finite Azumaya diagram, then any coproduct representation of A can be refined to an Azumaya diagram.*

(b) *If $A = \coprod_{j=1}^{m} B_j$ is another Azumaya diagram for A, then $m = n$, and there is a permutation p of $\{1, \ldots, n\}$ and a family $\{u(i): B_i \to A_{p(i)}\}_{i=1}^{n}$ of equivalences. In other words, there is an automorphism u of A which induces $u_i: B_i \approx A_{p(i)}$, $i = 1, \ldots, n$.*

(c) *If C is an abelian category, then any object A which has the b.c.c. has an AD.*

Proof. Induction on n; the case $n = 1$ is clear.

(a) Suppose $n > 1$. If $A \approx C_1 \oplus \cdots \oplus C_r$ then 18.17 implies that for some i, say, $i = 1$, we can write $C_1 \approx A_1 \oplus C_1'$, so that $A_2 \oplus \cdots \oplus A_n \approx C_1' \oplus C_2 \oplus \cdots \oplus C_r$. By induction we can refine the latter to an AD.

(b) If the C_i are indecomposable to begin with, then we must have $C_1' = 0$ and the uniqueness now follows also by induction. Then the coproduct $u = \coprod_{i=1}^{n} u_i$ is the desired automorphism.

(c) Clearly, A is a finite coproduct of indecomposable objects of C, so it remains to show that any indecomposable object A with the b.c.c. has local endomorphism ring. It suffices to show that a sum of endomorphisms $f + g = 1_A$ only if f or g is an equivalence. Let $A_n = \operatorname{im} f^n$. Then

$$A_n \xrightarrow[\text{inclusion}]{\overset{f|A_n}{\longleftarrow}} A_{n+1}$$

defines a bichain. Hence there exists an n_0 such that for $n \geq n_0$, $f|A_n: A_n \to A_{n+1}$ is an equivalence. Define $\rho: A \to A$ by $\rho = [(f|_{A_n})^n]^{-1} f^n$. Then $\rho|A_n = 1_{A_n}$ and $\operatorname{im} \rho = A_n$. Hence, $A = A_n \oplus \ker \rho$. Either $A_n = A$ and f is an equivalence or $A_n = 0$ and $f^n = 0$. But then $g = 1 - f$ has as inverse $1 + f + f^2 + \cdots + f^{n-1}$ and g is an equivalence. $\square$

18.19 *Exercise*

18.19.1 **(Cancellation Lemma).** Let C be an idemsplit category. Let A and A' be objects with AD. If B and B' are objects, then

$$A \approx A', \quad \text{and} \quad A \oplus B \approx A' \oplus B' \;\Rightarrow\; B \approx B'$$

(cf. Swan [68, p. 78, Theorem 2.6 and Lemma 2.9]).

18.19.2 If a ring R has an AD, then

$$R \oplus A \approx R \oplus B \;\Rightarrow\; A \approx B$$

for any modules A and B.

18.19.3 **Krull-Schmidt Theorem for Groups.** Any group G satisfying the d.c.c. chain condition on normal subgroups is a direct product of finitely many indecomposable groups. Moreover, if G satisfies the a.c.c. as well as the d.c.c. on normal subgroups, then any such decomposition is unique up to an automorphism of G (as stated in the last sentence of 18.18).

---

[4] Most frequently called the Krull-Schmidt theorem.

18.19.4 If R and S are local rings, then the full matrix rings R_n and S_m are isomorphic if and only if $n=m$, and R is isomorphic to S. Generalize to rings R and S having AD's. Show by example when this fails.

18.19.5 If M and M' are full sets of $n \times n$ matrix units in R_n, where R is a local ring, then there is a unit $x \in R_n$ such that $M' = xMx^{-1}$.

18.20 Corollary. *If R is a local ring, then every finitely generated projective module is free.* $\square$

In fact, every projective module over a local ring is free (Kaplansky [58a]).

Lifting Idempotents

If I is a subset of R, then an **idempotent** of R/I, or an **idempotent modulo** I, is an element $x \in R$ such that $x^2 - x \in I$. Thus, $x = x^2$ (modulo I). Such an idempotent is said to be **lifted** provided that there exists an idempotent $y \in R$ $y - x \in I$, and we say x has been lifted to y. If idempotents can be lifted modulo every right ideal, then R is called a **lifting** ring. We let $R \in$ lift denote this fact. If idempotents of $R/\mathrm{rad}\, R$ can be lifted, then we say that R is a **lift/rad ring**, or that $R \in$ lift/rad. Similarly, a direct sum decomposition $R/I = A \oplus B$ can be lifted, provided that $R = X \oplus Y$ for some right ideals X and Y of R containing I which map canonically onto A and B respective under the map $R \to R/I$. We next prove a theorem of Jacobson on lift/rad rings.

18.21 Proposition. *If I is a nil onesided ideal of R, then idempotents of R/I can be lifted.*

Proof. Let $u^2 - u \in I$. We shall find an $x \in R$ such that $e = u + x(1 - 2u)$ is an idempotent element of R that commutes with u. The equation $e^2 = e$ is equivalent to the equation

$$(x^2 - x)(1 + 4y) + y = 0$$

where $y = u^2 - u \in I$. This is quadratic in x, and application of the usual quadratic formula yields a (formal) solution $x = 1/2(1 - (1 + 4y)^{-\frac{1}{2}})$, or, upon expansion,

$$x = 1/2 \left(2y - \binom{4}{2} y^2 + \binom{6}{3} y^3 - \cdots \right).$$

Since $y \in I$, y is nilpotent, so this formula defines x as a polynomial in y with integer coefficients. Thus, this value for x commutes with u, belongs to I, and $e = u + x(1 - 2u)$ is an idempotent with the property that $e - u \in I$. Thus u lifts to e. $\square$

18.22 *Exercise*

18.22.1 If I is an ideal such that direct sum decompositions of R/I can be lifted, then idempotents of R/I can be lifted. (Hint: if $R/I = \bar{A} \oplus \bar{B}$, and $R = A \oplus B$ then any idempotent which generates A lifts any idempotent which generates $\bar{A}$.)

18.22.2   If I is an ideal of R contained in rad R, and if idempotents of R/I can be lifted, then any finite or countable set of orthogonal idempotents of R/I can be lifted to an orthogonal set of idempotents of R.

18.22.3   (Faith and Utumi [64a]). If R is a regular ring, then idempotents modulo any onesided ideal can be lifted, and any right selfinjective ring is a lift/rad ring.

Rings with Azumaya Diagrams

Earlier, *sup.* 18.15, we remarked that a module M has a finite Azumaya diagram iff the endomorphism ring R of M has a finite diagram iff R is a semilocal lift/rad ring. A part of this theorem is proved in the next proposition namely, 18.23.2 states that a ring R has an AD in mod-R (iff R has an AD in R-mod) iff R is semilocal lift/rad. Moreover, various important properties of semilocal lift/rad rings are displayed for future use.

A module M is said to be a **local module** provided that M has the equivalent properties:

LM 1. M has a unique maximal submodule

LM 2. $M/\mathrm{rad}\, M$ is simple.

Over a semilocal lift/rad ring, it is possible to describe all local modules (see 18.23.4).

18.23   Proposition and Definition

18.23.1   *A ring R with radical J is said to have an **Azumaya diagram** (AD) in case R contains finitely many sum-1 orthogonal nonzero idempotents $\{e_i\}_{i=1}^n$ such that $e_i R e_i$ is a local ring, $i = 1, \ldots, n$. Then $R = \sum_{i=1}^n \oplus e_i R$ is an AD in mod-R, and $R = \sum_{i=1}^n \oplus R e_i$ is an AD in R-mod.*

18.23.2   *A ring R has an AD if and only if R is a semilocal lift/rad ring.*

18.23.3   *In this case, for an idempotent $e \in R$, the following conditions are equivalent: (i) e is indecomposable (ii) $e R \approx e_i R$ for some i (iii) $e R/e J$ is simple (iv) $e J$ is the unique maximal submodule of $e R$. Then $e R$ is said to be a right and $R e$ a left, **principal indecomposable** module, or **prindec**.*

18.23.4   *A right R-module M is said to be **principal cyclic** if the equivalent conditions hold:*

(a) *M is a local module;*

(b) *M is an epic image of some right prindec. In this case there is an exact sequence $e_i R \to M \to 0$ if and only if $M e_i \neq 0$; and then M is simple if and only if $M \approx e_i R/e_i J$.*

18.23.5   *A complete set of sum-1 orthogonal prindecs is unique up to an inner automorphism of R. Thus, if*

$$\{e_i\}_{i=1}^n \quad and \quad \{f_j\}_{j=1}^m$$

are each sets of sum-1 orthogonal indecomposable idempotents, then $n=m$, and there exists a unit x of R and a permutation p of $\{1, \ldots, n\}$ such that

$$x e_i x^{-1} = f_{p(i)} \quad and \quad x (e_i R) x^{-1} = f_{p(i)} R = x e_i R,$$

$i = 1, \ldots, n$.

The *Proof of 18.23.1* is clear, since for any idempotent $e \in R$, there are canonical isomorphisms

$$\operatorname{End} e R_R \approx e R e \approx \operatorname{End}_R R e.$$

Proof of 18.23.2. Let $\bar{R} = R/J = \bar{A} \oplus \bar{B}$ be a direct decomposition of R/J, that is, $A + B = R$, and $A \cap B = J$ for right ideals A and B of R. By the replacement Lemma 18.17, $\{e_i\}_{i=1}^n$ can be numbered so that $\bar{R} = \bar{A} \oplus \bar{Y}$, where $\bar{Y} = \sum_{i=t}^n \oplus \bar{e}_i \bar{R}$, $Y = \sum_{i=t}^n \oplus e_i R$, and $1 \leq t \leq n$. The isomorphism $\bar{Y} \approx \bar{B}$ can be extended to an automorphism f of $\bar{R}$ such that $f(\bar{A}) = \bar{A}$. Then $f(\bar{1})$ is a unit of R/J which by 18.7 can be lifted to a unit a of R. Since $\bar{a} \bar{Y} = \bar{B}$, then $a Y + J = B$. Since

$$R = X \oplus Y = a X \oplus a Y = a R$$

this shows that a summand $a Y$ of R maps onto $\bar{B}$ under the canonical map h: $R \to \bar{R}$. Moreover, since $f(\bar{A}) = \bar{A}$, then $\bar{a} \bar{A} = \bar{A}$, so that $a A + J = A$. Since a is a unit, then $a J = J$, and so $a A \supseteq J$, and then $A = a A$. This shows that the summand $a X$ of R maps onto $\bar{A}$ under h. Thus, $R = a X \oplus a Y$ is the desired lifting of the decomposition of $\bar{R} = \bar{A} \oplus \bar{B}$. By Exercise 18.22.1 (or by 22.21), then idempotents of $\bar{R}$ lift. This proves that R is a lift/rad semilocal ring.

Proof of 18.23.3. Now

$$e R/e J = e R/e R \cap J \approx (e R + J)/J$$

and by end-of-chapter Exercise 5,

$$\operatorname{End}_R e R/e J \approx e R e/e J e.$$

Assume that R has an AD. Then, by the unique decomposition theorem, 18.18, (i) $\Leftrightarrow$ (ii). Since $e R e$ is then local, and $\operatorname{End}_R e R/e J$ is a field $\approx e R e/e J e$, (ii) $\Rightarrow$ (iii). This implies that $R/J \approx \sum_{i=1}^n \oplus e_i R/e_i J$ is semisimple. By 18.3.4, rad $e R = e J$, and then simplicity of $e R/e J$ implies that $e J$ is the unique maximal submodule of $e R$. Thus, (iii) $\Rightarrow$ (iv).

(iv) $\Rightarrow$ (i). If $e = g + h$ for two orthogonal idempotents then, since $e R$ is a B-object 18.3, $g R \neq e R$ and $h R \neq e R$ implies that $g R \subseteq e J$ and $h R \subseteq e J$, and then $e R \subseteq e J$. But then $e \in J$. Since 0 is the only idempotent of rad R, then $e = 0$, contradicting the assumptions on $\{e_i\}_{i=1}^n$, and therefore (iv) $\Rightarrow$ (i).

Proof of 18.23.4. (a) $\Rightarrow$ (b). If $M \neq 0$ is principal cyclic, then since $M = \sum_{i=1}^n M e_i$ properly contains rad $M = M J = \sum_{i=1}^n M e_i J$, then there is an i, such that $M e_i \nsubseteq M J$. Set $e = e_i$. Hence, let $m \in M$ be such that $m e R \nsubseteq M J$. Simplicity of M/MJ implies $M = m e R + M J$. Since $M J = $ rad M is superfluous 18.3, necessarily $m e R = M$, so there is an epimorphism $e R \to M$ sending $e a$ onto $m e a$. This proves that (a) $\Rightarrow$ (b), and (b) $\Rightarrow$ (a) is trivial since any prindec has property (a). The statement that $M/MJ \approx e_i R/e_i J$ if and only if $M e_i \neq 0$ is an exercise.

Proof of 18.23.5. If $\{e_i | i=1, \ldots, n\}$ and $\{f_j | j=1, \ldots, m\}$ are as stated in (1), then by the unique decomposition theorem 18.18, $n=m$, and there exists a permutation p of $\{1, \ldots, n\}$ and isomorphisms $g_i: e_i R \to f_{p(i)} R$, $i=1, \ldots, n$. Thus, the direct sum mapping $q = \sum_{i=1}^n \oplus g_i$ is an automorphism of R_R such that $g(e_i R) = f_{p(i)} R$, $i=1, \ldots, n$. If $x=g(1)$, then $g(r)=g(1) r = xr$ $\forall r \in R$. Since g is an automorphism, $1 = g(a) = x a$ for some $a \in R$. Then $\bar{1} = 1 = \bar{x} \bar{a}$ in $\bar{R}$. It is an easy exercise to show that the equation $\bar{1} = \bar{x} \bar{a}$ in a semisimple or Artinian ring implies $\bar{1} = \bar{a} \bar{x}$. Since ax is idempotent in R, necessarily $ax=1$, so x is a unit of R. Since $R x^{-1} = R$, we have

$$x e_i R x^{-1} = x e_i R = g(e_i R) = f_{p(i)} R \qquad\qquad (i=1, \ldots, n)$$

as stated. Since

$$1 = \sum_{i=1}^n e_i = x^{-1} 1 x = \sum_{i=1}^n x^{-1} e_i x = \sum_{j=1}^n f_j$$

and since $x e_i x^{-1} \in f_{p(i)} R$, it follows from the unique expressibility of 1 as a sum of elements in $f_j R$ that $x^{-1} e_i x = f_{p(i)}$, $i=1, \ldots, n$. $\qquad\square$

Basic Ring and Module

18.24 Proposition. *Let $R = e_1 R \oplus \cdots \oplus e_n R$ be an AD for a ring with radical J. Then R is said to be **basic** provided that the equivalent conditions (a) and (b) hold:*

(a) $e_i R \approx e_j R$ in mod-R, implies $i=j$;

(b) $e_i R/e_i J \approx e_j R/e_j J$, implies $i=j$.

*Renumber, if necessary, so that $\{e_i R\}_{i=1}^t$, $t \le n$, is an isomorphy class for $\{e_i R\}_{i=1}^n$. Then, $e_0 = e_1 + \cdots + e_t$ is called the **basic idempotent**, $e_0 R$ is the **basic module**, and $e_0 R e_0$ is the **basic ring** of R. A ring R is similar to its basic ring. Furthermore, the basic module is a summand of every generator of mod-R. A basic ring of a ring R is a basic ring, and any two basic rings of R are isomorphic by a map extendable to an inner automorphism of R. Thus, if e_0 and e_0' are basic idempotents of R, then there is a unit $x \in R$ such that $e_0' = x e_0 x^{-1}$, so $e_0' R e_0' = x e_0 R e_0 x^{-1}$ and $e_0' R = x e_0 R$.*

Proof. The equivalence of (a) and (b) is end-of-chapter Exercise 4. Then, by 18.23 a simple module $V \approx e_i R/e_i J$ if and only if $V e_i \neq 0$. By 3.31 this proves that the basic module $B = e_0 R$ is a generator of mod-R, so by Morita's theorem 4.29, R is similar to $e_0 R e_0 \approx \operatorname{End} B_R$. If G is a generator of mod-R, then R, hence B, is a summand of G^n, for some integer $n>0$. Then, every summand $e_i R$ of B is a summand of G^n, and the replacement lemma 18.17 then implies that every $e_i R$ is a summand of G, $i=1, \ldots, t$. Write $G = e_1 R \oplus X = e_2 R \oplus Y$. Since $e_1 R$ is not isomorphic to a summand of $e_2 R$, then the replacement lemma implies that $e_1 R$ is isomorphic to a summand of Y. Thus, $G = e_1 R \oplus e_2 R \oplus Z$, and by induction it follows that B is a summand of G.

By the proposition, any two basic modules $B = e_0 R$ and $B' = e_0' R$ are isomorphic to summands of each other, and, by the unique decomposition theorem 18.18, must be isomorphic. Using 18.18, one sees that the isomorphism is induced by a unit $x \in R \approx \operatorname{End} R_R$. Then $e_0' R e_0' = x e_0 R e_0 x^{-1}$, and so $e_0' = x e_0 x^{-1}$. The basic ring of R is a basic ring (exercise). $\qquad\square$

18.25 *Exercise*. Let R be lift/rad semilocal and let $J = \operatorname{rad} R$.

18.25.1 Every principal cyclic module is indecomposable.

18.25.2 Every module is generated by (is a sum of) principal cyclic modules.

18.25.3 A right R-module M is principal cyclic (local) if and only if M/MJ is simple.

18.25.4 For any local module M, $\operatorname{End} M_R$ is a local ring.

18.26 **Proposition.** *Let M be an object in an idemsplit (or abelian) category C, and let $R = \operatorname{End}_C M$. Then, the following conditions are equivalent:*

(a) *M has a finite Azumaya diagram,*

(b) *R has a finite Azumaya diagram,*

(c) *R is a lift/rad semilocal ring.*

Thus, (b) $\Leftrightarrow$ (c) for any ring R. When this is so, then sum-1 orthogonal indecomposable idempotents of R/J lift to sum-1 orthogonal indecomposable idempotents of R.

Proof. (a) $\Leftrightarrow$ (b). Let $R = \operatorname{End}_C M$ have radical J. By 18.15–18, a set of idempotents $\{e_i\}_{i=1}^n$ of R are orthogonal with sum $= 1_M$ if and only if M is a direct sum of the corresponding images, that is, the canonical morphism $\sum_{i=1}^n \oplus M_i \to M$ is an equivalence, where M_i is the image of e_i, $i = 1, \dots, n$ (cf. also (I, 3.34–35, pp. 152–3)). In this case, there is an induced isomorphism $e_i R e_i \approx \operatorname{End}_C M_i$, $i = 1, \dots, n$. Since $e_i R e_i \approx \operatorname{End} e_i R_R$, this proves $M = \sum_{i=1}^n \oplus M_i$ is an AD for M if and only if $R = \sum_{i=1}^n \oplus R e_i$ (resp. $R = \sum_{i=1}^n \oplus e_i R$) is an AD for R. Thus (a) $\Leftrightarrow$ (b), and (b) $\Leftrightarrow$ (c) is 18.23. $\square$

A **projective cover** of a module M is an exact sequence

$$0 \to K(M) \to P(M) \to M \to 0$$

such that $P(M)$ is projective, and $K(M)$ is a superfluous submodule of $P(M)$ (cf. Chapter 22). A ring R is **right (semi) perfect** provided that every (finitely generated) right R-module has a projective cover.

18.27 *Exercises* (Bass [60]).

18.27.1 A ring R is right semiperfect if and only if R is semilocal lift/rad and if and only if R is left semiperfect.

18.27.2* A ring R is right perfect if and only if the dual category $(\operatorname{mod-}R)^{\operatorname{op}}$ is a category with injective hulls.

18.27.3* A ring R is right perfect if and only if R satisfies any of the equivalent conditions: (a) R is semiperfect and the radical of R is left vanishing. (b) R has d.c.c. on principal left ideals. (c) R is semiperfect and every nonzero left module has nonzero socle. (d) Every flat right R-module is projective. (e) The direct limit of projective right R-modules is projective.

18.27.4* (Björk [69]). A left R-modules M satisfies the d.c.c. on cyclic submodules if and only if M satisfies the d.c.c. on finitely generated submodules.

18.27.5* (Björk [69]). Any right perfect ring satisfies the d.c.c. on finitely generated left ideals.

18.28   Proposition and Definition. *A lift/rad semilocal ring R is said to be* **selfbasic** *provided that R satisfies the equivalent conditions:*

(a) *R is its own basic ring.*

(b) *R/rad R is a product of fields.*

Proof. Let $J = \text{rad } R$, and $\bar{R} = R/J$. Since R is semilocal, then $\bar{R} = \prod_{i=1}^{n} \bar{S}_i$, is a product of simple full $n_i \times n_i$ matrix rings $\bar{S}_i$ over fields D_i, $i = 1, \ldots, n$. Then, $\bar{S}_i$ is the direct sum of n_i simple right ideals, and, moreover, $\bar{S}_i$ is the sum of all simple right ideals of $\bar{R}$ which are isomorphic to a simple right ideal of $\bar{R}$ contained in $\bar{S}_i$, $i = 1, \ldots, n$. Since prindecs $e_1 R$ and $e_2 R$ are isomorphic if and only if $e_1 R/e_1 J$ and $e_2 R/e_2 J$ are isomorphic, it follows that R is its own basic ring if and only if $n_i = 1$, that is, if and only if $\bar{S}_i = D_i$ is a field, $i = 1, \ldots, n$.   □

Thus, a selfbasic ring is basic, as stated in 18.24.

18.29 *Exercise*

18.29.1   Rings with AD are Morita invariant.

18.29.2   The basic ring of a full matrix ring R_n over a local ring R is canonically isomorphic to R.

18.29.3   The ring $T_n(R)$ of lower triangular matrices over a local ring R is a basic ring.

18.29.4   (Eisenbud and Griffith) If eR is a right prindec in a semiprimary ring R, and if X is a submodule of finite length, then $\text{End}_R eR/X$ is a local ring.

18.30   Chinese Remainder Theorem. *If $\{A_i\}_{i=1}^{n}$ are finitely many ideals of a ring R, then the following conditions are equivalent:*

(1) *The canonical map*

$$h \begin{cases} R/\bigcap_{i=1}^{n} A_i \to \prod_{i=1}^{n} R/A_i \\ a + \bigcap_{i=1}^{n} A_i \mapsto (a + A_1, \ldots, a + A_n) \end{cases}$$

is an isomorphism.

(2) *For any set $\{x_i\}_{i=1}^{n}$ of elements of R the system of congruences $X = x_i \pmod{A_i}$ has a solution $x \in R$.*

(3) *The ideals $\{A_i\}_{i=1}^{n}$ are comaximal in pairs, that is, $R = A_i + A_j$, whenever $i \neq j$.*

Proof. Evidently h is a ring monic. Moreover, h is surjective if and only if systems of congruences (2) have a simultaneous solution as stated. Then, if $y_j \equiv 1 \pmod{A_j}$ and $y_j \equiv 0 \pmod{A_i}$, $i \neq j$, then $a_j = 1 - y_j \in A_j$, and $1 = a_j + y_j \in A_i + A_j$, proving comaximality. Conversely, if there exists elements elements $y_1, \ldots, y_n$ satisfying the stated congruences, then the element $X = x_1 y_1 + \cdots + x_n y_n$ has the property (2).   □

18.31   Corollary. *If $\{A_i\}_{i=1}^{n}$ are finitely many pairwise comaximal ideals of a ring R, then for any module M, any system $\{y = y_i \pmod{MA_i}\}_{i=1}^{n}$ of congruences has a solution $y \in M$, and the canonical map $M/\bigcap_{i=1}^{n} MA_i \to \prod_{i=1}^{n} M/MA_i$ is an isomorphism.*

Proof. Apply the Chinese remainder theorem.   □

18.32   Corollary. *For a set $\{A_i\}$ of ideals of a ring R, and a permutation P of n, let $\mathbf{P} = A_{P(1)} A_{P(2)} \cdots A_{P(n)}$, let $\Lambda = A_1 \cap A_2 \cap \cdots \cap A_n$, and let*

(1) $$0 \to \Lambda/\mathbf{P} \to R/\mathbf{P} \xrightarrow{q(P)} R/\Lambda \xrightarrow{h} \prod_{i=1}^{n} R/A_i$$

denote the canonical exact sequence, and let $h(P) = h\,q(P)$. Then, the canonical map $h(P): R/\mathbf{P} \to \prod_{i=1}^{n} R/A_i$ is an isomorphism if and only if $\mathbf{P} = A_1 \cap A_2 \cap \cdots \cap A_n$, and every system $\{x \equiv x_i (\mathrm{mod}\, A_i)\}_{i=1}^{n}$ of congruences has a simultaneous solution in R. Moreover, $h(P)$ is an isomorphism for every permutation P of n if and only if the ideals $\{A_i\}_{i=1}^{n}$ are commutative and comaximal in pairs.

Proof. By the canonical exact sequence (1), $h(P)$ is an isomorphism if and only if both h and $q(P)$ is an isomorphism. By the lemma, this is equivalent to the equality $\mathbf{P} = \Lambda$, and the simultaneous solvability of congruences.

We prove the last assertion by induction on n. For $n = 2$, the last lemma implies that h is an isomorphism only if A_1 and A_2 are comaximal. Moreover, $\mathbf{P} = A_1 \cap A_2$ for every permutation P implies $A_1 A_2 = A_2 A_1$. Conversely, if A_1 and A_2 are comaximal, and commutative, then $1 = a_1 + a_2$, for $a_i \in A_i$, $i = 1, 2$. If $x_1, x_2 \in R$, then $x = x_2 a_1 + x_1 a_2$ is such that $x \equiv x_2 a_1 \equiv x_2 \pmod{A_2}$, and $x \equiv x_1 \pmod{A_1}$. This proves that h is an isomorphism. Furthermore, if $y \in A_1 \cap A_2$, then $y = y a_1 + y a_2 \in A_2 A_1 + A_1 A_2 = A_1 A_2$, so that $A_1 \cap A_2 = A_1 A_2$. This proves that $R/A_1 A_2 \approx R/A_1 \times R/A_2$ is an isomorphism, as desired.

For the general case, let $B_2 = A_2 \ldots A_n$. Assuming comaximality and commutativity of the ideals $\{A_i\}_{i=1}^{n}$, then A_1 and B_2 are comaximal and commutative, so $R/A_1 B_2 \approx R/A_1 \times R/B_2$ by the $n = 2$ case. Then, an inductive step yields $R/B_2 \approx \prod_{i=2}^{n} R/A_i$, and the sufficiency follows.

Conversely, if $h(P)$ is an isomorphism for every permutation P, then h is an isomorphism, so the A_i are comaximal in pairs by the Chinese remainder theorem. Moreover, there are canonical isomorphisms

(2) $$R/\Lambda = R/A_{P(1)} B \approx R/A_{P(1)} \times \prod_{j>1}^{n} R/A_{P(j)} \approx R/B A_{P(1)}$$

where $B = A_{P(2)} \ldots A_{P(n)}$. Since $\mathbf{P} = \Lambda$, for every P, then

$$A_{P(1)} B = A_1 \cap \cdots \cap A_n = B A_{P(1)}.$$

Since $A_{P(1)}$ and B are commutative and comaximal, then by the $n = 2$ case, there is a canonical isomorphism

(3) $$R/A_{P(1)} B \approx R/A_{P(1)} \times R/B$$

which by (2) implies that $R/B = \prod_{j=2}^{n} R/A_{P(j)}$. Then, by induction on n, the ideals A_j are commutative, for $j \neq P(1)$. But for $n > 2$, and for any pair $i \neq j$, there is a permutation P for which $i = P(2)$ and $j = P(3)$, proving that the ideals $\{A_i\}_{i=1}^{n}$ are commutative. $\square$

18.33   Lemma (Eilenberg, Nagao, and Nakayama [56]). *Let R be a semi-primary ring with radical J nilpotent of index $\leq m$. If A is any ideal of R, then $A^n = A^{n+1}$ for some integer $n \leq m$. Then $Q = A^n$ is an idempotent ideal, and $Q^k = Q + (Q \cap \mathrm{rad}\, R)^k$, for every k.*

Proof. Since R/J is a semilocal lift/rad ring, then there is an idempotent $e \in R$ such that

$$A = Re + B, \quad \text{and} \quad B = A \cap J.$$

Then, for every k,

(1) $A^k = Re A + B^k.$

For $k = 1$,

$$Re A + B = Re Re + Re B + B = Re + B = A$$

and by induction,

$$A^{k+1} = A A^k = Re Re A + Re B^k + B Re A + B^{k+1} = Re A + B^{k+1}.$$

This proves (1), which implies that $A^{k+1} = A^k$ if and only if $B^k = 0$. Since $B \subseteq J$, then one can assume that $k \leq m$. $\square$

18.34 A Lemma (Noether [21]). *Let R satisfy the a.c.c. on ideals. Then an ideal A of R is either prime, or else A contains a product of prime ideals. Thus, either R is prime, or a product of prime ideals is zero.*

Proof. If there is an ideal which does not contain a product of primes, then there is an ideal A maximal with respect to this property. Then A is not prime, so $A \supseteq BC$, for two ideals B and C containing A properly. By maximality of A, B and C are products of primes, hence A contains a product of primes. $\square$

18.34 B Lemma (Cohen [50], Ornstein [68]). *If R is a right Noetherian ring such that R/P is Artinian for each prime ideal $P \neq 0$, then either R is prime, or else R is right Artinian.*

Proof. If R is not prime, then there exist finitely many prime ideals $P_1, \ldots, P_n$ with product $= 0$. Consider the sequence

(1) $P_0 = R \supseteq P_1 \supseteq P_1 P_2 \supseteq \cdots \supseteq P_1 P_2 \ldots P_n = 0.$

We claim this chain can be refined to a composition series of R. For R/P_i is semisimple (Artinian), for each i, and therefore R/P_i has a composition series. Assume that $R/P_1 P_2 \ldots P_k$ has a composition series, let $B = P_1 P_2 \ldots P_k$, and $C = B P_{k+1}$. Possibly $B = C$, but in any case B/C is canonically an R/P_{k+1}-module, hence is semisimple. Then, the Noetherian condition implies that B/C has finite length. Thus, every factor in the sequence (1) is a semisimple module of finite length, and therefore (1) can be refined to a composition series. Then R is right Artinian. $\square$

A ring is said to be **Cohen** if R/P is right Artinian for each prime ideal $P \neq 0$ (or equivalently, R/P is semisimple). A ring R is said to satisfy the restricted **right minimum condition** (RRM) provided that R/I is right Artinian for any essential right ideal I. A ring is said to be RRA (RRN) provided that for any ideal $A \neq 0$, R/A is right Artinian (Noetherian).

18.35 *Exercise* [5]

18.35.1   A ring R is RRA if and only if R is RRN and R/P is Artinian for each prime ideal $P \neq 0$. (In this case, any nonzero prime ideal is maximal.)

18.35.2   Any RRN ring satisfies the a.c.c. on ideals. Any ring in which ideals are finitely generated right ideals satisfies the a.c.c. on ideals (but not conversely). Show that the latter condition can replace the Noetherian hypothesis in 18.34 B, and the conclusion still hold.

18.35.3   If every cyclic right module R/I, $I \neq 0$, is Artinian, and if R is not right Artinian, then R is a right Ore domain.

18.35.4   Any RRA ring is either prime, or contains just finitely many prime ideals $P_1, \ldots, P_n$. Then $P_1 \ldots P_n$ is nilpotent. Conclude that the radical of a non-prime RRA ring R is nilpotent, and every nil ideal is nilpotent.

18.35.5   A semiprime RRA ring which is not prime is semisimple.

18.35.6   A commutative RRA ring, which has zero divisors $\neq 0$, has just finitely many prime ideals. In this case, if R has no nonzero nilpotent elements, then R is a finite product of fields.

18.35.7   (Cohen) A commutative ring is Noetherian iff every prime ideal is finitely generated. This implies the a.c.c. on prime ideals. Show the converse is not true.

18.35.8   A right ideal I of a ring R is **(co-)irreducible** provided that R/I is a uniform right module. Show that a right Noetherian ring is hereditary iff every irreducible right ideal is projective.

18.35.9   (Zaks [71], Faith [75b]) (a) R is a **right bounded** ring if every essential right ideal contains an ideal $\neq 0$. If R is a right bounded Cohen ring with a.c.c. on ideals, and if every ideal is a projective right ideal, then R is right hereditary.

(b) Let R be a right bounded Cohen ring. Then every ideal is a finitely generated and projective right ideal iff R is right Noetherian and right hereditary.

(c) If B is an ideal of a ring R such that R/B is right Artinian, then every right ideal containing B is projective iff every ideal containing B is a projective right ideal. Conclude that a right Artinian ring is right hereditary iff every ideal is a projective right ideal.

(d) Any right bounded right Goldie prime Cohen ring in which every ideal is a principal right ideal is right Noetherian and right hereditary.

18.35.10   (Asano [49]) A Noetherian bounded Cohen prime ring in which every ideal is invertible is hereditary [Hint: apply 18.35.9 (a)]. Cf. Michler [69d] and Lenagan [71]. (Michler eliminates the Cohen hypothesis.)

18.36   Proposition and Definition. *A ring R is said to be **local-decomposable** provided that R is a finite product of local rings.*

---

[5] Exercises 1–7 are extracted from papers of Cohen [50] and Ornstein [68]. Other results on rings with restricted minimum conditions of Webber and Chatters are given in Chapter 20. See 20.29.

The following conditions on a ring R are equivalent:

(a) *R is similar to a finite product of local rings.*

(b) *R is a semilocal lift/rad ring, and the basic ring of R is a product of local rings.*

(c) *R is a finite product of full matrix rings over local rings.*

A local ring with nilpotent radical is said to be **completely primary**. *A* **primary ring** *is a full matrix ring R_n over a completely primary ring R. A primary ring is semiprimary. Any semiprimary ring R is a lift/rad ring, and is said to be* **primary-decomposable** *if it is similar to a local-decomposable ring.*

Proof. (a) $\Leftrightarrow$ (b) follows form 18.21 which asserts that any ring with AD is similar to its basic ring, and from 18.27 which states that rings with AD are Morita invariant. Also, since the basic ring of a matrix ring A_n over a local ring is A, then (c) $\Rightarrow$ (b). By 8.23, and the Morita theorem 4.29, assuming (a), then $R = \operatorname{End}_B P$, where B is a finite product of local rings, and P is finitely generated projective over B, say, $B^n = P \oplus X$, for some $n > 0$. If $B = \prod_{i=1}^t B_i$, where B_i is local, $i = 1, \ldots, t$, then this is an AD for B, and so, by the unique decomposition Theorem 18.18, $P = \prod_{j=1}^s P_j$, where P_j is a product of n_j copies of B_j, and $0 \le n_j < n$. Moreover, P_j is a fully invariant submodule of P, since $\operatorname{Hom}_B(P_i, P_j) = 0$ by virtue of the fact that $B_i B_j = 0$, $i \ne j$. Thus,

$$R = \operatorname{End}_B P = \operatorname{Hom}_B(\textstyle\prod_{i=1}^s P_i, \ \prod_{j=1}^s P_j) = \prod_{j=1}^s \operatorname{End}_B P_j.$$

Since $\operatorname{End}_B P_j = \operatorname{End}_{B_j} P_j$, and since $P_j \approx B_j^{n_j}$, then R is a finite product of $n_j \times n_j$ full matrix rings over B_j, $j = 1, \ldots, s$ (I, p. 152, 3.33.3). Thus, (b) $\Rightarrow$ (c).

Finally, any semiprimary ring is a semilocal lift/rad ring, and hence has an AD by 18.26. $\square$

For emphasis we note that a local ring is completely primary iff primary iff the radical is nilpotent.

18.37 Proposition (Asano [49]). *For a semiprimary ring R with radical J, the conditions are equivalent:*

(a) *R is primary-decomposable,*

(b) *R/J^2 is primary-decomposable,*

(c) *the prime ideals of R are commutative.*

Proof. Any semiprimary semiprime ring is semisimple, and the Wedderburn-Artin structure Theorem 8.8 proves the proposition for this case. Since J is nilpotent, and R/J is semisimple, 8.8 implies that J is the intersection of the prime ideals of R, every prime ideal of R is maximal, and there are only finitely many prime ideals $M_1, \ldots, M_r$. By 18.33, there is an integer n_i such that $Q_i = M_i^{n_i}$ is idempotent. If $Q_i = 0$ for some i, then M_i is nilpotent, so $M_i \subseteq J$, and $M_i = J$. This implies that R/J is simple, and therefore R is primary, and the theorem holds. Otherwise, no $Q_i = 0$. Moreover, the Q_i are then comaximal, since M_i is the only prime ideal containing $M_i^{n_i}$. Thus, if $i \ne j$, then $Q_i + Q_j$ is contained in no maximal ideal, and therefore equals R. Next, since we may assume that R

is not a prime ring, zero is a product of prime ideals, say

$$M_{P(1)}^{k_1}\dots M_{P(r)}^{k_r}=0, \qquad 0\leq k_i\leq n_i.$$

Hence, assuming (c) that the prime ideals of R commute, then

$$Q=M_1^{n_1}\dots M_r^{n_r}\subseteq M_{P(1)}^{k_1}\dots M_{P(r)}^{k_r}=0.$$

Hence, by the Chinese remainder theorem 18.30, then there is a canonical isomorphism

$$R=R/Q\approx\prod_{i=1}^{r}R/Q_i$$

of R into a finite product of (completely) primary rings R/Q_i, $i=1,\dots,r$. Thus, (c) $\Rightarrow$ (a), and (a) $\Rightarrow$ (b) is trivial.

(b) $\Rightarrow$ (c). Now (b) implies that M_i and M_j commute modulo J^2. However, since $M_i\supseteq J$, for every i, then $M_iM_j\supseteq J^2$, and

$$M_iM_j=M_iM_j+J^2=M_jM_i+J^2=M_jM_i.$$

Thus, (b) $\Rightarrow$ (c). $\square$

18.38   *Exercise*

18.38.1   (Nakayama [40], Asano [49]) If R is semilocal, and if I is an ideal, and if $I=R\,a=b\,R$, for $a,b\in R$, then $I=aR=Rb$. Furthermore, comaximal cyclic right ideals are commutative.

18.38.2   A completely primary ring R with radical J is a principal left ideal ring if and only if R has a unique composition series

$$R\supset J\supset J^2\supset\cdots\supset J^{n-1}\supset 0$$

in R-mod. This is equivalent to the requirement that J/J^2 be simple in R-mod. (See uniserial modules, Chapter 25.)

18.38.3   (Asano [49]) A semiprimary ring R is a principal left ideal ring if and only if R is a finite product of primary principal left ideal rings.

18.38.4   (Camillo) If R is a local ring, then the direct sum of simple modules has no proper essential extension in their direct product.

Tops and Bottoms

The **top** of a right R-module M is $M/\mathrm{rad}\,M$, and the **bottom** is the socle of M. Thus, $\mathrm{top}\,M$ is the quotient object of M modulo the intersection of the maximal submodules, and dually bottom $M=\mathrm{socle}\,M$ is the subobject which is the sum of the simple submodules.

18.39   Proposition. *If R is semilocal, and $J=\mathrm{rad}\,R$, then*

$$\mathrm{top}\,M=M/MJ$$

$$\mathrm{bottom}\,M=\mathrm{ann}_M\,J.$$

Proof. Exercise using (d) of 18.3. $\square$

When R has Azumaya diagram, then in 18.23 we have defined the terms:

prindec = *principal indecomposable right ideal (module) e R*

principal cyclic module = *a factor module e R/K of a prindec module.*

In case R is right Artinian we say that a prindec of maximal length is **dominant**. We now define a **uniserial module** to be a module with a linearly ordered lattice of submodules.

18.40   Proposition. *Let R be a semilocal ring with $J = \operatorname{rad} R$.*

18.40.1   *The conditions on a right R-module M are equivalent:*

(a) *M is uniserial of finite length n.*

(b) *M has finite length n, and $M J^k / M J^{k+1}$ is simple for every $0 \leq k < n-1$.*

(c) *M possesses a unique composition chain*

$$M \supset MJ \supset MJ^2 \supset \cdots \supset MJ^{n-1} \supset 0.$$

18.40.2   *Every finitely generated uniserial module is cyclic and local.*

Proof. Exercise. $\square$

18.41   *Exercise.* Let R be left and right Artinian.

18.41.1*   (Fuller [69a]) Then R is right selfinjective if and only if there exist pairings of right (left) prindecs such that the top of one is the bottom of the other (cf. QF rings, Chapter 24).

18.41.2*   (Boyle [73]) In this case, R is primary decomposable and every right or left prindec is uniserial if and only if every cyclic has the same top and bottom. This is equivalent to the assertion that R is a left and right PIR (cf. Exercise 18.38.3, 19.45 and also Chapter 25).

18.41   Let R be a right Artinian ring with radical J such that every right prindec eR is uniserial. Then, if n is the biggest integer for which $e J^n \neq 0$, then $M = eR + J^n$ is dominant in R/J^n. (Hint: $M = eR + J^n$ is faithful in R/J^n, and R/J^n embeds in a direct sum of $m < \infty$ copies of M. Cf. 19.13A and 25.4.2.)

Orders in Semilocal Rings

A theorem of Feller and Swokowski states sufficient conditions for a ring R to have a right quoring assuming that R contains an ideal I such that R/I has a right quoring. One condition is that I be reflective in the sense of the next definition. In this last section of Chapter 18, we use this theorem to characterize rings which have semilocal right quorings. These results are related to, and a continuation of Chapter 10 (Orders in Semilocal Matrix Rings).

18.42   Definition and Proposition

18.42.1   *An ideal I of a ring-1 R is said to be **closed** if the equivalent conditions hold:*

C 1. *I is a closed right, and a closed left ideal of R.*

C 2. $c \in R$ *is regular* $\Rightarrow [c+I]$ *is regular in* R/I.

18.42.2   *An ideal I of is* **reflective** *provided that the equivalence holds:*

$$c \in R \text{ is regular} \iff [c+I] \text{ is regular in } R/I.$$

In this case, R is said to be **I-reflective.** *Every reflective ideal is closed.*

18.42.3   *A right ideal I of R is said to be* **q-regular** *provided that*

$$x \in I \ \& \ \text{regular } c \in R \Rightarrow c+x \text{ regular in } R.$$

Every reflective ideal is q-regular.

18.42.4   *If R has a right quoring S, then a right ideal I of R is q-regular if and only if* $IS \subseteq \operatorname{rad} S$.

Proof. The equivalence of C 1 and C 2 is trivial, and so is the assertion 2. If I is reflective, and c regular in R, and if $x \in I$, then $[c+I] = [c+x+I]$ is regular in R/I, and so $c+x$ is regular in R. Thus, every reflective ideal is q-regular.

18.42.4 First, $IS = \{x\, c^{-1} \mid x \in I, \text{ regular } c \in S\}$ by 10.11. Let $u = 1 + x\, c^{-1}$. Then, the equivalence

$$u \in \text{units } S \iff u\, c = c + x \text{ is regular in } R$$

shows that every element $x\, c^{-1}$ of IS is quasi-regular, that is, that $IS \subseteq \operatorname{rad} S$, if and only if I is a q-regular right ideal of R. $\square$

18.43   Proposition. *Let S be a right quoring of R. Then any ideal I of R is reflective if and only if S/IS is canonically 10.16 the right quoring of R/I, and I is q-regular in R (equivalently* $IS \subseteq \operatorname{rad} S$).

Proof. Let I be a reflective ideal. Then, I is q-regular by Proposition 18.42.3. Since a reflective ideal is closed by 18.42.2, then $I = IS \cap R$, and so

$$\begin{cases} \bar{R} = R/I \to S/IS = \bar{S} \\ [r+I] \mapsto [r+IS] \end{cases}$$

is a canonically ring monomorphism. Since R is a right order in S, then $\bar{R}$ is a right order in $\bar{S}$. Furthermore, if $x \in R$ is such that $[x+I]$ is regular in R/I, then by reflectivity of I, x is regular in R, hence is a unit of S.

Therefore $[x+IS]$ is a unit of S/IS. This proves that S/IS is the right quoring of R/I and, moreover, $IS \subseteq \operatorname{rad} S$ by 18.42.4. Conversely, let $IS \subseteq \operatorname{rad} S$, and suppose that S/IS is the right quotient ring of R/I. Then, if $[c+IS]$ is a unit of S/IS, with $c \in R$, by the proof of 18.7, c is a unit of S, and conversely. This proves that I is reflective. $\square$

18.44   *Exercise*

18.44.1   Let S be a quoring, and let I be an ideal such that S/I is a quoring. (For example, S may be any ring with d.c.c. on principal right ideals, and then

I may be any ideal, since then S/I inherits the chain condition. Cf. perfect rings, Chapter 22.) Show that I is reflective if and only if $I \subseteq \mathrm{rad}\, S$. (If S is right Artinian, this is equivalent to the demand that I be nilpotent 18.6.) Conclude that in order for S/I to be the right quoring of S/I it is not necessary for I to be reflective (cf. 18.46).

18.44.2   (Feller and Swokowski [64]) Let M be a finitely generated torsion-free left module over a right and left Ore domain B with quofield D. Then $\mathrm{End}_D D \otimes_B M$ is the semisimple right and left quoring of the prime ring $R = \mathrm{End}_B M$. Thus, R satisfies (acc) $\oplus$ and (acc)$^{\perp}$. (Hint: M is contained in a finitely generate free left B-module generated by $r^{-1} a_1, \ldots, r^{-1} a_n$, where $a_1, \ldots, a_n$ generate M. Also see 19.37(z).)

18.44.3   (Zelmanowitz [67]) Show that 18.44.2 can not be generalized to one-sided Ore domains.

18.45   Definition and Proposition. *An ideal I of a ring R is said to be right* **quorite** *provided that for each pair $x \in I$, and regular $c \in R$, there correspond $x_1 \in I$, and regular $c_1 \in R$ such that $c x_1 = x c_1$.*

18.45.1   *If R has a right quoring S, then any closed ideal of R is right quorite.*

18.45.2   *If R has a right quoring S, and if I is a reflective ideal of R, then I is right quorite, and S/IS is canonically the right quotient ring of R/I.*

Proof. The proof of 18.45.1 is given in the proof of the fact stated in 10.11 that IS is an ideal of S when I is closed. Furthermore, 18.45.2 follows from the first statement in 3 of 18.42 inasmuch as a reflective ideal is closed by 18.42.2. $\square$

18.46   Proposition (Feller and Swokowski [61a, b]). *If R is a ring containing a reflective and right quorite ideal I such that R/I has a right quoring Q, then R has a right quoring S, and $S/IS \approx Q$ is canonically the right quoring of R/I. Thus, $IS = \{a c^{-1} \mid a \in I,\ c \in R \text{ regular}\}$ and $IS \cap R = I$ (cf. I, p. 409, 10.16).*

Proof. If R' denotes the set of regular elements of R, then the criterion for the existence of a right quoring S of R is the regularity condition 9.1 of Ore:

$$\forall a \in R,\ c \in R', \qquad \exists\, a_1 \in R,\ c_1 \in R' \quad \& \quad a c_1 = c a_1$$

or equivalently $a R' \cap c R \neq 0$.

Now assume the hypotheses of the proposition. We may suppose that $a \notin I$. Then, since R/I satisfies the regularity conditions, and since reflectivity of I implies that under the canonical map $R \to \bar{R} = R/I$ that R' maps onto $\bar{R}'$, then $\bar{a}\bar{R}' \cap \bar{c}\bar{R} \neq \bar{0}$. Hence, write $a c_1 = c a_1 + x$, with $c_1 \in R'$, $a_1 \in R$, $x \in I$. Since I is right quorite, there exist $d_1 \in R'$, $x_1 \in I$ such that $c x_1 = x d_1$. Then, $a(c_1 d_1) = c(a_1 d_1 + x_1)$ is the required element of $a R' \cap c R$. Thus, S exists, so we may apply 18.45. $\square$

18.47   Proposition (Faith [71b]). *A ring R has a semilocal right quoring S if and only if R contains a reflective, right quorite ideal T such that R/T is semi-prime right Goldie, and T is the sum of the q-regular right ideals of R. In this case, $TS = \mathrm{rad}\, S$, $T = R \cap \mathrm{rad}\, S$, and $S/\mathrm{rad}\, S$ is canonically the right quoring of R/T.*

Proof. Assume that S is a semilocal right quoring of R with radical J. Then, by 10.17, $\bar{S}=S/J$ is canonically the right quoring of $\bar{R}=R/T$, where $T=J\cap R$, and $TS=J$. Then T is q-regular by 18.42.4. If I is any q-regular right ideal of R, then the same result implies that $J\supseteq IS$, and so $T=J\cap R\supseteq IS\cap R\supseteq I$. Therefore, T is the sum of the q-regular right ideals. Since $T=J\cap R$ is a closed ideal of R, then T is right quorite by 18.45. Moreover, T is reflective by 18.43.

Sufficiency. Under the assumptions that T is reflective, right quorite, and R/T is semiprime right Goldie (with semisimple right quoring Q), then by 18.46, R has a right quoring S such that S/TS is canonically the right quoring of R/T. Since $S/TS\approx Q$ is semisimple, necessarily $TS\supseteq\operatorname{rad}S$. Then the assumption that T is q-regular implies by 18.42.4 that $TS=\operatorname{rad}S$. This implies that $S/\operatorname{rad}S\approx Q$, and that S is semilocal. $\square$

18.48   Corollary. *A ring R has a semilocal right quoring S which respectively is: (a) right Noetherian, (b) right Artinian, (c) a ring with right vanishing radical, (d) semiprimary if and only if R contains an ideal T with the properties stated in 18.47 and with the respective properties:*

(a′) *R satisfies the a.c.c. on closed right ideals.*

(b′) *T is nilpotent, and R satisfies the a.c.c. on closed right ideals. (Then R satisfies the d.c.c. on closed right ideals, and conversely.)*

(c′) *T is right vanishing.*

(d′) *T is right nilpotent.*

Proof. Since R has a semilocal right quoring S, then the lattice isomorphism 10.11 between the right ideals of S and the closed right ideals of R yields (a) $\Leftrightarrow$ (a′).

(a) Since S is semilocal, then S is semiprimary if and only if $\operatorname{rad}S$ is nilpotent. Since $TS=\operatorname{rad}S\supseteq T$ by 18.47, then T is nilpotent. Conversely, since $ST\subseteq TS=\operatorname{rad}S$, then $(TS)^n\subseteq T^nS$. Thus, if T is nilpotent of index $\leq n$, then $\operatorname{rad}S=TS$ is nilpotent of index $\leq n$.

(b) By 18.12, a semiprimary ring is Artinian if and only if it is Noetherian. Also, every Artinian ring is semiprimary. Thus, by (d) and (a), S is Artinian if and only if T is nilpotent and R satisfies the a.c.c. on closed right ideals.

(c) If $\operatorname{rad}S$ is right vanishing, so is T. Conversely, if

$$p_n=a_1\,c_1^{-1}\dots a_n\,c_n^{-1} \qquad\qquad a_n\in T,\ c_n\in R$$

is a sequence of products of elements of $\operatorname{rad}S$, then since T is right quorite, there exist $a_2'\in T$, $c_2'\in R$ such that,

$$c_1^{-1}a_2=a_2'\,c_1'^{-1}$$

and by induction, $p_n=a_1\,a_2'\dots a_n'\,c_n'^{-1}$ with $a_i'\in T$, $i\geq2$. This proves that $\operatorname{rad}S$ is right vanishing if T is $\square$

Note, (c′) is equivalent to the requirement that S is left perfect (see Chapter 22).

Exercises of Chapter 18

1. If $f\colon R \to S$ is a ring homomorphism, then

$$f(\operatorname{rad} R) \subseteq \operatorname{rad} f(R),$$

and the inclusion may be proper.

2. If P is a projective right R-module, then

$$\operatorname{rad}(\operatorname{End} P_R) = \{ f \in \operatorname{End} P_R \mid P \bigcirc \operatorname{im} f \}.$$

3. For any projective module $P \neq 0$, then

$$\operatorname{rad} P = PJ \neq P$$

where $J = \operatorname{rad} R$.

4. If P_i is a B-object of mod-R which is projective, and if $K_i \subseteq \operatorname{rad} P_i = P_i J$, $i = 1, 2$, then

$$P_1/K_1 \approx P_2/K_2 \;\Rightarrow\; P_1 \approx P_2.$$

5. If R is a right B-ring, then

$$\operatorname{rad}(\operatorname{End} P_R) = \operatorname{Hom}_R(P, P(\operatorname{rad} R))$$
$$= \{ f \in \operatorname{End} P_R \mid f(P) \subseteq P(\operatorname{rad} R) \}.$$

Moreover, abbreviating $\operatorname{End} P_R$ by $\operatorname{End} P$, then

$$\operatorname{End} P/\operatorname{rad}(\operatorname{End} P) = \operatorname{End}(P/\operatorname{rad} P).$$

6. (A) If R is a right B-ring, then $\operatorname{rad} R$ is left vanishing.

(B) If R satisfies the d.c.c. on principal left ideals, then R is a semilocal ring with left vanishing radical which contains every nil onesided ideal.

7. If R has a right Artinian right quotient ring, then the polynomial ring $R[x]$ has right Artinian right quotient ring.

8. The ring $R = Z[x, y]/(x^2 - xy)$ has a quoring S, since R is commutative, but S is not semilocal.

9*. (Cf. Robson [67, p. 605, Theorem 2.11]) A commutative Noetherian ring R has an Artinian quoring Q if and only if the maximum nilpotent ideal W is reflective if and only if the prime ideals belonging to 0 are minimal.

10. If e is an idempotent in a ring R, then $e \in \operatorname{rad} R$ if and only if $e = 0$.

11*. (Kaplansky [58a]) If R is a local ring, each projective R-module is free.

12*. (Amitsur [56]) If R is a ring, then $\operatorname{rad}(R[x]) = N[x]$, where N is a nil ideal of R, and $R[x]$ is the polynomial ring.

13*. (Snapper) If R is commutative, then $\operatorname{rad}(R[x]) = N[x]$, where N is the maximal nil ideal of R.

14*. (Goldman [51], Krull) If R is commutative, and if A is a commutative finitely generated algebra over R, then $\operatorname{rad} A$ is a nil ideal.

15. If R is semiprimary, then for every nonzero R-module M, $\operatorname{rad} M \neq M$ and $\operatorname{socle} M \neq 0$. Thus, every nonzero module has a maximal and a minimal submodule.

16. The conditions are equivalent on an ideal I of a ring R: (A) I/I^2 is Noetherian, (B) I^k/I^{k+1} is Noetherian, for every $k \geq 1$.

17. (Janusz) Let R be any ring such that R/J is semisimple and $J \neq J^2$, where $J = \operatorname{rad} R$. Then any simple right R-module M that is not injective can be embedded in J/J^2.

18. If M is an injective R-module, then M is an injective R/I-module for each ideal $I \subseteq \operatorname{ann}_R M$. Moreover, for every ideal A, $\operatorname{ann}_M A$ is an injective R/A-module.

19. (Converse of Exercise 17) Under the hypotheses of 17, show that if M is a nonzero submodule of J/J^2, then M is not an injective R-module.

20. If $R/\operatorname{rad} R$ is a simple ring, any right ideal $A \not\subseteq \operatorname{rad} R$ is a generator. If $A = eR$, where $e = e^2 \in R$ and $e \neq 0$, then $R \approx \operatorname{End}_{eRe} eR$ canonically (cf. 22.25).

21. Let R be a commutative entire ring with quotient field $K \neq R$. Show that $\operatorname{superfl} K_R = K$.

22. (Nagata [51a, b]) A semilocal commutative ring R with finitely many maximal ideals is a direct sum of local rings iff R is a subdirect sum of local rings.

Notes for Chapter 18

The origins of the concept of the Jacobson radical of a ring, defined by Jacobson [45a] and of the Perlis-Jacobson characterization, has been discussed in the preface to this volume. Moreover, the Jacobson radical will be compared to various other radicals in Chapter 26.

The concept of the radical of a module is a much older concept. Indeed, in groups, the intersection $\Phi(G)$ of the maximal subgroups of a group G was introduced by Frattini in 1885! (Thus, for abelian G, $\Phi(G) = \operatorname{rad} G$.) He proved that for a finite group G, $\Phi(G)$ is a nilpotent group. Moreover, a theorem of Wielandt states that G is nilpotent iff the derived group $[G, G] \subseteq \Phi(G)$ (see, for example, Huppert [67, pp. 268–271, esp. Satzes 3.6 and 3.11]).

As stated in the Preface, Artin [27] assumed the ascending chain condition in his structure theory for rings with the descending chain condition, a restriction which was removed, independently by Hopkins [39] and Levitzki [39].

The importance of semilocal rings stems from a vast number of applications from such diverse fields as algebraic geometry, commutative and noncommutative algebra, group theory, module theory, and category theory. In algebraic geometry, or commutative algebra, for example, one can consider the local ring at a point on an algebraic variety, or at a prime ideal of a ring.

According to Bourbaki [65, *Note Historique*, p. 131] the general idea of a local ring developed very slowly: Grell (1926), and Krull (1938), for domains, Chevalley (1944) for Noetherian rings, and the general case by Uzkov (1948).

Beginning about 1940, the local ring R_P at a prime ideal P of a domain R was consistently used by Krull (and his school), and in algebraic geometry by Chevalley and Zariski. Krull's term *stellenring* was superceded by Chevalley's terminology *local ring*, a ring associated with a point of a variety which gives "local properties" of the variety, for example, the ring of all functions regular at that point (also see Nagata [62, p. xi]). This chronology omits the important work of Hensel at the turn of the century on p-adic numbers; however, Hensel considered not R_P but the completion of R_P, that is, the p-adic completion (see Bourbaki, loc. cit. and also 21.7 A ff.).

Köthe [30a] studied noncommutative semilocal rings with a **Köthe radical**, that is, a nil ideal K containing every nil onesided ideal. (It is an open question if every ring has a Köthe radical.) Köthe (*loc. cit.*) proved that a semilocal ring R with Köthe radical K is isomorphic to a full matrix A_n over a local ring A iff R/K is simple. Moreover $R \approx \mathrm{End}_A E$, where E is one of the rows of A_n considered as a module over A. This theorem generalizes the Wedderburn-Artin theorem, the theorem of Noether [29] (for simple semisimple R), and is substantially the same theorem stated in 22.24. Moreover, Köthe [30a, p. 182, Satz 13] proved that in a semilocal lift/rad ring with radical J, and right prindecs eR and fR, that $eR \approx fR$ iff $eR/eJ \approx fR/fJ$. (Köthe's proof of this is for $J = $ Köthe radical.)

Köthe [30b] generalized two theorems of Shoda to a semilocal ring R with nil radical K:

(1) *The intersection of all the nil subrings of R is K.*

(2) *In case R is Artinian, any two maximal nilpotent subrings of R are conjugate, and every nilpotent subring is contained in a maximal nilpotent subring.*

(Shoda proved (1) for Artinian, and (2) for finite rings.) Also (1) has been generalized by Michler [66] to right Noetherian R. (See 17.31(b) and (c).)

(2) implies that *any nilpotent subring S of an $n \times n$ matrix ring k_n over a field k* can be placed simultaneously into triangular form, inasmuch as the ring $T_n(k)$ of upper triangular matrices is maximal nilpotent. In answering a conjecture of Köthe, Levitzki [30] (also [45a]) showed that every nilsubring of an Artinian (also Noetherian) ring is nilpotent, and thereby sharpened Köthe's theorem, that is, then nil subrings of k_n can be placed simultaneously in Δ-form (17.19). Levitzki [50] generalized nil $\Rightarrow$ nilpotence to multiplicatively closed systems (M-systems) of left and right Noetherian rings.

Fitting [33] established the relationship between the direct decomposition of a module M of finite length and endomorphism ring A (loc. cit., p. 528, Satz 4). In particular, M is indecomposable iff A is local (loc. cit. p. 533, Satz 8). This is called Fitting's "lemma" (see 17.17' and 17.30).

Noncommutative semilocal rings occur naturally even in the study of commutative algebra, since, as pointed out by 18.26, a module (or object of an abelian category) M satisfies the finite Krull-Schmidt (that is, has a finite Azumaya diagram) iff the endomorphism ring of M is semilocal lift/rad. These in turn are the semiperfect rings of Bass [60] (see Theorem 22.23), that is, rings over which every finitely generated module has a projective cover. In this paper, in characterizing rings over which every right module has a projective cover, Bass obtained a natural generalization of semiprimary rings, as the title suggests. (Instead of nilpotence of the radical, one requires transfinite nilpotency, which Bass called T-nilpotency, and we call left vanishing.)

Kaplansky [68, p. 4] expressed his doubt that Artinian rings are the natural generalization of finite dimensional algebras because "natural examples are not common," and suggests as an alternative rings which are finitely generated modules over Noetherian subrings of their centers (that is, PI-algebras). Be that as it may, semiprimary rings certainly do arise naturally as endomorphism rings of (Jordan-Holder) modules of finite lengths. (Exercise 18.14(b) which follows from 17.20 and 18.26.) Levitzki [44] characterized semiprimary rings as semilocal rings modulo the ideal N generated by all nilpotent one-sided ideals, and satisfying the d.c.c. on products of ideals in N; and Björk [70] characterizes semiprimary rings as follows: there exists an integer n such that R does not contain a strictly decreasing sequence of n principal left ideals.

Many of the other papers cited in this chapter also contribute to the structure of semiprimary rings, e.g. Hopkins [39] and Levitzki [39] (see 18.12), Eilenberg-Nagao-Nakayama [56] (see 18.33) and Asano [49] (see 18.37 and 18.38). Moreover, Chase [60] anticipated Björk's theorem (22.30) stating that right modules satisfy the d.c.c. on finitely generated submodules whenever R satisfies the d.c.c. on principal right ideals. Chase proved this holds (on both sides) for a semiprimary ring (18.14). (Some other theorems of Chase [60] are taken up in Chapter 20.)

Eilenberg-Nagao-Nakayama (*loc. cit.*) proved a hereditary semiprimary ring has the property that every factor ring has finite global dimension. Jans-Nakayama [57] and Chase [60] characterized semiprimary rings with the property that every factor ring has finite global dimension: they are triangular in the sense that there is a set $e_1, \ldots, e_n$ of sum-1 orthogonal indecomposable idempotents such that $e_i(\text{rad } R) e_j = 0$ for $i \geq j$. This happens iff $(R/(\text{rad } R)^2)$ has finite global dimension, and then gl.dim R is strictly less than the number r of simple rings in the Wedderburn-Artin decomposition of $R/\text{rad } R$. (Actually, gl.dim $R \leq$ gl.dim $R/(\text{rad } R)^2 < r$. See Chase [60, p. 22].) Moreover, when $R/\text{rad } R$ is "separable" then any semiprimary ring with $R/(\text{rad } R)^2$ of finite global is a factor ring of a unique hereditary semiprimary ring (Jans-Nakayama (*loc. cit.*), cited by Chase (*loc. cit.*)). This theory is generalized further by Harada [64], who completely determines the structure of a hereditary semiprimary ring R as a generalized triangular matrix ring over a semisimple ring. Moreover, for any ideal I,

$$\text{gl. dim } R/I \leq r - s + 1$$

where s is the number of simple ideals in a direct sum decomposition of I modulo rad R. Also, if $n =$ index of nilpotency of rad R, then

$$\text{gl. dim } R/I \leq n - 1.$$

Harada applies these results to give another proof of the main structure theorem for a hereditary order R over a rank 1 discrete valuation ring (Harada [63]). (See Reiner [75, p. 358] for this and other results on the structure of hereditary orders, including the theorem of Jacobinski [71] stating that hereditary orders are "extremal".)

Zaks [68] carried out a similar theory for T-rings, that is, semiprimary rings in which every indecomposable direct factor has a least ideal $\neq 0$, and every minimal ideal is a projective module. The factor rings have finite global dimension, so they are generalized triangular matrices over semisimple rings in Harada's sense, a fact which Levy describes (*loc. cit.*, p. 76) (without reference to Harada).

Going back for a moment to semilocal rings, many papers have been written on orders in semilocal rings, beginning with those of Goldie and Lesieur-Croisot (an account of which is given in Chapter 9 (Volume I)). Theorem 18.47 gives a characterization of these rings in terms of a maximal q-regular (reflective, quorite, etc) ideal which is reminiscent of the Perlis-Jacobson characterization of the radical of a ring. For a survey of the literature on orders in semilocal rings, consult Elizarov [69] and Faith [71b]; and for orders in QF rings, consult Notes for Chapter 24.

In concluding this chapter on semilocal rings, a warning: in commutative algebra, semilocal has a more precise meaning: a commutative ring R is quasi-semilocal if R has just finitely many maximal ideals, and then it is called semi-local if it is also Noetherian (Nagata [62, p. 13]). The subject of commutative semilocal and local rings is well nigh inexhaustible, as a glance at Nagata [62], or Krull [48], readily substantiates. (Also refer to Nagata [50, 51a, b, 60].) Precise results on local rings of global dimension 2, in fact a characterization, are contained in Vasconcelos [72] and Greenberg [74].

I ought to add a word about the origins of 18.24, the part stating that a semi-local lift/rad ring R is similar to its basic ring in its protean form owes to Wedderburn who never formalized it in categorical language; if $R = D_n$ is a $n \times n$ matrix ring over a ring D, then $X \mapsto X e_{11}$ defines an equivalence of categories mod-$R \approx$ mod-D in the natural way; since, if e_{11} is the obvious matrix unit, then $D \approx e_{11} R e_{11}$. Similarity was also used informally in many other instances, and in its present form 18.24 is a trivial consequence of a theorem Morita [58].

References

Amitsur [56], Asano [49], Azumaya [50], Baer [43], Bass [60, 68], Björk [69, 70], Bourbaki [65], Boyle [73], Chase [60, 61], Cohen [50], Eilenberg, Nagao, and Nakayama [56], Elizarov [69], Faith [71b], Faith and Utumi [64], Feller and Swokowski [61a, b], Fitting [33], Fuller [69a], Harada [63, 64], Hopkins [39], Huppert [67], Jacobson [43, 45a, b, 64], Jans and Nakayama [57], Kaplansky [46, 68], Köthe [30a, b], Krull [25, 38, 48], Levitzki [30, 39, 44, 45a], Nagata [50, 51a, b, 60, 62], Nakayama [40], Noether [29], Ornstein [68], Perlis [42], Posner [60a, b], Robson [67], Schmidt [28], Swan [68], Zaks [68], Zelmanowitz [67], Greenberg [74], Vasconcelos [72], Artin [27], Ginn and Moss [75], Jacobinski [71], Michler [66], Morita [58], Reiner [75].

Chapter 19. Quasinjective Modules and Selfinjective Rings

If M is a module such that every map $f: S \to M$ of a submodule S is induced by an endomorphism of M, then M is said to be **quasinjective,** or QI, for short. Every module which is injective modulo annihilator, and every semisimple module, is QI (see 19.2). The QI modules coincide with the class of fully invariant submodules of injective modules 19.3. A module which is finitely generated over endomorphism ring is said to be **finendo.** Any finendo QI module is injective modulo annihilator 19.14A. Over a right Artinian ring, any QI right module is finendo and conversely 19.16. Thus, every faithful quasinjective over a right Artinian ring is injective 19.15, a result which holds for finitely generated faithful modules over commutative rings 19.17.

The ideas behind the proof of this are best expressed by Beachy's theorem 19.13A which states that an "essentially right Artinian ring R" may be characterized by the fact that every faithful right R-module M is compactly faithful in the sense that R embeds in a finite product of copies of M. (This also provides the "essentially Hopkins-Levitzki theorem": an essentially right Artinian ring is essentially right Noetherian!)

An example of a ring with simple right modules which are not injective modulo annihilator is a left full linear ring $\operatorname{End}_D M = L$ over an infinite dimensional left vector space M over D. The canonical right L-module M is not injective 19.46. (In fact, if every quasinjective module is injective, then 20.5 the ring R is necessarily right Noetherian.)

Right selfinjectivity of a ring R is describable by a finendo condition: every finendo faithful right R-module generates mod-R (cf. 19.20; also 25.7B).

Other main topics in this chapter are (1) the quasinjective hull of a module 19.7; (2) the double annihilator condition 19.10 for finitely generated $\operatorname{End}_R M$ submodules of a QI module M which yields (3) the density theorem 19.22 for primitive rings; (4) von Neumann regular rings 19.25; (5) the radical and ring structure of the endomorphism ring of a QI-module 19.27; (6) the Utumi characterization of the radical of a right selfinjective ring 19.28; (7) the Findlay-Lambek rational extensions 19.32–3; (8) the Johnson-Utumi maximal right quotient ring 19.34; (9) Dedekind finite rings 19.39–43.

Included in this chapter is a characterization of prime and semiprime right Goldie rings: A ring R is prime right Goldie iff R possesses an indecomposable finendo faithful injective right module E with the two properties: (1) E has no nontrivial fully invariant submodules; (2) End E_R is a field 19.54. (Semiprime Goldie rings are characterized similarly 19.55B.) When this is so, then $Q = \operatorname{Biend} E_R$ is the right quotient ring of R.

The last two sections of this chapter are devoted to Goodearl's theorems on the prime ideal structure of a regular right selfinjective ring R: the ideals of R containing a prime ideal P are linearly ordered by inclusion, and are all prime. Moreover, if R is not the injective hull of P, then P is a primitive ideal, the ideals containing P are then well ordered, and are all primitive 19.64. Similar theorems hold in any right selfinjective ring for any prime ideal containing the Jacobson radical; in particular, the ideals containing any primitive ideal are linearly ordered 19.69.

19.1 *Examples*

19.1.1   A right A-module M is injective in case the exactitude of

(a) $0 \to X \to Y \to Z \to 0$

in mod-A implies that of

(b) $0 \to \operatorname{Hom}_A(Z, M) \to \operatorname{Hom}_A(Y, M) \to \operatorname{Hom}_A(X, M) \to 0.$

Furthermore, M is quasiinjective, or QI, when (a) $\Rightarrow$ (b) whenever $Y = M$. If M is a module which is injective modulo annihilator, that is, if M is injective as an $A/\operatorname{ann}_A M$ module, then M is a QI canonical right A-module.

By Baer's criterion for injectivity (I, p. 157), a QI module M is injective if there is an embedding $R \hookrightarrow M^n$ of R into some finite power of M. This happens iff M finendo and faithful 19.15.

19.1.2   Any semisimple module M is quasiinjective since every submodule N splits (I, 8.2 p. 366).

19.1.3   Any cyclic p-group $\mathbb{Z}_{p^n}$ is quasiinjective in mod-$\mathbb{Z}$ (Proof?).

19.1.4   A ring R is said to be **right selfinjective** if R is injective in mod-R. As stated in 19.1.1, Baer's criterion implies that R is right selfinjective if and only if R is quasiinjective in mod-R.

19.1.5   A module M is quasiinjective if and only if M is a fully invariant submodule of its injective hull (19.3). (The sufficiency, which is trivial, implies Example 19.1.3.) If E is injective, I is an ideal of $B = \operatorname{End} E_R$, and A is an ideal of R, then IE, EA, $\operatorname{ann}_E I$, and $\operatorname{ann}_E A$ are all quasiinjective and, moreover, $\operatorname{ann}_E A$ is injective in mod-R/A.

19.1.6   Quasiprojectives are defined dually to quasiinjectives, and the duals to 19.1.5 hold. Thus, if K is any fully invariant submodule of a projective module P, then P/K is quasiprojective; and if I is any ideal of $\operatorname{End} P_R$, or if A is any ideal of R, then P/K is quasiprojective, for $K = IP$, or $K = PA$.

19.1.7   A direct sum of two quasiinjectives need not be quasi, but M^n is quasi, for any quasiinjective M, and integer $n > 0$ (see 19.5). Moreover, M^c is then QI for any cardinal number c iff M is injective over $R/\operatorname{ann}_R M$ (Exercise 19.21(p)). In particular, by 19.14A, this holds for any finendo quasiinjective.

19.1.8   If M is a right R-module, and if $\hat{R}$ is the injective hull of R, then $\hat{R} \oplus M$ is quasiinjective if and only if M is injective in mod-R. (This follows from the last statement in 19.1.1.)

19.1.9   If every direct sum of any two QI modules is QI, then R is a right Noetherian V-ring in which every quasinjective module is injective (cf., 20.4B).

19.1.10   The unique simple module in the $R = K[y, D]$ ring of differential polynomials in a single derivation D over a universal (Kolchin) field K is injective, faithful but, of course, is not finite dimensional 7.41, hence not finendo.

Quasinjectives Characterized

Let $\hat{M} = E(M)$ denote the injective hull of any module M. Recall that $M \triangledown N$ denotes that N is an essential submodule.

19.2   Proposition (Johnson-Wong [61]). *If M is any right R-module, if $\hat{M} = E(M)$, and if $\Lambda = \operatorname{End} \hat{M}_R$, then:*

(a) *ΛM is the intersection of all quasinjective submodules of $\hat{M}$ containing M;*

(b) *ΛM is quasinjective;*

(c) *M is quasinjective if and only if $M = \Lambda M$.*

Proof. (b) If $f: N \to \Lambda M$ is any map of a submodule N of ΛM into ΛM, then f is induced by some $\lambda \in \Lambda$. Since $\lambda(\Lambda M) \subset \Lambda M$, λ induces $\bar{\lambda} \in \operatorname{Hom}_R(\Lambda M, \Lambda M)$, and $\bar{\lambda}$ also induces f, showing that ΛM is quasinjective.

(a) Let P be any quasinjective submodule of $\hat{M}$ containing M. We wish to show that $P \subseteq \Lambda M$, so it is sufficient to show that $\alpha P \subseteq P \ \forall \alpha \in \Lambda$. To do this, we note that $Q(\alpha) = \{x \in P \mid \alpha x \in P\}$ is a submodule of P, and we have only to show that $Q(\alpha) = P \ \forall \alpha \in \Lambda$. Since $q \to \alpha q$, $q \in Q = Q(\alpha)$ a map of Q into P, and since P is quasinjective, there exists $\alpha_1 \in \operatorname{Hom}_R(P, P)$ such that $\alpha_1 q = \alpha q \ \forall q \in Q$. Since $\hat{M}$ is injective, there exists $\alpha' \in \Lambda$ such that $\alpha' x = \alpha_1 x \ \forall x \in P$. Since $\alpha' P \subseteq P$, if $(\alpha' - \alpha) P = 0$, we have $\alpha P \subseteq P$. Thus, if $Q(\alpha) \neq P$, then $(\alpha' - \alpha) P \neq 0$. Since $\hat{M} \triangledown M$, necessarily $\hat{M} \triangledown P$, and consequently $(\alpha' - \alpha) P \cap P \neq 0$. But if x, $0 \neq y \in P$ are such that $y = (\alpha' - \alpha) x \in (\alpha' - \alpha) P \cap P$, then since $\alpha' x = \alpha_1 x \in P$, we have that $\alpha x = \alpha_1 x - y \in P$. But then $x \in Q(\alpha)$, so that $\alpha x = \alpha' x$, and so $y = 0$, a contradiction which establishes (a).

(c) is an immediate consequence of (a) and (b).   □

19.3   Corollary. *A module M is quasinjective if and only if M is a fully invariant submodule of its injective hull.*   □

A submodule N of M is **closed** in case each submodule of M which contains N and is essential over N coincides with N

19.4   Proposition (Faith-Utumi [64a]). *Let M be quasinjective in mod-R, and let N be a closed submodule. Then, any map w of a submodule K of M into N can be extended to a map u of M into N.*

Proof. By Zorn's lemma we can assume that K is such that w cannot be extended to a map of T into N for any submodule T of M which properly contains K. Since M_R is quasinjective, w is induced by a map $u: M \to M$. Suppose $u(M) \nsubseteq N$, and let L be a complement of N in M. Since N is closed, N is a complement of L.

Therefore, since $u(M)+N \supset N$, we see that $(u(M)+N) \cap L \neq 0$. Let $0 \neq x = a+b \in (u(M)+N) \cap L$, $a \in u(M)$, $b \in N$. If $a \in N$, then $x \in N \cap L = 0$, a contradiction. Thereofore, $a \notin N$, and $a = x - b \in L \oplus N$. Now $T = \{y \in M \mid u(y) \in L \oplus N\}$ is a submodule of M containing K. If $y \in M$ is such that $u(y) = a$, then $y \in T$, but $y \notin K$, since $a \notin N$. Let π denote the projection of $L \oplus N$ on N. Then πu is a map of T in N, and $\pi u(y) = u(y) = w(y) \ \forall \ y \in K$. Thus πu is a proper extension of w, a contradiction. Therefore, $u(M) \subseteq N$, and u is the desired extension of w. $\square$

Analogously to the corresponding result for injective modules, one proves that if a product $\prod_{i \in I} M_i$ of R-modules $\{M_i \mid i \in I\}$ is quasinjective then M_i is quasinjective $\forall \ i \in I$. Unlike the case for injectives, however, the converse does not hold.

19.5 *Example.* Let $M = \mathbb{Q} \oplus \mathbb{Z}_p$, where p is a prime. Then the canonical epimorphism $\mathbb{Z} \to \mathbb{Z}_p$ is a map of the subgroup $\mathbb{Z}$ of $\mathbb{Q}$ onto $\mathbb{Z}_p$ which can not be extended to $\mathrm{Hom}_{\mathbb{Z}}(\mathbb{Q}, \mathbb{Z}_p)$, and therefore cannot be extended to $\mathrm{Hom}_{\mathbb{Z}}(M, M)$. Thus a sum of quasinjective modules need not be quasinjective. Cf. 19.9(c).

Quasinjective Hull

(P, M, f) denotes a monomorphism $f: M \to P$, and is called an **extension** of M.

An extension (P, M, f) of a module M is a **minimal quasinjective extension** in case P is quasinjective and the following condition is satisfied:

If (A, M, g) is any quasinjective extension of M, then there exists a monomorphism $\varphi: P \to A$ such that

commutes.

19.6 Corollary. *Let M be quasinjective. (a) If N is any closed submodule of M, then N is a summand of M, and N is quasinjective. (b) If P is any submodule of M, then there exists a quasinjective essential extension of P contained in M. (c) Each minimal quasinjective extension of a module K is an essential extension of K.*

Proof. (a) If $e: M \to N$ is the extension given by the theorem of the injection map $N \to N$, then $M = N \oplus \ker(e)$, so that N is a summand of M; N is quasinjective by the remark preceding the corollary.

(b) By Zorn's lemma, P is contained in a closed submodule N which is an essential extension of P, and N is quasinjective by (a).

(c) is an immediate consequence of (b). $\square$

19.7 Proposition. *In the notation of* 19.2, *ΛM is a minimal quasinjective extension of M. Any two minimal quasinjective extensions are equivalent.*

Proof. Let (A, M, g) be any quasinjective extension of M, let $\hat{A} = E(A)$, and let $\Omega = \mathrm{Hom}_R(\hat{A}, \hat{A})$. Then, by 19.2 $\Omega A \subseteq A$. Since $M_0 = \Lambda M$ is an essential extension of M, the monomorphism $g: M \to \hat{A}$ can be extended to a monomorphism (also denoted by g) of M_0 in A. Since $g(M_0)$ is quasinjective, then $\Omega(g(M_0)) \subseteq g(M_0)$, and we conclude that $\Omega(B) \subseteq B$, where $B = A \cap g(M_0)$. Then, by 19.2 B is quasinjective. It follows that $g^{-1}B$ is a quasinjective extension of $M \subseteq M_0 = \Lambda M$. Since ΛM is the smallest quasinjective extension of M contained in $\hat{M}$, we conclude that $g^{-1}B = M_0$, so $B = g(M_0) \subseteq A$. This establishes that $M_0 = \Lambda M$ is a minimal quasinjective extension. It follows immediately that if (A, M, g) is also a minimal quasinjective extension of M, that (A, M, g) is equivalent to ΛM. $\square$

In the future $Q(M)$ will denote any of the equivalent minimal quasinjective extensions of M. By 19.6 and 19.7 we have:

19.8 Corollary. *Let M be a quasinjective module and let N be a submodule. Then $M = Q(N)$ if and only if $M \,\nabla\, N$.* $\square$

19.9 *Exercise.* (a) Let M be a module, let $\{M_i \,|\, i \in I\}$ be a family of independent submodules, and let Q_i be an essential extension of M_i in M $\forall i \in I$. Then $\{Q_i \,|\, i \in I\}$ is an independent family of submodules, and $\sum_{i \in I} Q_i$ is an essential extension of $\sum_{i \in I} M_i$.

(b) A QI-module M is indecomposable if and only if $\hat{M}$ is indecomposable.

(c) For any ring R, and module M, the direct sum $R \oplus M$ is quasinjective iff R is selfinjective, and M is injective. Moreover, $\hat{R} \oplus M$ is quasinjective iff M is injective.

(d) (Natasescu and Popescu) For any ring, any projective simple module is injective.

Double Annihilator Condition

If M is a right R-module with endomorphism ring S, then an additive subgroup X of M such that $sx \in X$ $\forall s \in S$ is an S-module, and a submodule of the canonical left S-module M. Thus, we may refer to S-submodules of M which are not necessarily R-submodules. An R-submodule which is also an S-submodule is called a **fully invariant** submodule (I, p. 178).

For any nonempty subset X of M,

$$\mathrm{ann}_R X = \{r \in R \,|\, X r = 0\} = X^\perp$$

is a right ideal of R, called the **annihilator** of X. If $M = R$, then $X^\perp$ is a annihilator right ideal, or right annulet. For a nonempty subset A of R, the set

$$\mathrm{ann}_M A = \{m \in M \,|\, m A = 0\} = {}^\perp A$$

is an S-submodule of M, called the **annihilator** of A in M. Then, $\mathrm{ann}_M \mathrm{ann}_R X$ is called the **double annihilator** of X, and $\mathrm{ann}_R \mathrm{ann}_M A$ is called the **double annihilator** of A. Moreover, A is an annihilator of a subset of M if and only if $A = \mathrm{ann}_R \mathrm{ann}_M A$, in which case A is said to satisfy the **double annihilator condition** with respect to M. This is abbreviated by M-d.a.c., or simply d.a.c., when M is a fixed module in a discussion. Dually, for annihilators in M of subsets of R. A collection of subsets (of M or R) is said to satisfy the d.a.c. provided that every set in the collection satisfies the d.a.c., that is, is an annihilator. For example, the left and right annulets of R satisfy the R-d a.c.

The proposition below comes from Wedderburn-Artin [50]—Tate for (semi) simple modules, and Jacobson [56, 64] (see the lemma on p. 27), and Johnson and Wong [61].

19.10 Proposition. *Let M be a quasiinjective R-module with endomorphism ring S.*

(a) *Every finitely generated S-submodule of M satisfies the double annihilator condition.*

(b) *If F is a finitely generated S-submodule, and if N is an S-submodule that satisfies the double annihilator condition, then so does $N+F$.*

Proof. (a) is the $N=0$ case of (b). (b) is proved by induction on the number of generators of F. It suffices to prove (b) for the case $F=Sx$. For any subset X, ${}^{\perp}(X^{\perp}) \supseteq X$, so we must show that ${}^{\perp}((N+Sx)^{\perp}) \subseteq N+Sx$. Now

$$(N+Sx)^{\perp}=N^{\perp}\cap(Sx)^{\perp}=N^{\perp}\cap x^{\perp}.$$

Let $y \in {}^{\perp}((N+Sx)^{\perp})={}^{\perp}(N^{\perp}\cap x^{\perp})$, so $y(N^{\perp}\cap x^{\perp})=0$.

Consider the correspondence:

$$\theta: xa \mapsto ya, \qquad\qquad a \in N^{\perp}.$$

If $a, b \in N^{\perp}$ are such that $xa=xb$, then $a-b \in x^{\perp}=(Sx)^{\perp}$, hence $(a-b)\in N^{\perp}\cap x^{\perp}$, and therefore $y(a-b)=0$; that is, $ya=yb$. This shows that θ is a mapping $xN^{\perp} \to yN^{\perp}$. Since $\theta(xar)=\theta(xa)r \ \forall r \in R$, θ is a map of the R-submodules $xN^{\perp}$ and $yN^{\perp}$. Since M is quasiinjective, θ is induced by an element of Λ, which we also designate θ. Since $(\theta x - y) N^{\perp}=0$, then $z=\theta x-y$ is an element of $N={}^{\perp}N^{\perp}$, and hence $y=-z+\theta x \in N+Sx$, proving

$$^{\perp}((N+Sx)^{\perp})=N+Sx. \quad \square$$

R is right **selfinjective** provided that R is injective in mod-R.

19.11 Corollary. *Any finitely generated left ideal of a right selfinjective ring is a left annulet.* $\square$

19.12 Proposition. *If E is any injective right A-module, then for any ideal B of A, $\mathrm{ann}_E B$ is an injective A/B-module.*

Proof. Let F be the injective hull of $X=\mathrm{ann}_E B$ taken in mod-A/B. Then F is a canonical A-module, and is essential over X. Therefore, F can be embedded in the injective hull $\hat{X}$ of X, hence $X \subseteq F \subseteq E$. But $FB=0$, so $F \subseteq X=\mathrm{ann}_E B$, and $F=X$ as asserted. $\square$

Finendo and Compact Faithful Modules

In this section in addition to the results on finendo modules mentioned in the chapter introduction, we introduce the concept of compact faithful modules, and characterize rings over which every faithful module is compact faithful.

An object M of mod-R will be said to be **compactly faithful**, or by abuse, **compact faithful** (CF), provided that there is an embedding $0 \to R \to M^n$ for some finite integer $n > 0$. Thus, any CF module is faithful. (Compare faithful and cofaithful modules defined in Volume I, p. 143, and also 3.28 on p. 146.)[1]

19.13A Theorem (Beachy [71b]). *The following conditions on a ring R are equivalent; in which case R is said to be **essentially right Artinian**.*

(a) *R has an essential Artinian right ideal A.*

(b) *R has a finitely generated essential right socle S.*

(c) *Every faithful right R-module M is compactly faithful.*

Proof. (a) $\Rightarrow$ (b) follows from (I, 8.3, p. 367) which states that $A \supseteq S$, so that $S = \text{socle} A$. However, any Artinian module manifestly has finite essential socle, so (b) follows.

(b) $\Rightarrow$ (c) Embed R in M^a for some cardinal a. (This can be done as stated on p. 143 of Volume I.) Then, (b) implies that there is a finite subset of the set of projections $\{f_i : M^a \to M\}_{i \in a}$ such that the intersection of their kernels intersected with S is zero. Say $\{f_i\}_{i \in b}$, with b a finite cardinal $\leq a$, has this property. Then $R \cap (\cap_{i \in b} f_i) = 0$, so that R embeds in M^b.

(c) $\Rightarrow$ (b) If E is the smallest cogenerator of mod-R, (I, 3.55, p. 167), then E is faithful by the dual of (I, 3.26, p. 144), so that R embeds in E^n for an integer $n > 0$. But, E^n is the direct sum of injective hulls E_i of simple modules V_i, for a set $\{V_i\}_{i \in I}$, and thus, R is contained in a finite direct sum of these, since R is finitely generated (by 1). Adjust notation so that $R \subseteq \bigoplus_{i=1}^n E_i$. Since the right side of the containment has finite essential socle, then so does the left, that is, so does R. This completes the proof since (b) $\Rightarrow$ (a) is trivial. □

We remark that if a ring R is defined to be **essentially right Noetherian** analogously to (a), then the theorem of Hopkins-Levitzki 18.13 still holds:

19.13B Corollary. *Any essentially right Artinian ring is essentially right Noetherian.*

Proof. By (b), R has an essential Noetherian right ideal S. □

19.13C *Exercise and Definition.* An ideal N of a ring is said to be **right essentially nilpotent** iff N contains a nilpotent right ideal which is an essential submodule.

1. (Shock [71c]) Show that a nil ideal N is right essentially nilpotent iff N contains a right vanishing right ideal essential in N. Also conclude that any nil ideal of an essentially right Artinian ring is right essentially nilpotent.

2. A ring R is right Artinian iff every factor ring is essentially right Artinian. □

---

[1] Beachy [71a, b, c] and Beachy-Blair [75] call a CF module cofaithful.

19.14 A   Theorem and Definition. *Any right A-module M which is finitely generated over* $\operatorname{End} M_A$ *is said to be* **finendo**. *A finendo right A-module M is* QI *if and only if M is an injective canonical* $A/\operatorname{ann}_A M$ *module.*

The proof will follow 19.15.

19.14 B   Proposition. *For any right R-module M, set* $B = \operatorname{End} M_R$, *and*

$$A = \operatorname{End}_B M = \operatorname{Biend} M_R.$$

(a) (*M finitely generated in* mod-*R* $\Rightarrow$ *compactly faithful in B-mod.*)

$$R^n \to M \to 0 \ exact \ in \ \text{mod-}R \ \Rightarrow \ 0 \to B \to M^n \ exact \ in \ B\text{-mod.}$$

(b) (*M is finendo* $\Rightarrow$ *compactly faithful in* mod-Biend M_R.)

$$B^n \to M \to 0 \ exact \ in \ B\text{-mod} \ \Rightarrow \ 0 \to A \to M^n \ exact \ in \ \text{mod-}A;$$

$$0 \to A \to M^n \ exact \ in \ \text{mod-}A \ \Rightarrow \ 0 \to R/\operatorname{ann}_R M \to M^n \ exact \ in \ \text{mod-}R.$$

(c) *A finendo faithful module is compactly faithful.*

Proof. (a) Apply the left exact functor $(\,,M) = \operatorname{Hom}_R(\,,M)$ to the exact sequence on the left, and use the natural isomorphisms

$$(B^n, M) \approx (B, M)^n \approx M^n.$$

(Consult (I, 3.38, p. 155) for the first, and (I, 3.6.3, p. 121) for the second.) (a) then follows forthwith, and the first implication of (b) is the same as (a), while the second then follows from the fact that $R/\operatorname{ann}_R M$ embeds in A canonically. Moreover, (c) is then trivial.   □

19.15   Proposition. *Let M be a* QI *faithful right R-module. The following conditions are equivalent:*

(a) *M is finendo.*

(b) *M is compactly faithful.*

When this is so, then M is injective in mod-*R*.

Proof. (a) $\Rightarrow$ (b) by (c) of the last proposition. Moreover, applying $(\,,M)$ to the exact sequence in (b), and using the fact that M is QI yields $B^n \to M \to 0$ exact in B-mod. Thus, (b) $\Rightarrow$ (a).

Baer's criterion (I, 3.41, p. 157) for injectivity requires only that every map $I \to M$ of a right ideal I of R can be extended to a map $R \to M$. This is indeed the case for any QI module containing R. Thus, (b) implies that M^n, whence M, is injective.   □

Proof of Theorem 19.14 A. If M is any QI right R-module, then M is a QI right $R/\operatorname{ann}_R M$ module, and $\operatorname{End}_R M$ coincides with the endomorphism ring of M over $R/\operatorname{ann}_R M$. Thus, 19.15 implies that M is injective over $R/\operatorname{ann}_R M$.   □

The class of right Artinian rings satisfies the condition of 19.14 A, as observed by Fuller [69]. Moreover, these are the only rings with the property that the quasinjective modules are finendo, as the next theorem shows.

19.16 A   **Theorem** (Vámos [68], Beachy [71b], Faith [72b]). *The following conditions on a ring A are equivalent:*

(a) *A is right Artinian.*

(b) *Every quasinjective right module is finendo.*

(c) *Every factor ring of A is essentially right Artinian.*

When this is so, then a right module M is quasinjective iff M is injective modulo $\mathrm{ann}_A M$.

Proof. (a) $\Rightarrow$ (b). For any right module M, the factor ring $\bar{A} = A/\mathrm{ann}_A M$ is right Artinian, so M is compactly faithful over $\bar{A}$ by 19.13. Thus, if M is quasinjective, then M is finendo by 19.14 A.

(b) $\Rightarrow$ (c) Let M denote a direct sum of a complete set of representatives of simple right A-modules. Then, M is a QI faithful module over $\bar{A} = A/\mathrm{rad}\, A$, so 19.15 implies an embedding $\bar{A} \hookrightarrow M^n$, for some integer n. Thus, $\bar{A}$ is semisimple, so that M is a finite direct sum. Let E denote its injective hull. Since E is faithful (as stated in the proof of (c) $\Rightarrow$ (b) of 19.13 A), then A embeds in E^m, for an integer $m > 0$. Then, A has a finite essential socle (along with E^m), so A is essentially right Artinian by 19.13 A.

Before giving the proof of (c) $\Rightarrow$ (a), we make a definition. A module M is said to be **essentially Artinian**, or **finitely embedded** provided that the socle of M is an essential submodule of finite length. It is an easy exercise to show that this condition holds iff the submodules of M have the **finite intersection property**; namely: if $\{M_i\}_{i \in I}$ is any collection of submodules, then $\bigcap_{i \in I} M_i = 0$ iff $\bigcap_{j \in J} M_j = 0$ for all finite subsets J of I. (Cf. proof of 19.13 A.)

(c) $\Rightarrow$ (a) follows from the following corollary. $\square$

19.16 B   **Corollary** (Vámos [68]). *A module M is Artinian iff every factor module is essentially right Artinian.*

Proof. In effect, if $M_1 \supseteq \cdots \supseteq M_n \supseteq \cdots$ is any descending sequence with intersection Q, the fact that M/Q has the finite intersection property implies the existence of an integer $n > 0$ such that $M_{n+i} = Q$ for all $i \geq 0$. Therefore M is Artinian. $\square$

19.17   **Theorem.** *Over a commutative ring A, any finitely generated module M is a compactly faithful* $A/\mathrm{ann}_A M$-*module.*

Proof. We may assume that M is faithful, and generated by $x_1, \ldots, x_n$. Then, the mapping $A \to M^n$ sending $a \mapsto (x_1 a, \ldots, x_n a)$ is an embedding. $\square$

19.18   **Proposition.** (a) *A right R-module M is quasinjective iff* M^n *is quasinjective, for any integer* $n > 0$.

(b) *Any finendo quasinjective right R-module M is injective over* $\mathrm{Biend}\, M_R$.

Proof. Exercise. $\square$

19.19   **Lemma.** *If R is a ring, then* $\hat{R}$ *is a faithful injective right R-module which is finite, in fact, cyclic over endomorphism ring.*

Proof. Exercise.

A right *R*-module *E* is **projectivendo** provided that *E* is a projective module over End E_R. Then, the Morita theorem 4.13 (I, p. 190) asserts:

An object E generates mod-*R if E is finendo, projectivendo, and balanced.*

The proofs of the foregoing results contain an idea which permits selfinjective rings to be characterized thus:

19.20 Proposition. *The conditions on a ring R are equivalent:*

(a) *R is right selfinjective.*

(b) *Every faithful finendo module E generates* mod-*R.*

(c) *Every faithful finendo injective module E generates* mod-*R.*

(In this case, then E is projectivendo.)

Proof. A ring *R* is right selfinjective if and only if mod-*R* has an injective generator. For any ring *R*, the injective hull of *R* in mod-*R* is cyclic over endomorphism ring 19.19, and hence, (b) [or (c)] implies that *R* is a generator which is injective.

Conversely, assume that *R* is injective in mod-*R*, and let $M \in$ mod-*R* be a faithful module finite over $S = \text{End} M_R$, say

$$S^n \to M \to 0$$

is exact in *S*-mod. Then, 19.14 implies that $0 \to R \to M^n$ is exact in mod-*R*, and injectivity of *R* implies an isomorphism $M^n \approx R \oplus X$. Therefore, *M* is a generator. In this case, Morita's theorem 4.1 implies that *M* is projective over endomorphism ring. $\square$

A submodule *N* of the left *R*-module *M* is (Cohn) **pure** if for all right *R*-modules *U*, the homomorphism $U \otimes_R N \to U \otimes_R M$ induced by the inclusion $N \to M$ is monic. This is equivalent to the following condition: if

$$(1) \qquad \sum_{j=1}^{n} r_{ij} x_j = a_i \in N \qquad\qquad (i = 1, 2, \ldots, m)$$

is a finite set of equations in the unknowns $x_1, \ldots, x_n$ where $r_{ij} \in R$, and if this system has a solution in *M*, then it has a solution in *N* too. Any summand of *M* is pure.

An *R*-module *A* is called **algebraically compact** if it is a summand in every *R*-module in which it is a pure submodule, or, equivalently, if

$$(2) \qquad \sum_j r_{ij} x_j = a_i \in A \qquad\qquad i \in I$$

is an arbitrary set of equations in the unknowns $x_j, j \in J$, where *I* and *J* are arbitrary index sets and each equation contains but a finite number of non-zero $r_{ij} \in R$, and if every finite set of equations in (2) has a solution in *A*, then the entire system (2) is solvable in *A*. Compare pure-injectivity 20.40.3.

A module *M* is said to be **bounded** provided that there exists an element $x \in M$ such that $\text{ann}_R x = \text{ann}_R M$. Then, if $A = \text{ann}_R M$, *M* is said to be *A*-**bounded.** Any module *M* which is cyclic over an endomorphism ring *S* is bounded.

An R-module M is said to have the **exchange property** if for every R-module A containing M and for submodules N and A_i $(i \in I)$ of A, the direct decomposition

(3) $A = M \oplus N = \sum_{i \in I} \oplus A_i$ $(I = \text{arbitrary index set})$

implies the existence of R-submodules B_i of A_i $(i \in I)$ satisfying

(4) $A = M \oplus (\sum_{i \in I} \oplus B_i)$

(cf. Exchange lemma 18.17).

19.21 *Exercise.* (a) The direct limit of any directed family of pure subobjects of a module M is a pure submodule.

(b)* (Fuchs [69], and Warfield [69 a], and Crawley-Jonsson [64]) A quasinjective module has the exchange property.

(c)* (Cf. Fuchs [70]) Assume that

$$A = M_1 \oplus \cdots \oplus M_m = \sum_{i \in I} \oplus N_i$$

where every M_j and every N_i is a quasinjective R-module, and I is an arbitrary index set. Then there exist isomorphic refinements, that is, there exist R-modules A_{ji}, $j = 1, \ldots, m$, $i \in I$, such that

$$M_j \approx \sum_{i \in I} \oplus A_{ji} \quad \text{and} \quad N_i \approx A_{1i} \oplus \cdots \oplus A_{mi}$$

for every j and i, respectively.

(d) If $x \in M$, and if $\text{ann}_R x = \text{ann}_R M$, then for every $y \in M$, there is a map $f: x R \to y R$ such that $f(x) = y$. Conclude that a quasinjective module M bounded if and only M is cyclic over endomorphism ring.

(e) (Fuchs [69]) An A-bounded R-module M is quasinjective if and only if M is injective canonical R/A-module.

(f)* (Fuchs [69]) Any A-bounded quasinjective module is algebraically compact. Furthermore, any summand of a product of bounded quasinjective modules is algebraic compact.

(g)* (Kaplansky, Fuchs, Hulanicki, and Harrison. Cf. Kaplansky [69, pp. 84–86]) Any algebraically compact abelian group is a product of bounded quasinjective modules.

(h) (Fuchs, l.c.) Compare the definition of N as a pure submodule of M with the definition requiring that $IN = N \cap IM$, for any right ideal I of R. Call this $\cap$-**pure**. Show that any bounded quasinjective module is algebraically compact using this notion of purity. Prove that if M is flat, then N is a $\cap$-pure submodule iff M/N is flat (cf. I, 11.21, p. 433).

(i) (Fuchs and Rangaswamy [70]) An abelian group M is quasiprojective if and only if M is either free, or a torsion group such that the p-subgroup of M is a direct sum of groups isomorphic to $\mathbb{Z}_{p^n}$, for fixed $n = n(p)$.

(k)* (Azumaya [66], Utumi [67]) A ring R is an injective cogenerator in mod-R if and only if every faithful module is a generator (cf. 19.20).

(l)* (Osofsky [66 b]) If R is an injective cogenerator, then R is semilocal, and has finitely generated essential right socle.

(m)* (Faith-Walker [67]) If R is a cogenerator in mod-R, and if R is semilocal, then R is right selfinjective.

(n)* (Warfield [72 b]) (1) A right R-module M has the exchange property iff End M_R has the exchange property. (2) If A is a lift/rad ring, and if $A/\text{rad}\,A$ is regular (see 19.25), then A has the exchange property. [Apply to exercise (b). Cf. 18.22.]

(o)* R is an injective cogenerator in mod-R iff R is right selfinjective and every faithful right R-module is compact faithful.

(p) (Fuller [69 b]) Any injective R-module E is $\prod$-injective in the sense that any product E^c is injective, for any cardinal number c. Prove that a QI module M is $\prod$-QI iff M is an injective $R/\text{ann}_R M$ module. [Hint: By 19.15 a QI-module E is injective if R embeds in E, and R embeds in some power M^c of any faithful module M (I, *sup.* 3.24 p. 143).]

(q) If a prime ideal P is not an essential right ideal, then P is a right annulet. Moreover, if $(^{\perp}P)^2 \neq 0$, then P is a minimal prime ideal.

Dense Rings of Linear Transformations and Primitive Rings

If M is a left vector space over a field D, then by our convention of writing homomorphisms opposite scalars (I, p. 119), if a is any element of $L = \text{End}_D M$, and $x \in M$, then we let $xa = (x)a$. A subring A of L is said to be **dense** provided that for each finite subset $y_1, \ldots, y_n$ of elements of M, with $n \leq \dim_D M$, and any set $x_1, \ldots, x_n$ of n linearly independent vectors, there is an element a of A such that $x_i a = y_i, i = 1, \ldots, n$. (Intuitively, this means that A has "enough" l.t.'s (cf. Exercise 1 in 19.23B).) We also say that A is **dense** in L, or a **dense ring of l.t.'s in** (or **on**) M.

As defined in Chapter 18 (following 18.0), a ring R is **right primitive** if R has a faithful simple right R-module. Any dense ring A of l.t.'s in a left vector space $M \neq 0$ is right primitive since M is faithful over A, and simple since given any nonzero A-submodule M', we must have $M' = M$ by density of A. (To wit: if $y \in M$, and if $x \neq 0$ in M', then $y = xa \in M'A = M'$ for some $a \in A$.) The converse is the:

19.22   Chevally-Jacobson Density Theorem. *Let M be a simple R-module with endomorphism ring S. If $x_1, \ldots, x_n$ are finitely many elements of M that are linearly independent over S, and if $y_1, \ldots, y_n$ are corresponding arbitrary elements of M, then there exists $a \in R$ such that $x_i a = y_i, i = 1, \ldots, n$.*

Proof. Let A denote the S-submodule of M generated by $x_1, \ldots, x_n$, and, for each i between 1 and n, let $N_i = A - S x_i$. By 19.10, $^{\perp}(N_i^{\perp}) = N_i, i = 1, \ldots, n$. Since $x_i \notin N_i$, then $x_i N_i^{\perp} \neq 0$, and since M_R is simple, then $x_i N_i^{\perp} = M, i = 1, \ldots, n$. Then $y_i = x_i a_i$, with $a_i \in N_i^{\perp}, i = 1, \ldots, n$, and $a = a_1 + \cdots + a_n$ has the desired property. $\square$

19.23 A   Corollary. *A ring R is primitive if and only if R is isomorphic to a dense ring of linear transformations in a vector space.* $\square$

19.23B *Exercises.* 1. Let M be a left vector space over D, and let A be a subring of L. Then, A is dense iff for every finitely generated vector subspace V of M every l.t. of V is induced by an element of A. Thus, if $\dim_D V = t$, then there is a subring $A_V = \{a \in A \mid Va \subseteq V\}$ and an ideal $K_V = \{a \in A \mid Va = 0\}$ of A_V such that A_V / K_V is isomorphic to the full matrix ring $D_t \approx \operatorname{End}_D V$ (cf. Jacobson [64, p. 33, Theorem 3]).

2. If A is a dense subring of L, then so is any subring $B \supseteq A$. If $\dim_D V = a$ is infinite, then a l.t. t has **rank** b if $\dim_D Vt = b$. The only proper ideals of L are those of the form

$$L_b = \{t \in L \mid \operatorname{rank} t < b\}$$

for an infinite cardinal not exceeding $a + 1$. (The units of L are those l.t.'s of rank a.)

Thus, the ideal $L_{\aleph_0}$ consisting of all l.t.'s of finite rank is the unique minimal ideal $\neq 0$, and is dense in L. Conclude that any subring of L containing $L_{\aleph_0}$ is dense. Show that $L_{\aleph_0}$ consists of all row and column finite matrices relative to some well ordered basis of M. (Also see problem 6.)

3. Every factor ring of L is a primitive ring, or 0. If J is a right (left) ideal, then J induces a dense set of l.t.'s in $M/\operatorname{ann}_M J$ (resp. in MJ).

4. Show that every cyclic left ideal is generated by an idempotent. (So L is regular in the sense of 19.24.)

5. If D is any field which is transcendental over center, then the polynomial ring $D[x]$ is primitive (see, I, Exs. 13.11.9–11 pp. 467–9). Conclude that an integral domain may be primitive. (Also see examples of simple integral domains 7.28. Recall that any simple ring is primitive.)

6. Next consider M as a *right* vector space over D of dimension d, and recall from (I, 3.19.9, p. 136) that L is isomorphic to the ring D_d of all row finite matrices over D, relative to a well ordered basis $\{x_i\}_{i \in d}$ of M. If B is a subring of D, and if A is a subring which contains all matrices of the form

$$\begin{pmatrix} M & & & 0 \\ & b & & \\ & & b & \\ & & & \ddots \\ 0 & & & & \ddots \end{pmatrix}$$

where M is an arbitrary $n \times n$ matrix in D_n, for any integer n, and $b \in B$, then A is dense in L. (This includes the case where $B = 0$.) If B is contained in the center of D, the center of A consists of all diagonal matrices

$$\operatorname{diag}\{d, d, d, \ldots\}$$

and is isomorphic to B. Conclude that an arbitrary commutative integral domain can be isomorphic to the center of a primitive ring, that the center of a primitive ring need not be primitive, and may have nonzero radical (I. Kaplansky). (See also end-of-chapter problem 21.)

7. If in 6, we let B be a subring of the matrix ring D_n, for a fixed integer n, and let M be any $kn \times kn$ matrix over D for increasing $k = 1, 2, \ldots$, then A is still primitive, and moreover, B is an epimorphic image of A. Thus, any subring of a simple Artinian ring is a homomorphic image of a primitive ring.

Regular Rings

For convenience in references, we repeat a proposition and definition previously introduced in Chapter 11 of Volume I.

19.24 Proposition and Definition. *A ring R is said to be **regular** provided that the following equivalent properties hold:*

(a) *Every cyclic right R-module is flat.*

(b) *Every right R-module is flat.*

(c) *If $x \in R$, then there exists $a \in R$ with $x a x = x$. (Then $x a$ is an idempotent which generates $x R$.)*

(d) *Every cyclic right ideal is generated by an idempotent.*

(e) *Every finitely generated right ideal is generated by an idempotent.*

(i') *The right-left symmetry of* (i), $i = a, b, c, d, or$ e.

Proof. Part of this is 11.24, and its proof, and also the following 19.25. □

19.25 Proposition (Von Neumann [36]). *Let R be a ring such that every cyclic (resp. simple) right ideal is generated by an idempotent. Then, every finitely generated right ideal (resp. every sum of finitely many simple right ideals) is generated by an idempotent.*

Proof. Let $I = x_1 R + \cdots + x_n R$ be any finitely generated right ideal, and set $A = x_1 R + \cdots + x_{n-1} R$, $B = x_n R$. An induction hypothesis provides idempotents e and f such that $A = e R$ and $B = f R$. Let $B_1 = (1 - e) B = (1 - e) f R$. Now $A + B = \{e u + f v \mid u, v \in R\}$, and

$$A + B_1 = \{e u' + (1 - e) f v' \mid u', v' \in R\}$$
$$= \{e(u' - f v') + f v' \mid u', v' \in R\},$$

so we see that $A + B_1 = A + B$.

Write $B_1 = f_1 R$, with $f_1^2 = f_1 \in R$. Since $f_1 \in B_1$, then $e f_1 = 0$. Next, put $f' = f_1(1 - e)$. Then

$$f' f_1 = f_1(1 - e) f_1 = f_1(f_1 - e f_1) = f_1 f_1 = f_1$$

so $(f')^2 = f' f_1(1 - e) = f_1(1 - e) = f'$. Since $f' = f_1(1 - e) \in f_1 R$, and since $f_1 = f' f_1 \in f' R$, it follows that $f' R = f_1 R = B_1$. Hence $A + B = e R + f' R$. Since

$$e f' = e f_1(1 - e) = 0, \quad f' e = f_1(1 - e) e = 0,$$

then e and f' are orthogonal, and the idempotent $e + f'$ generates I:

$$I = A + B = e R + f' R = (e + f') R.$$

The proof also works for the case when $x_i R$ is a simple right ideal, $i = 1, \ldots, n$. Then $B = x_n R$, and $B_1 = (1 - e) B$ are simple right ideals, or zero, and therefore there exists an idempotent f_1 such that $B_1 = f_1 R$, and the rest of the proof is unchanged. $\square$

19.26 A   Theorem. *Any semisimple ring R is regular. Any union of a chain of regular rings is regular, and the product of any family of regular rings is regular. Moreover, for any regular ring R the following are equivalent:*

(a) *R is semisimple;*

(b) *R is right Artinian;*

(c) *R is right Noetherian;*

(d) *R satisfies* (acc)$^{\perp}$;

(e) *R satisfies* (acc) $\oplus$;

(f) *R contains no infinite set of orthogonal idempotents.*

Proof. A semisimple ring R has the property that any right ideal is a direct summand, hence generated by an idempotent, so 19.24(e) applies.

The characterization of 19.24(c) implies the theorem for products, and unions of chains, of regular rings.

The equivalence of (a)–(e) is an exercise. $\square$

19.26 B   *Examples.* (a) A union of a chain of fields is a field, but a union of a chain of semisimple rings is not always semisimple. For example, we may construct the chain of total matrix algebras $\{k_n\}$, $n = 2^t$, $t = 1, 2, \ldots$. Thus,

$$k_2 \subset k_4 \subset k_8 \subset \cdots$$

and the union A is a regular ring not satisfying 19.26(e), hence not semisimple. (The algebra A is infinite dimensional yet locally finite dimensional, and simple. Compare Köthe [31] for a characterization of countable simple algebraic algebras which are locally finite dimensional and central over k. As Köthe's title suggests, one similarly constructs noncommutative algebraic division algebras of infinite dimensions.)

(b) As is to be shown, any full linear ring is a regular ring, and hence, so is any product of the same. (Any infinite product of fields (or regular rings) is a nonsemisimple regular ring.)

(c) Certain of the examples of primitive rings in Exercise 19.23 B are regular, e.g., the ring $S + D$ generated by the scalar l.t.'s $d \in D$, and the ideal $S = L_{\aleph_0}$ consisting of all l.t.'s of finite rank in the full linear ring L of a vector space V of uncountable dimension over a field D. This follows from the regularity of D and L, and the fact that S is an ideal. (Similarly, $I + D$ is regular for any ideal I of L.) Also, the ring A of Exercise 6 in 19.23 B can be chosen to be regular, by the judicious choice of B. (Take $B = 0$, etc.)

19.26 C   Theorem (Faith [72 b]). *Let R be a regular ring. Then, a module M which satisfies the d.c.c. on annihilators of right ideals of R is semisimple and injective, and $R/\mathrm{ann}_R M$ is a semisimple ring. (Moreover, M is also $\sum$-injective in the sense that any direct sum of copies of M is injective. Cf. 20.3 A)*

Proof. For any idempotent e of R, the set $M(1-e)$ is an annihilator of the right ideal eR. Since M is faithful over $\bar{R} = R/\text{ann}_R M$, the d.c.c. on annihilators in M implies the a.c.c. on right ideals of $\bar{R}$ generated by idempotents. (Obviously, any idempotent of $\bar{R}$ lifts to one in R.) Then, by 19.26A, $\bar{R}$ is semisimple, and hence any $\bar{R}$-module M' is injective over $\bar{R}$, by (I, 8.12, p. 370). But, since $\bar{R}$ (and indeed, any R-module) is a flat R-module (I, 11.34, p. 434), then M' is an injective R-module by (I, 11.35, p. 440). In particular, any direct sum of copies of M is injective over R, that is, M is $\sum$-injective. □

19.26 D   *Exercise.* 1. If $R = \text{End}_D V$, where V is a left vector space over D, then R is a regular ring (and left selfinjective).

2. Any regular ring is right and left nonsingular (I, p. 396). If a regular ring R contains a subring A, and if R is an essential extension of A in mod-A, then A is right nonsingular. Thus, any right order in a semisimple (or regular) ring is right nonsingular (I, 9.13, p. 397).

Endomorphism Rings of Quasinjective Modules

As in (I, p. 168), $M \bigtriangledown N$ signifies that N is an essential submodule of M. Also recall from (I, p. 394) that the **singular submodule** of a right R-module M is defined by

$$\text{sing } M = \{x \in M \,|\, R \bigtriangledown \text{ann}_R x\}$$

Two important properties are: (i) sing M is a fully invariant submodule; and (ii) $M \supseteq S \supseteq N$ and $S \bigtriangledown N \Rightarrow \text{sing } M/N \supseteq S/N$; and conversely if $N = \text{sing } M$. [As before, these are exercises.] Thus, the right singular submodule of R is an ideal, denoted sing R. This may not coincide with the left singular ideal, but the two are used so infrequently, that it will not be necessary to employ notation to distinguish them. As a matter of fact, it is the disappearance of the singular submodule which is most frequently referred to: a module M is **nonsingular** (also called neat) if sing $M = 0$. When sing $R = 0$, we say that R is right nonsingular, and then $\text{sing}(M/\text{sing } M) = 0$ for any $M \in \text{mod-}R$ (see 19.46(a)). In this terminology, then 19.29 states that the endomorphism ring of any quasinjective nonsingular module is selfinjective and von Neumann regular. In particular, 19.27 states that any selfinjective ring R is von Neumann regular and selfinjective modulo radical (and rad $R = \text{sing } R$).

(a) of the next theorem was discovered by Utumi [56]; (b) by Wong and Johnson [59], Utumi [67], and Osofsky [68a]; and (c) by Faith and Utumi [64a].

19.27   Theorem. *Let $S = \text{End } M_R$, where M is a quasinjective right R-module. Then:*

(a) rad $S = \{s \in S \,|\, M \bigtriangledown \ker s\} = \text{sing } S$.

Moreover, $S/\text{rad } S$ is von Neumann regular; and

(b) *$S/\text{rad } S$ is a right selfinjective; and*

(c) *S is a lift/rad ring; indeed, idempotents modulo any subset I containing* rad R *lift.*

Proof. (a) Let $I = \{s \in S \mid M \nabla \ker s\}$. If $s \in S$, and if $u, v \in I$, then

$$\ker(u+v) \supseteq \ker u \cap \ker v, \qquad \ker(su) \supseteq \ker u.$$

Since $\ker u \cap \ker v$ is an essential submodule, it follows that I is a left ideal of S; however, if $s \in I$, $\ker(1+s) = 0$, since then $\ker s \cap \ker(1+s) = 0$. Thus, $(1+s)$ has a left inverse $\forall s \in I$, so that each $s \in I$ is a left quasiregular. This establishes that $I \subseteq \operatorname{rad} S = J$.

Next let s be an arbitrary element of S, let L be a relative complement submodule of M corresponding to $K = \ker s$, and consider the mapping $sx \mapsto x \; \forall x \in L$. (This is indeed a mapping, since if $sx = sy$, with $x, y \in L$, then $s(x-y) = 0$, and then $x - y \in K \cap L = 0$.) Since M is quainjective, the map $sx \mapsto x$ of sL in L is induced by some $t \in S$. If $u = x + y \in L + K$, $x \in L$, $y \in K$, then

$$(s - sts)(u) = s(x) - sts(x) = s(x) - s(x) = 0.$$

Since $M \nabla (L+K)$, and since $\ker(s-sts) \supseteq L+K$, we conclude that $s - sts \in I$. Although we have not yet shown that I is an ideal (it would not be too hard to do so), roughly speaking, we have schown that S is a regular ring modulo I.

Now we show that $J = I$. If $s \in J$, and if $t \in S$ is chosen so that $u = s - sts \in I$, then $-st \in J$ (since J is an ideal) and $(1-st)^{-1}$ exists (since J is quasiregular 18.6). Therefore $(1-st)^{-1}u = s$, and $s \in I$ (since I is a left ideal). Thus $J = I$ as asserted. Moreover, S/J is a regular ring, proving (a).

To prove (b), we first prove (c), and more generally that idempotents of S modulo any *subset* I containing J lift. Let $s^2 - s \in I$. Then, $\ker s \cap \ker(1-s) = 0$, so that

$$y \in \ker(s^2 - s) \; \Rightarrow \; (1-s)\, y \in \ker s \; \& \; s\, y \in \ker(1-s)$$

proving that

$$M \nabla [\ker s \oplus \ker(1-s)].$$

Now let $M \supseteq N_i$ be such that $N_1 \nabla \ker s$ and $N_2 \nabla \ker(1-s)$, $i = 1, 2$, are maximal essential extensions. Then, $N_1 \oplus N_2 = M$ by 19.6(a). Let e and $1-e$ be the projection idempotents, $eM = N_2$, and $e N_1 = 0$. Then:

$$(e-s)[\ker s \oplus \ker(1-s)] = 0$$

so that $e - s \in J \subseteq I$, that is, idempotents modulo I lift.

(b) In order to prove that $\bar{S} = S/J$ is right selfinjective, we first establish (b$_1$) and (b$_2$) below.

(b$_1$) If e and f are idempotents of S, with $eM \cap fM = 0$, then since eM and fM have no proper essential extensions in M, neither does $eM \oplus fM$, and so, by 19.6(a), $eM \oplus fM$ is a direct summand of M, say $eM = fM = gM$, with $g = g^2 \in S$.

(b$_2$) On the other hand, if $eM \cap fM \neq 0$, then $\bar{e}\bar{S} \cap \bar{f}\bar{S} \neq 0$. For let K be a complement submodule of $eM \cap fM$ in M, that is, K maximal with respect to $eM \cap fM \cap K = 0$. Then, $eM \cap fM \oplus K$ is an essential submodule of M, and K

is a direct summand of M since K has no essential extensions. Let N be a maximal essential extension of $eM \cap fM$ in eM. Since eM is QI, then N is a direct summand of eM, and hence $N \oplus K$ is a direct summand of M, and hence there is a projection idempotent $g \in S$ with kernel K, such that $N = gM$. Then $eg = g$, and by symmetry, $fh = h$ for some idempotent h with kernel K where hM is a maximal essential extension of $eM \cap fM$ in fM. Then, $g - h$ annihilates an essential submodule, so $g - h \in I$. But, $\bar{g} \neq 0$ lies in $\bar{e}\bar{S} \cap \bar{f}\bar{S}$.

Now let I be any right ideal of S, and let $\{\bar{e}_i\}_{i \in A}$ be a maximal independent set of idempotents of I, that is, maximal with respect to the directness of the sum $T = \sum_{i \in A} \bar{e}_i \bar{S}$. By regularity of $\bar{S}$, T is an essential right ideal, so any homomorphism from I to $\bar{S}$ is determined by its effect on T. For suppose $g: I \to \bar{S}$ agrees with f on T. If $x \in I$, let Q be an essential right ideal such that $xQ \subseteq T$. Then Q annihilates $f(x) - g(x)$, so $f(x) = g(x)$ as asserted.

Let $e_i = e_i^2 \in S$ be a lifting of $\bar{e}_i$ (as the notation implies), and let $\bar{x}_i = f(e_i) \ \forall \ i \in A$. Then, by (b$_1$) and (b$_2$), the sum T is direct, and the direct sum of the mappings $x_i e_i : e_i M \to M$ is a map $g: T \to M$ which extends by quasiinjectivity to a map $m: M \to M$. Then, $f(x) = \bar{m} x \ \forall \ x \in I$, which by Baer's criterion (I, 3,41, p. 157) shows that $\bar{S}$ is right selfinjective. $\square$

For the proof of (b), I have followed Osofsky [71, p. 44], where other results of interest may be found, notably Osofsky's theorem stating that a *ring R is semisimple Artinian iff every cyclic right module is injective*. A corollary of her proof is another theorem of hers stating that a *hereditary ring cannot contain an infinite product of subrings*. A corollary of her theorem is that a *right selfinjective ring R is right hereditary iff R is semisimple*. (For then, every factor module R/I would be injective by a result proved in Cartan-Eilenberg so every cyclic module is injective.)

19.28   Corollary (Utumi [56, 67]). *Let R be a right selfinjective ring. Then:* (a)

$$\text{rad } R = \text{sing } R$$

and $R/\text{rad } R$ is a right selfinjective regular ring.

(b) *If every nonzero right ideal contains a minimal right ideal then*

$$\text{rad } R = {}^{\perp}(\text{socle } R),$$

the left annihilator of socle R.

Proof. (a) corresponds to (a) of the Theorem 19.27 since $R \approx \text{End } R_R$ canonically. (b) Socle R is the intersection of the essential submodules by 8.3, so (a) implies (b). $\square$

19.29   Corollary. *Let M be a nonsingular quasiinjective right R-module. Then* End M_R *is right selfinjective and von Neumann regular.*

Proof. If $f \in \text{rad } S$, where $S = \text{End } M_R$, then by 19.27, M is essential over $\ker f$. Since $\text{sing } M = 0$, by 19.32, M is rational over $\ker f$. This implies that $fM = 0$, that is, $f = 0$. Since therefore rad $S = 0$, so we can apply 19.27. $\square$

Parts of the next theorem are from Faith-Utumi [64a], Harada [65] and Sandomierski [67].

19.30   Theorem. *Let M be any quasinjective.*

(a) *If $S_2 \supseteq S_1 = \operatorname{sing} M$, and $S_2/S_1 = \operatorname{sing} M/S_1$, then S_2 is a direct summand of M.*

(b) *Any mapping $f: M \to N$ into a nonsingular module N splits.*

Proof. (b) Let $K = \ker f$. Then, $\operatorname{sing} M/K = 0$ hence by (I, p. 393, 9.5.1), K is a closed submodule of M, hence 19.6(a) suffices for (b).

(a) Using 9.5.1 again, we see that S_2 contains the maximal essential extension $P = Q(S_1)$ of S_1 in M. Since P splits by 19.6, then $S_2 = P$ is closed, so that (b) $\Rightarrow$ (a). □

19.31   *Exercise.* Prove any of the Osofsky theorems stated directly preceding 19.28.

19.31.2   An infinite dimensional full right linear ring is right but not left selfinjective.

19.31.3   (Utumi) A primitive ring with nonzero socle is a full right linear ring if and only if it is right selfinjective.

19.31.4   A right nonsingular ring R is a right full linear ring if and only if it is a right selfinjective prime ring containing uniform right ideals.

19.31.5   The injective hull of any domain D is a simple, hence primitive ring, which is a full linear ring if and only if D is a right Ore domain. Conclude that primitivity (hence, density) and selfinjectivity does not imply fullness, by finding an example of a domain which is not Ore.

The Maximal Rational Extension of a Module and Ring

We finally come to the important concept of a rational extension M of a module N of G. Findlay and J. Lambek [58], which we introduced in (I, 4.5, p. 193). It is one of the basic ideas implicitly used in the study of quotient rings in Volume I, and in fact, the quotient ring Q of a semiprime right Goldie ring R is a maximal rational extension of the right R-module R. (This may be verified directly using the definitions, or obtained as a consequence of the next theorem which shows that an essential extension of a nonsingular module is a rational extension.) Moreover, the maximal quotient ring of a ring (19.34, following) is defined via rationality. (This was previously discussed for nonsingular rings in (I, pp. 531–2, 16.12–4).)

While we repeat the definition of a rational extension for ready referencing in Volume II, the proof of the equivalence of the two stated definitions is delegated as an exercise.

19.32   Proposition and Definition (Findlay-Lambek [58]). *A module P is a **rational extension** of a submodule M provided that (1) $\operatorname{Hom}_R(K/M, P) = 0$ for every between module $P \supseteq K \supseteq M$, or equivalently, (2) provided for any pair $a, b \in P$, and $b \neq 0$, there corresponds an element $r \in R$ such that $ar \in M$ and $br \neq 0$.*

Let $\hat{M}$ be the injective hull of M in mod-R, *and let* $S = \operatorname{End} \hat{M}_R$.

(a) *For any submodule N of M, the double annihilator*

$$\bar{N} = \operatorname{ann}_{\hat{M}} \operatorname{ann}_S N$$

of N (first in S, then back in $\hat{M}$) is a maximal rational extension of N, and $\bar{N}$ contains every rational extension of N in $\hat{M}$. (Thus, $\bar{N}$ is the unique maximal rational extension of N in $\hat{M}$.)

(b) *Every rational extension of M can be embedded in $\bar{M}$ by a map which induces the identity map on M.*

(c) *Any two maximal rational extensions of M are isomorphic by a map which induces the identity map of M.*

(d) *Every rational extension of M is an essential extension, and conversely when* $\operatorname{sing} M = 0$. *(Thus,* $\operatorname{sing} M = 0$ *implies that* $\bar{M} = \hat{M}$.)

Proof. If $\bar{N} \supseteq K \supseteq N$ is a submodule, and if $f : K/N \to \bar{N}$, then f is induced by $f' : K \to N$ for some f' with $\ker f' \supseteq N$. Since $\hat{M}$ is injective, then f' can be assumed to be an element of $S = \operatorname{End} \hat{M}_R$. By the definition of $\bar{N}$, $\ker f' \supseteq \bar{N}$, and so $f' = 0$, hence $f = 0$. This proves that $\operatorname{Hom}_R(K/N, \bar{N}) = 0$, so $\bar{N}$ is a rational extension of N.

If $x \in \hat{M}$, and $t \in \operatorname{ann}_S N$, is such that $t(x) \neq 0$, that is, if $x \notin \bar{N}$, there is no element $a \in R$ such that $xa \in N$ and $t(x)a \neq 0$. This shows that any rational extension of N contained in $\hat{M}$ is contained in $\bar{N}$. This proves (a), and (c) is a consequence, since by the definition, any rational extension is essential and therefore any rational extension of M embeds in $\bar{M}$ (I, 3.58 A, p. 169). (d) is left for the reader to prove. $\quad\square$

Rationally Closed Modules

A submodule N of M is **rationally closed** provided that N is the only rational extension of N contained in M. Any closed submodule N (closed in the sense of 19.4) is rationally closed inasmuch as every rational extension is essential.

19.33   Corollary. (a) *Any maximal rational extension P of a module M contains a unique maximal rational extension of any submodule N of M.*

(b) *If P is any rational extension of M, then the contraction map $I \to I \cap M$ induces an isomorphism of the respective lattices of rationally closed submodules of P and M.*

(c) *If* $\operatorname{sing} M = 0$, *then any maximal rational extension P of M contains a unique injective hull of any submodule N of M.*

(d) *If $P \supseteq K \supseteq M$ are modules, then P is rational over M iff P is rational over K, and K is rational over M.*

(e) *If P is rational over M, then P is a maximal rational extension iff P is rationally closed in every over-module.*

Proof. By 19.32, we may assume that $\hat{M} \supseteq P = \bar{M}$, and then, for any submodule N of M, $P \supseteq \bar{N}$. Thus, (a) follows from (a) of 19.32, and (c) then follows from (d) of 19.32.

(b) We may assume that $\bar{M} \supseteq P$. Then, $\bar{M} \supseteq \bar{P}$. The second definition of rational extension (2) implies that $\bar{M}$ is rational over $\bar{P}$, so we have $\bar{M} = \bar{P}$. Therefore, it is necessary and sufficient to prove (b) for the case $P = M$. If Q is rationally closed in M, then by the definition, Q is rationally closed in $\bar{M}$, and since $\bar{Q} \cap M$ is rational over Q, then $\bar{Q} \cap M = Q$. Finally, if I is rationally closed in M, and if $I \cap M = Q$, then, since I is patently rational over Q, we have $I \subseteq \bar{Q}$ by (a). Since I is rationally closed, then $I = \bar{Q}$. This completes the proof of (b).

The necessity of (d) is immediate, the sufficiency is an exercise, and (e) is a corollary. ☐

19.34 Proposition (Johnson [51], Utumi [56]). *The maximal rational extension $\bar{R}$ of R in mod-R is a ring, and the inclusion $R \hookrightarrow \bar{R}$ in mod-R is a ring monomorphism. Then there is a canonical isomorphism of rings*

(1)
$$\operatorname{Biend} \hat{R}_R \to \bar{R}$$
$$f \mapsto (1)f$$

and an embedding of rings

(2)
$$\bar{R} \to \operatorname{End} \hat{R}_R$$
$$x \mapsto t_x$$

such that $x = t_x(1)$, where $\hat{R}$ denotes the injective hull of R in mod-R. The ring $\bar{R}$ is called the (Johnson-Utumi) maximal right quotient ring of R.

Proof. First, if $t \in S = \operatorname{End} \hat{R}_R$, then

(3)
$$t((1)f) = (t(1)f) \qquad \qquad \forall f \in \operatorname{Biend} \hat{R}_R.$$

Now $\operatorname{ann}_S R = \operatorname{ann}_S 1$, and so

(4)
$$\bar{R} = \{x \in R \mid t(x) = 0 \text{ whenever } t(1) = 0\}.$$

Thus (3) implies that the image of the mapping (1) is contained in $\bar{R}$. Next, since R embeds in $\hat{R}$, then $\hat{R}$ is cyclic over S, and $\hat{R} = S1$ is generated over S by the identity element 1 of R. If $x \in \bar{R}$, then $x = t_x(1)$, for a unique $t_x \in S$. This follows since $t(1) = t'(1)$ implies that $(t - t')(1) = 0$, whence $(t - t')(x) = 0 \ \forall x \in \bar{R}$. Hence, if $y \in \bar{R}$, then $y = (1)f$, where $f \in \operatorname{Biend} \hat{R}_R$ satisfies

$$(x)f = (t_x(1))f = t_x(y).$$

This proves that (1) is an isomorphism in mod-R, and that (2) is an embedding in mod-R. Then, the ring operation of $\operatorname{Biend} \hat{R}_R$ defines via (1) a ring operation in $\hat{R}$ with the stated properties. ☐

The isomorphism (1) and (2) is a theorem of Lambek [63].

19.35 Corollary (Johnson [51] and Wong-Johnson [59]). *If R is a ring, then $\bar{R}$ is a regular ring if and only if R is right nonsingular. In this case $\bar{R}$ is the injective hull of R in mod-R, and is a right selfinjective regular ring, denoted $\hat{R}$.*

Moreover, there is an isomorphism of lattices

cyclic right ideals of $\hat{R} \rightleftarrows$ rationally closed right ideals of R

$$I \longmapsto I \cap R$$

$$\hat{J} \longleftarrow J$$

where $\hat{J}$ denotes the injective hull in $\hat{R}$ of any right ideal J of R.

Proof. By (d) of 19.32, $\operatorname{sing} R = 0$ implies that $\bar{R} = \hat{R}$, and then $S = \operatorname{End} R_R$ is a right selfinjective and regular by 19.27. Furthermore, by the last proposition, the canonical embedding of $\bar{R}$ in $\operatorname{End} \hat{R}_R$ is an isomorphism since $\bar{R} = \hat{R}$. Thus, $\bar{R}$ has the desired property. It remains only to describe the lattice isomorphism between the respective lattices of rationally closed submodules induced by the contraction map $I \to I \cap R$. Since $\bar{R} \approx \operatorname{Biend} \bar{R}_R$ canonically, then $\operatorname{End} \bar{R}_R = \operatorname{End} \bar{R}_{\bar{R}}$ canonically.

This shows that the R-summands of $\bar{R}$ are the same as the $\bar{R}$-summands, and thus are right ideals of $\bar{R}$. Since $\bar{I} = \hat{I}$ is the injective hull of I, and therefore an $\bar{R}$-summand of $\bar{R}$, then $\bar{I}$ is a cyclic right ideal of $\bar{R}$. Moreover, since $\bar{R}$ is right selfinjective, any rationally closed right ideal is injective by 19.32(d), and hence is a cyclic right ideal. Finally, since $\bar{R}$ is a regular ring, every cyclic right ideal is generated by an idempotent, and hence is a summand of $\bar{R}$, whence rationally closed.

Finally, if $\bar{R}$ is regular, then $\operatorname{sing} \bar{R} = 0$ in mod-$\bar{R}$ because the right annihilator of any element x is generated by an idempotent $1 - e$, where $\bar{R}e = \bar{R}x$. Then, the fact that $\bar{R}$ is an essential extension of R in mod-R implies that $\operatorname{sing} R = 0$. $\square$

19.36 Corollary. *If the maximal right quotient ring $\bar{R}$ of R is injective in mod-R, then $\bar{R}$ right selfinjective.*

Proof. The fact that $\bar{R}$ is rational over R can be used to prove that the injective hull E of $\bar{R}$ in mod-$\bar{R}$ is an essential extension of R in mod-R, and hence $E = \bar{R}$ as claimed. To do this, if $y \in E$, and $y \neq 0$, then $0 \neq ya \in \bar{R}$, for some $a \in \bar{R}$. By rationality of $\bar{R}$, there is an element $r \in R$ such that $ar \in R$, and $yar \neq 0$. Then, there is an element $t \in R$ such that $yart \in R$ and $yart \neq 0$. Since $art \in R$, this proves that E is essential over R as asserted. $\square$

19.37 *Exercise.* (a) (Faith and Chase [64]) Let R be a semiprime ring with the property that every rationally closed left ideal of R contains a minimal rationally closed left ideal. Then, R is left nonsingular, and R is a product $\prod_{i \in I} S_i$ of full left linear rings $\{S_i\}_{i \in I}$. Moreover, if e_i is the identity element of S_i, then $R_i = e_i R$ is a prime ring $\forall i \in I$, and the canonical map $R \to \prod_{i \in I} R_i$ is a representation of R as an essential, irredundant subdirect product. (By which we mean that if $J_i = \ker(R \to R_i) \ \forall i \in I$, then $\bigcap_{i \in I} J_i = 0$, and $\sum_{i \in I} J_i$ is an essential left ideal of R. Cf. subdirect products, Chapter 26. This subdirect product also is irredundant in the sense of Levy [63a].)

(b) Converse of (a). If R is a left nonsingular ring that is an essential, irredundant subdirect product of a family $\{R_i\}$ of rings such that each R_i has a maximal left quotient ring which is a full left linear ring S_i, then the canonical map $R \to \prod_{i \in I} S_i$ embeds R in its maximal left quotient ring. Moreover, R is a semiprime ring in which each rationally closed left ideal contains a minimal rationally closed left ideal.

(c) A ring R is isomorphic to a product of full left linear rings iff R is a semiprime left selfinjective ring in which every nonzero left ideal (resp. annulet) contains a minimal nonzero left ideal (resp. annulet). (Cf. Chapter Exercise 23.)

(d) Let R be a domain, not Ore, embedded in a field D (see, e.g. Cohn [71]). Show that $\hat{R}$ does not embed in $D = \hat{D}$. (Hint: $\hat{R}$ contains nontrivial nilpotent elements!) However, for any right Ore ring $A \hookrightarrow D_n$, the quotient ring $Q(A) \hookrightarrow D_n$ (L. Small).

19.38 *Exercise.* (a) Show that the maximal right quotient ring of a maximal right quotient ring S of a ring R is S.

(b) (Chase-Faith [64]) Let M be a quasinjective left R-module such that $S = \text{End}_R M$ is a regular ring. (For example, M can be any QI module with $\text{sing } M = 0$.) Then S is a product of full left linear rings if and only if every direct summand of M contains a minimal direct summand.

(c) The following conditions are equivalent: (1) $\bar{R}$ is a semisimple ring. (2) $\bar{R}$ is right Artinian (Noetherian) and $\text{sing } R = 0$. (3) R satisfies (acc) $\oplus$ and $\text{sing } R = 0$. (4) R satisfies (dcc) $\oplus$ and $\text{sing } R = 0$.

(d) The conditions are equivalent: (1) R has a semisimple classical right quotient ring; (2) R is a semiprime ring with (acc) $\oplus$ and $(\text{acc})^\perp$; (3) R is a semiprime ring with $\text{sing } R = 0$, and (acc) $\oplus$. In this case, $\bar{R}$ is the classical right quotient ring of R, and then R is prime if and only if $\bar{R}$ is simple.

(e) The matrix ring R_n is the maximal right quotient ring of the ring $T_n(R)$ of lower triangular matrices over R. Conclude that even in the case R is a field that $A = T_n(R)$ may have nilpotent ideals $\neq 0$, while A is a right nonsingular ring with maximal right quotient ring $\hat{A} = R_n$ which is simple Artinian (and both injective and projective over A). (Any subring S of R_n containing all first column matrices $K e_{11} + \cdots + K e_{1n}$, where K is a right order in R, has maximal right quotient ring $S = R_n$. Cf. Faith [67a], p. 129, Problem 13.)

(f) Let Q be a ring, and R a subring such that $Q = R_S$ is the right quotient ring of R with respect to a multiplicative set S. Then, Q is a flat left S-module (I, pp. 529–530).

(g) (Cateforis [69]) Show for a right nonsingular ring R, that $\hat{R}$ is a flat left R-module iff every finitely generated right ideal I of R is essentially finitely related in the sense that there is a presentation $0 \to K \to F \to I \to 0$ of I with F finitely generated free such that K contains an essential finitely generated submodule. Then, every von Neumann regular ring T between R and $\hat{R}$ is left flat over R.

(h) (Sandomierski [67]) If $\hat{R}$ is a semisimple ring, then every left $\hat{R}$ module is left flat over R.

(i) For any ring A and subring R it is true that A is left flat over R iff every injective right A-module is an injective R-module (I, 11.35.3, p. 440).

(j) (Cateforis [69]) For any ring R, and maximal right quotient ring $Q = \bar{R}$ the two sets of conditions (A) and (B) are equivalent: (A) R is a right nonsingular ring over which every finitely generated nonsingular right module is projective; (B) R is right semihereditary, Q is a flat right [sic!] module, and $Q \otimes_R Q$ is a right nonsingular module.

(k) (Silver [67]) An inclusion $R \hookrightarrow A$ in RINGS is an epic in RINGS iff the canonical map $A \otimes_R A \to A$ is an isomorphism.

(l) For any essential extension M of a module N, if $x \neq 0$ in M, then $\text{ann}_R x/N$ is an essential right ideal. Use this to prove for any essential ring extension A of R that the kernel of the canonical map $A \otimes_R A \to A$ is contained in the singular R-submodule of $A \otimes_R A$. Conclude that if $A \otimes_R A$ is a nonsingular right R-module, then $R \hookrightarrow A$ is a ring epic (see (j) and (k)). Moreover, if A is a right nonsingular R-module, then this condition is necessary and sufficient (cf. (i)).

(m) If $R \subseteq A$ are rings, and if $R \hookrightarrow A$ is a flat epic, (that is, $R \hookrightarrow A$ is an epic in RINGS, and A is a flat left R-module), then a right A-module E is injective as a right R-module iff injective as a right A-module. Conclude that r.gl.dim $A \leq$ r.gl.dim R.

(n) (Goodearl [71]) If R is right nonsingular, and $Q = \hat{R}$, then every finitely generated nonsingular right R-module embeds in a free right R-module iff Q is a flat right R-module and $R \hookrightarrow Q$ is a ring epic (see (k)). When this is so, then Q is also a left quotient ring of R. (And when R has finite right Goldie dimension, then all that is required is that Q be right flat.)

(o) If E is an injective module, and if M is nonsingular then any map $f: E \to M$ splits.

(p) Let R be a right nonsingular ring, and M a right R-module. Then, $\text{sing}(M/\text{sing} M) = 0$. Conclude that the singular submodule splits in any epic image of an injective module over R (cf. Sandomierski [67, p. 120]). Thus, any indecomposable injective module E is either nonsingular, or $\text{sing} E = E$.

(q) Let R be right nonsingular. Any nonsingular right R-module M embeds in a product of copies of $\hat{R}$, and if M is finitely generated, M embeds in a finite product of copies of $\hat{R}$ (see, for example, Goodearl [72, p. 19]). Any finitely generated uniform nonsingular module embeds in $\hat{R}$. Conclude that any indecomposable nonsingular injective right R-module embeds in $\hat{R}$.

(r) If R is right nonsingular, and if M is a nonsingular right R-module, then the kernel of the canonical map $M \otimes_R \hat{R} \to \hat{M}$ is contained in the singular submodule of $M \otimes_R \hat{R}$.

(s) (Zelmanowitz [71]) Every finitely generated right R-submodule of a direct product $\bar{R}^a$ of copies of the right quotient ring $\bar{R}$ of R is torsionless over R iff $\bar{R}$ is also a left quotient ring of R. If $\bar{R}$ is right Artinian, then every finitely generated torsionless right R-module embeds in a free R-module.

(t) (Zelmanowitz [71]) For any ring R, any nonsingular right module M is an essential extension of a direct sum of closed right ideals of R, and hence embeds in a direct product of copies of the injective hull $\hat{R}$ of R. If M is countably generated, then M embeds in the direct sum of copies of $\hat{R}$. If R is right self-injective, then any finitely generated nonsingular right module is isomorphic to a finite direct sum of injective hulls of right ideals of R (and hence is injective and projective (Sandomierski [68])). Finally, if R is semilocal and right selfinjective (making R a semilocal lift/rad ring), then every nonsingular module isomorphic to a direct sum of right ideals (cf. (u)).

(u) If R is right Noetherian, then any nonsingular right module M embeds in a direct sum of copies of $\hat{R}$.

(v) Over a right Goldie semiprime ring R, any indecomposable injective module E is either torsion or torsionfree (see (q)).

(x) (Faith [66a]) If R is right nonsingular of finite Goldie dimension (or equivalently, if $\hat{R}$ is semisimple Artinian, see (c)), then $\hat{R}$ is $\sum$-injective.

(y) (Sandomierski [67]) Any direct sum of a family $\{E_i\}$ of nonsingular injective right R-modules of the ring R of Exercise (x) is injective. (Hint: any E_i is a summand of $Q = \hat{R}$, and is a canonical injective Q-module. See (m) and (p).) Moreover, any epic image of an injective right R-module has splitting singular submodule.

(z) (Zelmanowitz [67]) Suppose that R has a semisimple (classical) left quotient ring Q. If M is any torsionless left R-module of finite Goldie dimension, then $\mathrm{End}_R M = E$ has semisimple (classical) left quotient ring $\mathrm{End}_R(Q \otimes_R M) = S$. Moreover, if Q is a twosided quotient ring of R then S is a twosided quotient ring of E. (In particular, M can be any finitely generated nonsingular module over Q in this case.)

Dedekind Finite Rings

19.39 Propositon and Definition. *A ring R is said to be **Dedekind finite** provided the implication*

$$x y = 1 \;\Rightarrow\; y x = 1$$

is valid $\forall\, x, y \in R$. A ring R is Dedekind finite if and only if $R/\mathrm{rad}\,R$ is. A sufficient condition for this is that $y^{\perp} = 0$ implies $^{\perp}y = 0$ $\forall\, y \in R$, a condition expressed by saying that every right regular element of R is regular.

Proof. If $x y = 1$, then $g = 1 - y x$ is an idempotent of R such that $x g = g y = 0$. Then, $y x = 1$ if and only if $g = 0$. Now $y^{\perp} = 0$, and so $^{\perp}y = 0$ implies $g = 0$.

Let $\bar{a}$ denote the image in R/J of any $x \in R$ under the canonical map $R \to R/J$, where $J = \mathrm{rad}\,R$. If R/J is Dedekind finite, then $x y = 1$ implies that $\bar{y}\,\bar{x} = 1$, that is, that $g = y x - 1 \in J$. But the zero element is the only idempotent contained in $\mathrm{rad}\,R$. Thus, $y x = 1$, and R is Dedekind finite. Conversely, if $\bar{x}\,\bar{y} = 1$, then $u = x y - 1 \in J$, so that $x y = 1 + u = v$ is a unit of R. Assuming that R is Dedekind finite, then $x y v - 1 = 1$ implies that x is a unit of R, and hence that $\bar{x}$ is a unit of R/J, and so $\bar{y} = \bar{x}^{-1}$. Thus, $\bar{y}\,\bar{x} = 1$, and R/J is Dedekind finite. $\square$

19.40 Proposition (Jacobson [50]). (a) *If a ring A is Dedekind infinite, then there is an infinite set $\{e_{ij}\}_{i,j \in \omega}$ of matrix units of A, that is,*

$$e_{ij} e_{kl} = \delta_{jk} e_{il} \qquad\qquad\qquad i, j, k, l \in \omega,$$

where δ is the Kronecker-δ.

(b) *A ring R is Dedekind finite provided that either R, or $R/\mathrm{rad}\,R$, satisfies any of the following chain conditions:* (1) $(\mathrm{acc})^{\perp}$, (2) $(\mathrm{dcc})^{\perp}$, (3) $(\mathrm{acc})\oplus$, $(\mathrm{dcc})\oplus$, (5) *either the a.c.c., or the d.c.c., on principal right ideals generated by idempotents.*

Proof. (a) If $xy=1$, then $g=yx$ is an idempotent such that

$$x(1-g)=(1-g)\,y=0$$

and

$$e_{ij}=y^{i}(1-g)\,x^{j}\qquad\qquad i,j\in\omega$$

are orthogonal idempotents. Furthermore, $e_{ii}=0$, for any i, implies that $g=1$. This proves (a), and (b) is evident.

(b) Given the infinite set of matrix units, then $\{e_{ii}R\}_{i\in\omega}$ is an infinite independent set of annihilator right ideals of R, and if $f_{n}=\sum_{i=1}^{n}e_{ii}$, then $\{f_{i}R\}_{i\in\omega}$ is a strictly ascending sequence of right ideals generated by idempotents. This suffices for (b). $\square$

19.41 Proposition and Definition (Utumi [65]). *A right ideal I of S is said to* **essentially generate** *an idempotent $e\in S$ provided that eS is an essential extension of I. Every ring S in which every right, and every left, ideal essentially generates an idempotent, in particular, every right and left selfinjective ring, is Dedekind finite.*

Proof. By Jacobson's theorem 19.40, if S is not Dedekind finite, then S contains an infinite set $\{e_{ij}\}_{i,j\in\omega}$ of matrix units. Let $Se,e=e^{2}$, be an essential extension of $\sum Se_{ii}$. We may suppose that $e\,e_{ij}=e_{ij}$ for every i,j taking $e\,e_{ij}$ instead of e_{ij} if necessary. Let fS be an essential extension of $\sum_{j>1}(e_{1j}+e_{jj})S$, f being an idempotent.

First assume that $e\neq fe$. Then $0\neq(1-f)e\in Se$, and hence there is x such that $0\neq x(1-f)e\in\sum_{i=1}^{n}Se_{ii}$ for some finite n. Then $x(1-f)e\,e_{n+1,n+1}=0$ by the orthogonality of (e_{ii}). On the other hand, $(1-f)(e_{1j}+e_{jj})=0$ for each $j>1$ by the definition of f. Hence

$$x(1-f)\,e_{1,n+1}=x(1-f)(e_{1,n+1}+e_{n+1,n+1})-x(1-f)\,e\,e_{n+1,n+1}=0.$$

Thus, $x(1-f)\,e_{1,n+1}=0$, and so $x(1-f)\,e_{1,i}=0$ for every i. In particular, $x(1-f)\,e_{1}=0$. For any $j>1$ we have

$$x(1-f)\,e_{jj}=x(1-f)(e_{1j}+e_{jj})-x(1-f)\,e_{1j}=0.$$

Therefore $x(1-f)\,e_{ii}=0$ for every i. Now denote by A the set of elements y in Se such that $y\,e_{ii}=0$ for every i. Then $x(1-f)\,e\in A$. A is a left ideal contained in Se, and is disjoint to $\sum Se_{ii}$. Since Se is essential over $\sum Se_{ii}$, it follows that $A=0$, whence $x(1-f)\,e=0$, a contradiction.

Next, suppose that $e=fe$. Then $e_{11}=e\,e_{11}=fe\,e_{11}\in fS$. Hence

$$0\neq e_{11}z\in\sum_{j>1}(e_{1j}+e_{jj})S$$

for some z. Let $e_{11}z=\sum_{j=2}^{m}(e_{1j}+e_{jj})z_{j}$. Since the sum $\sum e_{ii}S$ is direct, we have $e_{11}z-\sum_{j=2}^{m}e_{1j}z_{j}=0$ and $e_{jj}z_{j}=0$ for $j=2,\ldots,m$, and therefore $e_{11}z=0$, a contradiction. $\square$

19.42 *Exercise.* (a) (Utumi [65]) Under the assumptions of the proposition, if $eS \approx fS$, for idempotents e and $f \in S$, and if $eS \subseteq fS$ (that is, if $fe = e$), then $eS = fS$.

(b) If S is left selfinjective, and Dedekind finite, and if $f: I \to J$ is an isomorphism of left ideals, then there is a unit $u \in S$ such that $(x)f = xu \ \forall x \in I$.

(c) Any semilocal ring is Dedekind finite.

(d) (Utumi [63b]) Let R be right nonsingular. Then every closed right ideal is a right annihilator iff every nonzero inessential right ideal has nonzero left annihilator. Moreover, R is also left nonsingular. Then every closed right or left ideal is an annulet iff the maximal right quotient ring of R is also the maximal left quotient ring.

A ring R is **completely Dedekind finite** if every factor ring is Dedekind finite 19.39. Any semilocal ring is Dedekind finite, and hence completely Dedekind finite.

Two right ideals A and B of a ring R are **comaximal** (or **relatively prime**) provided that $R = A + B$. Two right ideals X and Y are **commutative**, or **commute**, if $XY = YX$.

19.43 **Proposition** (Nakayama [40], Asano [49]). *Let R be completely Dedekind finite.*

(a) *Any two comaximal ideals $A = aR$, and $B = bR$, which are cyclic right ideals, are commutative.*

(b) *If C is any ideal which is cyclic as a right and left ideal, say $C = Rc = aR$, then $C = Ra$.*

Proof. (a) Write $1 = ax + by$, with $x, y \in R$. Then the image of 1 in $\bar{R} = R/B$ is $1 = \bar{a}\bar{x}$. Since R/B is Dedekind finite, then $1 = \bar{x}\bar{a}$. Since $ba \in A$, then $ba = ab'$, for some $b' \in R$. Then, $\bar{b}' = \bar{x}\bar{a}\bar{b}' = \bar{x}\bar{b}\bar{a} = 0$, that is, $b' \in B$. Therefore,

$$BA = bRaR \subseteq baR \subseteq ab'R \subseteq AB,$$

and, by symmetry, $AB \subseteq BA$, so $AB = BA$.

(b) Then $c = av$, $a = uc$, with $u, v \in R$. Then, $c = ucv$. Write $cv = v'c$, with $v' \in R$. Then, $c = uv'c$ so that $1 - uv'$ lies in the left annihilator ideal I of the ideal Rc. Since R/I is Dedekind finite, then $1 - v'u$ lies in I, that is, $c = v'uc = v'a$ lies in Ra. This proves that $A = Rc = Ra$. $\square$

19.44 **Corollary.** *If R is a semiprimary ring, and if the prime ideals of R are principal right ideals, then R is primary decomposable.*

Proof. 19.43 and 18.37. $\square$

19.45 **Propositon** (Asano [49]). *The conditions on a right Artinian ring R are equivalent:*

(1) *R is a PRIR.*

(2) *R is primary decomposable, and every right prindec is uniserial.*

(3) *R is a finite product of primary PRIR's.*

(4) *R is a finite product of matrix rings over completely primary* PRIR's.

(4 + *n*) *The condition* (*n*) *on* $R/(\operatorname{rad} R)^2$, $n = 1, 2, 3, 4$.

Proof. By 18.38, if R is local Artinian, then R is PRIR if and only if R is a uniserial module in mod-R if and only if J/J^2 is simple, so (2) $\Leftrightarrow$ (3) $\Leftrightarrow$ (4) $\Leftrightarrow$ (1). Assuming (1), then R is primary decomposable by 19.44. Thus, (1) $\Rightarrow$ (3). This also establishes the equivalence of (5)–(8). Since R is primary decomposable if and only if R/J^2 is 18.37, the proof is complete. $\square$

Injectives over Regular Rings

We begin this section with a quite general theorem for injectives over non-singular (eg. regular) rings, and then study special injectives over some special regular rings.

19.46 A   Theorem (Sandomierski [67]). *If R is right nonsingular, then* $\operatorname{sing}(M/\operatorname{sing} M) = 0$ *for every* $M \in$ *mod-R. Moreover, then* $\operatorname{sing} M$ *splits in any right module M which is an epic image of an injective module E.*

Proof. If $x \in S_2 \smallsetminus S_1$, then by the definition of S_2, there is an essential right ideal J such that $0 \neq xJ \subseteq S_1$. Now J contains the annihilator K of x. Moreover, if $0 \neq j \in J$, then, since $xj \in S_1$, there is an essential right ideal Q such that $xjQ = 0$, that is, $jQ \subseteq K$. Now $\operatorname{sing} R = 0 \Rightarrow jQ \neq 0$, hence $J \triangledown K$, proving that K is essential along with J. But then $x \in S_1$ contrary to choice. Thus, $\operatorname{sing}(M/S_1) = 0$.

Next let $M = E/K$ be an epic image of an injective E, and write $\operatorname{sing} M = S/K$. Then, $\operatorname{sing}(M/\operatorname{sing} M) = 0 \Rightarrow \operatorname{sing}(E/K)/(S/K) = 0$ so therefore S/K splits in $M = E/K$ by 19.30. $\square$

A full left linear ring $L = \operatorname{End}_D V$, where V is a left vector space over D, is left selfinjective by 19.46 B, and hence every principal left ideal K of L is injective, (since K is a summand of L by 19.25). There is no reason to suppose, in general, that any principal right ideal $\neq 0$ of L is injective, and, indeed, this happens iff V is finite dimensional 19.48 B.

19.46 B   Proposition. *Let R be a ring satisfying the two equivalent conditions:*

(a) $R \approx \operatorname{End}_D V$, *where V is a left vector space over a field D.*

(b) *R is a prime left selfinjective ring with a simple right ideal $\approx V$, and* $D = \operatorname{End} V_R$.

Then, V is injective in mod-R *if and only if V is finite dimensional over the field* $D = \operatorname{End} V_R$.

Proof. Assuming (a), then the canonical right R-module V is simple by the density theorem 19.23 and R is left selfinjective by the theorem of Utumi 19.27. Since $D = \operatorname{End} V_R$ is a field, then D embeds in V in D-mod, and so by (6) of 4.11, V embeds canonically in R in mod-R. Thus, (a) $\Rightarrow$ (b).

Conversely, assuming (b), then R embeds canonically in $L = \operatorname{Biend} V_R$. If $a \in L$, and $a \neq 0$, then $Va \neq 0$, and so $Ra \cap R \neq 0$. This implies that L is contained in the maximal left quotient ring $\hat{R}$ of R. Since R is left selfinjective, and regular, then $R = \hat{R}$, which proves (a) that $L = R = \operatorname{End}_D V$.

If V is finite dimensional, then R is semisimple Artinian, and every R-module is injective. Conversely, assume (a) and V injective. If $\hat{R}$ is the right quotient ring of R, then by 19.35 the right ideal $V\hat{R}$ of $\hat{R}$ generated by V is the injective hull of V. Then, injectivity of V implies $V=V\hat{R}$, which shows that $\hat{R}$ embeds canonically in $R=\operatorname{End}_D V$, and hence that $R=\hat{R}$ is right selfinjective. Since R is also left selfinjective, then Utumi's theorem 19.41 implies that R is Dedekind finite, which is possible if and only if V is finite dimensional. $\square$

The equivalence (a) $\Leftrightarrow$ (b) is a theorem of Utumi [56].

19.47 *Example. A regular right V-ring may have infinite dimensional simple modules.*

The example, surprisingly perhaps, is one which the author cited in his *Lectures* [67] of a regular ring which is not a left V-ring! Let $L=\operatorname{End} V_F$ be the full right linear ring over an infinite dimensional right vector space V over a field F, let S be the ideal consisting of l.t.'s of finite dimensions, and let $R=S+F$ be the subring generated by S and the subring F consisting of scalar transformations (sending every $v\mapsto v\,a$ for some $a\in F$). Clearly, R is a regular ring. Moreover, V is canonically isomorphic to a simple left ideal of L, say $V=L\,e$, for $e=e^2\in L$. Now L is injective in mod-L, hence $W=eL$ (the L-dual of V) is a simple injective in mod-L, but, by 19.46 B, V is not injective in L-mod, and hence neither is L. (Thus, L is not a left V-ring.)

First we show that R is a right V ring. Let W be a simple right (or left ideal) of R, and let $f: I \to W$ be a mapping of a right (or left) ideal of R. In order to determine if W is injective, it suffices to assume that I is an essential one-sided ideal, and thus, that I contains the socle S of R. Since R/S is simple, either $I=S$, or $I=R$. If $I=S$, and f is a morphism of mod-R, then f is a morphism of mod-L, since

$$[f(x\,a)-f(x)\,a]\,s=0 \qquad\qquad \forall\ s, x\in S,\ a\in L.$$

Since W is a summand of L in mod-L, then W is injective along with L, and so f has an extension to a mapping $L\to W$ which induces a mapping $R \to W$ extending f. This proves that W is injective in mod-R. If W is a simple right R-module not contained in S, then $W\approx R/S$ is injective over R/S, whence W is injective over R (cf. 19.52 ff.). This proves that R is a right V-ring.

In order to prove that R is not a left V-ring, assume for the moment that V is injective in R-mod, and that $f: I \to V$ is a morphism in L-mod, where I is a left ideal of L. In order to extend f, we may assume that I is an essential left ideal, that is, that $I\supseteq S$. Then, f has an extension f' in R-mod, and f' is a morphism of L-mod, since

$$s[(a\,x)\,f'-a(x)\,f']=0 \qquad\qquad s\in S,\ x\in I,\ a\in L.$$

But this implies that V is injective in L-mod, a contradiction. Thus, R is not a left V-ring. (This fact is also a consequence of the author's [67a, p. 103, Theorem 3.1] which implies that the maximal left quotient ring of R is also a right quotient ring. This involves a contradiction.)

Let Q denote the maximal left quotient ring of L. Then the injective hull in L-mod of V is $QV = Qe$, and

$$\text{End}_L\, Qe = \text{End}_Q\, Qe = eQe = eLe$$

(the last equality since eQe is the quotient ring of eLe). This shows that V is a fully invariant L-submodule of its injective hull [cf. (c) of the following theorem 19.54].

19.48 A   Corollary. *If $L = \text{End}_D V$ is a left linear ring then L has a nonzero injective right ideal iff V is finite dimensional.*

Proof. If V is finite dimensional, then every L-module is injective. Conversely, let $\dim_D V \geq \aleph_0$. Any right ideal $I \neq 0$ contains a simple (minimal) right ideal W. Since $W \approx V$ in mod-L, then 19.46 B states that W is not injective. But, W is a summand of L, hence of I, so I can not be injective.   □

19.48 B   Corollary. *A right nonsingular ring R is canonically isomorphic to the biendomorphism ring of an injective right ideal $V \neq 0$ if and only if R is right selfinjective and regular.*

Proof. Any regular ring R is nonsingular (Johnson [51]) and $R \approx \text{End}_R R = \text{Biend}\, R_R$. The converse is a corollary of the proof of the last part of 19.46 B which shows that $R = \hat{R}$ is right selfinjective.   □

QI Rings

A ring R is said to be a **right QI ring** ("Quasi" $\Rightarrow$ "Injective") provided that every quasiinjective right module is injective. This implies that every simple, and every semisimple, right module is injective, and hence such a ring is a right V-ring (cf. 7.32). [Moreover, any right QI ring is right Noetherian (proof?).]

19.49   Example. (a) Let $R = k[y, D]$ denote the ring of differential polynomials over a universal field k, and let V be the unique simple right R-module. Then, by Cozzen's theorem 7.40, V is injective in mod-R. Nevertheless, by 19.46 B, V is not injective in $L = \text{Biend}\, V_R$, since V is not finite dimensional over $D = \text{End}\, V_R$.

(b) R *is a right QI-ring.* Let Q be any QI right R-module, and let E denote the injective hull. By a result proved in the next chapter (20.6), we can write $E = \sum_{i \in I} \oplus E_i$, where E_i is indecomposable for every $i \in I$. Since $Q \cap E_i \neq 0$, for every $i \in I$, it suffices to assume that E is indecomposable. But by Cozzen's results, either $E = Q$ is simple, or else E is the right quotient field of R. Since Q is by 19.3 a fully invariant R-submodule of E, then Q is a left ideal of E. Since E is a field, then $Q = E$ in this case too.

(c) *Every QI module over R is $\coprod$-QI.* Thus, R affords an example of a Noetherian ring not Artinian having the stated property.

A right QI ring A has the property for any ideal B:

(I) *Every injective right A/B module M is injective as a canonical right A-module.*

This is equivalent to the requirement on ideals B:

(II) *The inclusion functor* mod-$A/B \rightsquigarrow$ mod-A *preserve injectives.*

(III) has been discussed in the setting of categories, namely:

19.50 Proposition. *If* $T: C \rightsquigarrow D$ *is a functor of abelian categories with left adjoint* $S: D \rightsquigarrow C$, *and if* C *has enough injectives, then* T *preserves injectives if and only if* S *is exact.*

Proof. This is 6.28.

19.51 A Corollary. (I) *and* (II) *are each equivalent to the requirement that* A/B *is a flat left* A-*module. A necessary and sufficient condition for this is that the inclusion functor* mod-$A/B \rightsquigarrow$ mod-A *preserves injective hulls of simple right* A/B-*modules.*

Proof. The first stated equivalence follows from 19.50 as does the necessity of the second statement. Now suppose that the inclusion functor mod-A/B preserves injective hulls of simple right A/B-modules. If M is an injective right A/B-module, then M embeds in some product E^I of any injective cogenerator E. By injectivity of M, $E^I = M \oplus X$, for some right A/B-module. Choose E to be the product $\prod_{j \in J} \hat{V}_j$, where $\{V_j\}_{j \in J}$ is a representative class of simple right A/B-modules. Now E is a product in mod-A of injective right A-modules, so E is injective in mod-A, and hence so is E^I. Then, M, a summand of an injective, is injective, and therefore the inclusion functor preserves injectives. $\square$

19.51 B Corollary. *If* A *is a right* V-*ring, then, for any ideal* B, A/B *is a flat left* A-*module.* $\square$

19.52 Proposition. *If* R *is a regular ring, then every injective right* R/A-*module is canonically an injective right* R-*module, for any ideal* A. *If* R *is a commutative ring, then conversely.*

Proof. If R is regular, then every R-module is flat, so Corollary 19.51 A applies. Conversely, if R is commutative, and if R/A is flat, for every ideal A, that is, if every cyclic module is flat, then R is regular (19.24). $\square$

19.53 Corollary (Kaplansky). *A commutative ring* A *is regular iff* A *is a* V-*ring.*

Proof. Assume A is regular, and let $V = A/M$ be a simple A-module. Then, V is injective over A/M, hence by 19.52, V is injective over A. Thus, A is a V-ring. Conversely, assume that A is a V-ring, and let B be any ideal. If V is any simple A/B-module, then V is injective over A, and hence over A/B. This proves that the inclusion functor mod-$A/B \rightsquigarrow$ mod-A preserves the injective hulls of simple A/B-modules, so A/B is flat over A by 19.51 A, and hence A is regular by 19.52. $\square$

A Characterization of Goldie Prime Rings

Recall from Chapter 9, a ring R is right Goldie if R satisfies the a.c.c. on complement right ideals, that is, (acc) $\oplus$, and the a.c.c. on annihilator right ideals, that is,

$(acc)^\perp$. The theorem of Goldie and Lesieur-Croisot characterizes rings having simple Artinian right quotient rings as right Goldie prime rings (Theorem 9.9). In Chapter 16, we gave Gabriel's proof. Now we give another proof, and another characterization of these rings.

19.54 Theorem (Faith [72b]). *The following conditions on a ring R are equivalent:*

(a) *R is a prime right Goldie ring.*

(b) *R has simple Artinian right quotient ring $Q(R)$.*

(c) *R has a faithful finendo indecomposable injective right R-module E, having no nontrivial fully invariant submodules, and $D = \operatorname{End} E_R$ is a field.*

When this is so, then $Q(R) \approx \operatorname{End}_D E$, and E is isomorphic in mod-$Q(R)$ to a minimal right ideal of $Q(R)$.

Proof. We shall prove the theorem without recourse to the theorems of Goldie and Lesieur-Croisot.

(c) $\Rightarrow$ (b) Let $n = \dim_D E$. Then $A = \operatorname{End}_D E$ is a full matrix ring D_n over the field D. Moreover, there is a canonical embedding of the right A-module E in A. Since E is a simple right A-module, then E is a minimal right ideal of A. Since E is faithful in mod-R, then R embeds in A canonically. Now $A \approx E^n$ in mod-A. Write $E^n = E_1 \oplus \cdots \oplus E_n$, a direct sum of n isomorphic A-submodules of A. If, for example, $R \cap E_1 = 0$, then the projection $E^n \to E_2 \oplus \cdots \oplus E_n$ maps R isomorphically in mod-R into E^{n-1}, which, by the proof of 19.15, implies that $D^{n-1} \to E \to 0$ is exact in D-mod, contrary to the assumption that $\dim_D E = n$. Thus, if $U_i = R \cap E_i$, $i = 1, \ldots, n$, then R contains an essential R-submodule $U_1 \oplus \cdots \oplus U_n$ of $A = E^n$. This proves that the right R-module A is the injective hull of R in mod-R. By the Krull-Schmidt theorem applied to E^n, this proves that R satisfies the a.c.c. on direct sums of right ideals contained in R. Moreover, any subring of a right Noetherian ring satisfies the a.c.c. on right annulets, so we have proved that R is a right Goldie ring, which together with primeness of R (see below), implies (a).

Write $E = eA$, for some idempotent $e \in A$, and let $D = eAe$. We invoke for the first (and last) time the hypothesis on fully invariant submodules: If I is any non-zero right ideal of R, then DI is a fully invariant submodule of E, so that $DI = E$, and there exists a basis of E over D consisting of n elements $x_1, \ldots, x_n$ of I. Then, $E = \sum_{i=1}^n D x_i$, and faithfulness of E implies that $\bigcap_{i=1}^n x_i^\perp = 0$. *Thus, ($R$ is prime and) R embeds as a right R-module in a direct sum I^n* [under a map $r \mapsto (x_1 r, \ldots, x_n r)$] *of n copies of any nonzero right ideal.* Now, if H is any essential right ideal of R, then $V_i = E_i \cap H \neq 0$, $i = 1, \ldots, n$, so that H contains the direct sum $V_1 \oplus \cdots \oplus V_n$ of uniform right ideals. Moreover, since R is prime, $V_i V_1 \neq 0$, for any i. If $y_i \in V_i$ is such that $y_i V_1 \neq 0$, then V_1 embeds in V_i under the map $v \mapsto y_i v$, since $y_i v = 0$ for some nonzero $v \in V_1$ would imply that $y_i E = 0$, whence $y_i V_1 = 0$, by virtue of the fact that E_i is a minimal right ideal of A. This proves that H contains a direct sum V^n of right ideals isomorphic to $V = V_1$, and hence that R embeds in H as a right R-module. Therefore, any essential right ideal H of R contains an element x which has zero right annihilator in R, and, by the fact that R is essential in A, has zero right annihilator in A. But in an Artinian ring A, this implies by 19.40 that x is a unit in A (and hence that x is a regular element of R). Similarly, any regular element of R is a unit of A.

The proof that A is the right quotient ring of R is immediate from this. If q is any element of A, then, since R is an essential R-submodule of A, the right ideal $(q:R)$ is essential in R, and therefore contains a regular element x. Thus, $q = a x^{-1}$, where $a = q x \in R$, proving that A is the right quotient ring $Q(R)$ of R.

(a) $\Rightarrow$ (c). By the a.c.c. on complement right ideals (or equivalently, the a.c.c. on direct sums contained in R), R contains a uniform right ideal, which can be chosen to be a right complement U of any maximal right complement right ideal. (Then, U is a minimal complement right ideal. Cf. 7.17)

Assertion: The injective hull $E = \hat{U}$, of any uniform right ideal U has the property stated in (c).

Proof of the Assertion. We first show that the right singular ideal Z of R is nil. If $x \in Z$, then there is a finite integer n such that $y = x^n, x^{n+1}, \ldots, x^{2n}$, all have the same right annihilator. It follows that $y^{\perp} \cap y R = 0$, whence $y = x^n = 0$. Next, if A is a nil left ideal contained in Z, then by 9.7, an element $a \in A$ which is maximal in $\{x^{\perp} | 0 \neq x \in A\}$ generates a nilpotent ideal of index ≤ 3. Then (semi)primeness of R forces $A = 0$, and $Z = 0$.

Next, there is a finite integer n such that R contains an essential right ideal $W = U_1 \oplus \cdots \oplus U_n$, with $U_i \approx U$, $i = 1, \ldots, n$. Thus, the injective hull E^n of U^n is the injective hull $\hat{R}$ of R. Write End $E^n_R = B_n$, where $B =$ End E_R. Since E^n is injective, the radical J of B_n consists of those b with essential kernels (theorem of Utumi 19.28). Since $R \subseteq E^n$, and $b \in J$, then $x = b(1) \in E^n$ belongs to the singular submodule S of E^n. But $S = 0$, since R is an essential submodule with zero singular submodule. Then, $\ker b \supseteq R$, and if $y \in E^n$, there is an essential right ideal I with $y I \subseteq R$, and $b(y) I = 0$, so that $b(y) \in S = 0$. Thus, $b = 0$, and therefore, B_n, hence B, has zero radical. Since B is a local ring, then B must be a field. Then, the fact that $R \subseteq E^n$ implies that $n = \dim_B E$ (19.15), and hence, $A = \text{End}_B E \approx B_n$ canonically. Moreover, $E^n \approx A$ in mod-A is an essential extension of R in mod-R, and then it follows from the proof of the next proposition that E has no nontrivial fully invariant submodules. This completes the proof of the assertion, and of (a) $\Rightarrow$ (c).

Now (c) $\Rightarrow$ (a) [parenthetically in the proof of (c) $\Rightarrow$ (b)], and (b) $\Rightarrow$ (c) by the next proposition, so the equivalence of (a), (b) and (c) is completed by the next proposition. $\square$

19.55 A **Proposition.** *Let R be a ring with simple Artinian right quotient ring Q. Let E denote a minimal right ideal of Q. Then, E is the (up to isomorphism) unique finendo indecomposable injective faithful right R-module. Moreover, E has no nontrivial fully invariant R-submodules.*

Proof. By the Wedderburn theorem, $Q \approx \text{Biend } E_Q$. Let $D = \text{End } E_Q$, and let $n = \dim_D E$. Since Q is Artinian, then n is finite. Since $D = \text{End } E_R$ (in fact, $Q = \text{End } Q_R$), it follows that the right R-module E is finendo. Since Q is the injective hull of R, and E is a summand of Q, then E is an injective, faithful, right R-module. Moreover, E is indecomposable in mod-R, since any R-summand of E is a Q-summand.

Let H be a nonzero fully invariant R-submodule of E, and write $E = H \oplus G$, for a D-submodule G of E. Let $g \in Q$ be the idempotent which induces the projection $E \to H$ parallel to G in D-mod. Then, $Eg = H$, and $H(1-g) = 0$. Since H is an R-submodule, then $gr = grg$, for any $r \in R$, that is, $gR(1-g) = 0$. But, then $(gR \cap R)((1-g)R \cap R) = 0$. Since $H \neq 0$, then $g \neq 0$, and $gR \cap R \neq 0$, so that primeness of R implies that $g = 1$. Thus, $H = E$, so E has no fully invariant nontrivial submodules.

That any finendo faithful indecomposable injective right R-module F is isomorphic to E follows from the fact that by 19.15, $R \hookrightarrow F^m$ for a finite integer $m > 0$, and hence E embeds as a direct summand in F^m, whence $E \approx F$ by the $K - S$ Theorem 18.18. $\square$

The next theorem may be proved similarly as 19.34.

19.55 B   Theorem. *A ring R has a semisimple right quotient ring if and only if R is a semiprime ring containing finite faithful set $\{E_1, \ldots, E_n\}$ of indecomposable finendo injective right R-modules, none of which have nontrivial fully invariant submodules, and the endomorphism ring of each is a field.*

The condition of the theorem is equivalent to the existence of a *faithful finendo injective module F such that* End F *is a finite product of n fields, and F has precisely n fully invariant indecomposable submodules $\neq 0$.*

A prime ring may have nonfaithful finendo indecomposable injective modules, yet not be right Goldie, as the example, Example 19.47, of a regular V-ring R shows, since the simple module $W = R/S$ is injective, and 1-dimensional over its endomorphism ring R/S.

A right ideal I of R is **co-irreducible** if R/I is a uniform right module.

19.56 A   Corollary. *If R is a right QI-ring, and if I is any co-irreducible right ideal such that $\widehat{R/I}$ is finendo, then I is a maximal complement right ideal.*

Proof. Any right QI ring is right Noetherian, and right V, hence semiprime, 7.32 C with semisimple classical right quotient ring Q. Moreover, by 19.55 A, every finendo indecomposable injective right R-module E embeds in Q. Thus, $E = \widehat{R/I}$ embeds in Q under a map f such that I is the right annihilator in R of $x = f[1+I]$. Since Q is semisimple, the right annihilator of x in Q is a principal right ideal eQ, that is, a complement right ideal of Q, and hence $I = eQ \cap R$ is a complement right ideal of R. Clearly, I is a maximal right complement. $\square$

19.56 B   Remarks. (a) More generally, if R is any right n.s. ring with quotient ring $\hat{R}$, then a cyclic module R/I embeds in $\hat{R}$ if and only if I is a complement right ideal.

(b) If R is semiprime right Goldie, then any torsion free module M is right QI if and only if it is injective. This follows since then M is canonically a right $Q = Q(R)$ module, and hence $\hat{M}$ is a direct sum of finendo indecomposable injective right R-modules, each of which is canonically isomorphic to a principal indecomposable right ideal (right prindec) of Q. But a right prindec E of Q has no fully invariant submodules, as 19.55 A shows. $\square$

Prime Ideals in Selfinjective Regular Rings

In this section, we prove an important result of K.R.Goodearl [73 b] on the prime ideal structure of the rings of the title. The simplest examples of nonartinian, that is, nonsemisimple regular selfinjective rings are: the full right linear rings on right vector space V over a field D such that $\dim_D V = \infty$; and the maximal right quotient rings of domains which are not Ore. In either case, the lattice of ideals are linearly ordered. Goodearl's theorem states *that the lattice of ideals of regular right selfinjective rings containing a prime ideal is linearly ordered.* A sharpened version of this theorem which strengthens this to a well ordering for closed prime ideals is proved in the following section. As a corollary, it follows that a conjecture of Kaplansky to the effect that prime regular rings are primitive holds good for selfinjective rings.

For convenience, we say that an ideal I of a right selfinjective ring R is **closed (essential)** if it is a closed (essential) right ideal.

19.57 A Lemma. *Let R be right selfinjective and regular. Then, any nonzero closed ideal I is a ring and a direct factor of R, that is, $R = I \times K$, for an ideal K.*

Proof. By (I, 3.59, p. 170), R contains an injective hull of I, which coincides with I since I is closed. Write $R = I \oplus K$, for a right ideal K. Then $KI \subseteq K \cap I = 0$, so that $KI = 0$, and $(IK)^2 = 0$, which in a semiprime ring implies $IK = 0$. Then, K is an ideal, and $R = I \times K$. (Cf. e.g. the Chinese remainder theorem 18.31.) □

19.57 B Corollary. *A right selfinjective regular ring R is directly indecomposable in RINGS iff R is a prime ring.*

Proof. Any prime ring is directly indecomposable. Conversely, if the right selfinjective ring R is directly indecomposable in RINGS, then by the lemma, every nonzero ideal I of R must be an essential right ideal, for otherwise the closure I' of I in R would be a direct factor of R, and $I' \neq R$, a contradiction. □

As note in Exercise 19.21(p), in any ring R, a prime ideal P is either essential, or a right annulet. In a right selfinjective ring, right annulets are not in general closed. However:

19.58 *In a right selfinjective regular ring R, a prime ideal P is either essential or closed.*

Proof. We first note that the closure H of P in R (= the maximal essential extension of P in R = the unique injective hull of P in R 19.35) is an ideal. Let x be an element of R, It suffices to show that $xH + P$ is an essential extension of P, since by 19.35, P has a unique injective hull in R. Let $y = xh + p$ be any element of $xH + P$, $h \in H$, $p \in P$. We must show that $y \neq 0 \Rightarrow yR \cap P \neq 0$. If $xh = 0$, then done. Otherwise, $h \neq 0$, and so there is an essential right ideal J such that $hJ \subseteq P$. Then, $yJ \subseteq xhJ + pJ \subseteq P$, since P is an ideal. Moreover, $yJ \neq 0$, since sing $R = 0$ by 19.35. This proves that $xH + P \subseteq H$, and $xH \subseteq H$, which proves that H is an ideal. Then, by 19.57A, $R = H \times K$, for an ideal K, and then $HK = 0$ and primeness of P imply that $H \subseteq P$, that is, $P = H$ is closed. □

The next two lemmas are more general than needed for the theorem on linearly ordered ideals, but (as Goodearl remarks) are useable in the proof of his following theorem on well ordered ideals 19.64.

First a remark: any finitely generated right ideal of R is generated by an idempotent 19.24, hence is a summand of R, so is projective *and* injective.

19.59   Definition and Proposition. *We say that M is **subisomorphic** to a module A, or **submorphic**, for short, provided that M embeds in A. If M embeds in a finite product A^n, then M is **n-submorphic** to A.*

If A is a right ideal of a right selfinjective regular ring R, then for $x \in R$, the right ideal xR is n-submorphic to A iff x belongs to the ideal RA generated by A.

Proof. If $x \in RA$, then $x \in r_1 A + \cdots + r_n A$ for some elements $r_i \in R$. Letting B be the direct sum of n copies of A, there is an epimorphism of B onto $r_1 A + \cdots + r_n A$. Since xR is projective, it follows that xR embeds in B.

Conversely, assume that xR embeds in $A_1 \oplus \cdots \oplus A_n$, where each $A_i = A$. Inasmuch as xR is injective, there must exist an epimorphism $f: A_1 \oplus \cdots \oplus A_n \to xR$. From the injectivity of R we infer that each of the maps $A_i \to A_1 \oplus \cdots \oplus A_n \to xR$ is given by left multiplication by some $r_i \in R$. Thus there exist elements $a_i \in A$ such that $x = f(a_1, \ldots, a_n) = r_1 a_1 + \cdots + r_n a_n$. $\square$

19.60   Lemma. *Any finitely generated nonsingular module M module is injective and projective.*

Proof. Any finitely generated right ideal has this property, as we observed. Let x be one of the generators of M. The right annihilator ideal I of x is contained as an essential submodule of a right ideal eR, $e = e^2 \in R$. Now,

$$0 \to eR/I \to R/I \to R/eR \to 0$$

is exact, and $R/eR \approx (1-e)R$ is projective, so that the sequence splits, and eR/I is a summand of the nonsingular module $R/I \approx xR$. Since eR/I is a singular module, then $I = eR$, so that xR is projective and injective. By induction on the number of generators of M, so is M. $\square$

19.61   Lemma. *Let R be regular and right selfinjective. Given nonsingular injective right R-modules A and B, there exists a central idempotent $e \in R$ such that Ae is submorphic to Be and $B(1-e)$ is submorphic to $A(1-e)$.*

Proof. Let $\mathscr{A}$ denote the collection of all pairs (C, D), where $C \leq A$, $D \leq B$, and $C \approx D$. Say that a family $\{(C_i, D_i)\} \subseteq \mathscr{A}$ is *independent* whenever $\{C_i\}$ is an independent family of submodules of A and $\{D_i\}$ is an independent family of submodules of B. Choosing a maximal independent family $\{(C_i, D_i)\} \subseteq \mathscr{A}$, we obtain decompositions $A = A_1 \oplus A_2$ and $B = B_1 \oplus B_2$ such that A_1 is an injective hull for $\bigoplus C_i$ and B_1 is an injective hull for $\bigoplus D_i$. Inasmuch as $C_i \approx D_i$ for all i, we obtain $A_1 \approx B_1$. In view of the maximality of $\{(C_i, D_i)\}$, we note that it is not possible for any nonzero submodule of A_2 to be submorphic to B_2.

Since A is nonsingular, the annihilator ideal $H = \{r \in R | A_2 r = 0\}$ is closed. In view of 19.57A, we must have $H = eR$ for some central idempotent $e \in R$. Then $Ae = A_1 e \approx B_1 e \leq Be$, hence Ae is submorphic to Be.

We claim that $B_2(1-e)=0$. Suppose, on the contrary, that there is a nonzero element $x \in B_2(1-e)$. Now xR is projective by 19.60, hence $xR \approx tR$ for some idempotent $t \in R$, and we note that $te=0$. Inasmuch as $t \neq 0$ (because $x \neq 0$), we see that $t \notin H$, hence $yt \neq 0$ for some $y \in A_2$. Now ytR is projective (by 19.60), and there is an epimorphism $xR \approx tR \to ytR$, hence there must be a monomorphism $ytR \to xR$. But then ytR is a nonzero submodule of A_2 which is submorphic to B_2, which is impossible.

Thus $B_2(1-e)=0$, whence $B(1-e)=B_1(1-e) \approx A_1(1-e) \leq A(1-e)$, and therefore $B(1-e)$ is submorphic to $A(1-e)$. $\square$

19.62 Theorem. *Let R be regular and right selfinjective. If P is a proper ideal of R, then the following conditions are equivalent:*

(a) *P is prime.*

(b) *For any central idempotent $e \in R$, either $e \in P$ or $1-e \in P$.*

(c) *The ideals of R containing P are linearly ordered under inclusion.*

Proof. (a) $\Rightarrow$ (b) is clear.

(b) $\Rightarrow$ (c). If not, then we can choose ideals H and K in R, both containing P, along with elements $x \in H-K$ and $y \in K-H$. According to 19.61, there exists a central idempotent $e \in R$ such that xRe is submorphic to yRe and $yR(1-e)$ is submorphic to $xR(1-e)$. If $e \in P$, then $yRe \subseteq P \subseteq H$, and $xR(1-e) \subseteq H$ anyhow, hence yR is 2-submorphic to H. However, 19.59 then says that $y \in H$, which is false. Likewise, if $1-e \in P$, we obtain $x \in K$, which is also false. Thus neither e nor $1+e$ belongs to P, which contradicts (b).

(c) $\Rightarrow$ (a). If I, J are ideals of R which properly contain P, then by (c) there is no loss of generality in assuming that $I \subseteq J$. Since R is regular, $I^2 = I \nsubseteq P$, and consequently $IJ \nsubseteq P$. $\square$

19.63 Corollary. *If P is any prime ideal in R, then every proper ideal in R which contains P is also prime.* $\square$

Well Ordered Ideal Lattices

As in the Correspondence theorem (I, p. 195), ideals R will denote the lattice of ideals of R. If R is a full linear ring on a vector space V over a field D and if $\dim_D V$ is an infinite cardinal a, then for the ideals of R are 0 and

$$\{t \in R \mid \operatorname{rank} t < b\}$$

for infinite cardinals b not exceeding $a+1$. Thus, ideals R embeds in the lattice of cardinals $\leq a+1$, hence, is well ordered. In this section, this result is extended to include:

19.64 Theorem (Goodearl [73b]). *For any closed prime ideal P of a right selfinjective regular ring R, ideals R/P is well ordered. If R is prime, that is, $P=0$, then each ideal of R is isomorphic to $H(\alpha)$ for some infinite cardinal α, where*

$$H(\alpha) = \{0\} \cup \{x \in R \mid xR \approx E[\alpha(xR)]\}$$

and $E[A]$ is the injective hull; $\alpha(A)$ is the direct sum of α copies of any R-module A.

We postpone the proof of Theorem 19.64 in order to develop some intermediate results. First, by 19.57A, there is a ring decomposition $R = P \times K$, so R/P is a regular, right self-injective ring; hence we may assume (without loss of generality) that $P = 0$. This assumption that R is a prime ring will be in force until Theorem 19.64 is proved.

Note that $H(\alpha) \subset H(\beta)$ whenever $\alpha \leq \beta$. For if $x \in R \smallsetminus H(\beta)$, then

$$xR \approx E[\beta(xR)] = E[\alpha\beta(xR)] \approx E[\alpha(E[\beta(xR)])] \approx E[\alpha(xR)]$$

and so $x \notin H(\alpha)$.

In view of the assumption that R is a prime ring, 19.61 takes on the following strengthened form:

19.64A *Given any nonsingular injective modules A and B, either A is submorphic to B or B is submorphic to A.*

For convenient referencing, we restate Proposition 3.60 (I, p. 171).

19.64B *Any two injective modules which are submorphic to each other must be isomorphic.*

19.64C *Let B be a nonsingular injective module, α any cardinal. If B has a nonzero submodule C such that $C \approx E[\alpha C]$, then $B \approx E[\alpha B]$.*

Proof. We obviously may assume that $\alpha > 1$. Set $\mathscr{A} = \{A \leq B \mid A \approx E[\alpha A]\}$, and expand $\{C\}$ to a maximal independent family $\{A_i\} \subseteq \mathscr{A}$. We have $B = E[\bigoplus A_i] \oplus D$ for some D, and the maximality of $\{A_i\}$ implies that D has no nonzero submodules which belong to $\mathscr{A}$. Inasmuch as C is a nonzero submodule of $E[\bigoplus A_i]$ which belongs to $\mathscr{A}$, it follows that $E[\bigoplus A_i]$ cannot be submorphic to D. According to 19.64A, D must thus be submorphic to $E[\bigoplus A_i]$, whence B is submorphic to $2E[\bigoplus A_i]$. Since $\alpha > 1$ we infer that $A_i \approx 2A_i$ for all i, hence $E[\bigoplus A_i] \approx 2E[\bigoplus A_i]$. Thus B is submorphic to $E[\bigoplus A_i]$, hence in view of 19.64B we obtain $B \approx E[\bigoplus A_i]$, from which we conclude that $B \approx E[\alpha B]$.

19.64D *Let A be a nonsingular injective module which is isomorphic to a proper submodule of itself. Then $A \approx 2A \approx E[\aleph_0 A]$.*

Proof. There exists a monomorphism $g: A \to A$ which is not epic. Since gA is isomorphic to A and so is injective, we obtain $A = gA \oplus B$ for some nonzero B, and then $g^n A = g^{n+1} A \oplus g^n B$ for all positive integers n. Now $\{gB, g^2 B, \ldots\}$ is a countably infinite independent sequence of pairwise isomorphic submodules of A, hence the submodule $G = \bigoplus g^n B$ satisfies $G \approx 2G \approx \aleph_0 G$. The module A must contain an injective hull C for G, and we see that $C \approx 2C \approx E[\aleph_0 C]$, hence from 19.64C we obtain $A \approx 2A \approx E[\aleph_0 A]$.

Given any nonzero nonsingular injective module A, it follows easily from 1964.D that $A \approx 2A$ if and only if $A \approx E[\aleph_0 A]$. Therefore

$$H(\aleph_0) = \{0\} \cup \{x \in R \mid xR \not\approx 2(xR)\}.$$

19.64E *If A, B, C, D are nonsingular injective modules with $A \oplus B \approx C \oplus D$, then there exist decompositions $A = A_1 \oplus A_2$ and $B = B_1 \oplus B_2$ such that $A_1 \oplus B_1 \approx C$ and $A_2 \oplus B_2 \approx D$.*

Proof. We may obviously assume that C is a closed submodule of $A \oplus B$ and that $D = (A \oplus B)/C$.

Let $p_1: A \oplus B \to A$ and $p_2: A \oplus B \to B$ be the projections. Since $A_1 = A \cap C$ is a closed submodule of A, A_1 is injective and $A = A_1 \oplus A_2$ for some A_2. Also, we have an exact sequence $0 \to A_1 \to C \to p_2 C \to 0$, which now splits because A_1 is injective. Setting $B_1 = p_2 C$, we obtain $C \approx A_1 \oplus B_1$. Since B_1 is now injective, we must have $B = B_1 \oplus B_2$ for some B_2.

Note that $B_1 = (1 - p_1) C \leq C + A$, whence $B \leq C + A + B_2$, and thus $C + A + B_2 = A \oplus B$. We also have $A = A_1 + A_2 \leq C + A_2$, hence $C + A = C + A_2$, and consequently $A \oplus B = C + A_2 + B_2$.

Now set $F = C \cap (A_2 + B_2)$. Since $p_2 C = B_1$ and $p_2(A_2 + B_2) = B_2$, we find that $p_2 F \leq B_1 \cap B_2 = 0$, and consequently $F \leq A$. As a result, $F \leq A \cap C = A_1$, so that $p_1 F \leq A_1$. We also have $p_1 F \leq p_1(A_2 + B_2) = A_2$, whence $p_1 F \leq A_1 \cap A_2 = 0$. Since $F \leq A$, we conclude that $F = 0$.

Thus $A \oplus B = C \oplus (A_2 \oplus B_2)$, and therefore $D = (A \oplus B)/C \approx A_2 \oplus B_2$.

19.64F *Let A, B, C be nonsingular injective modules such that $A \not\approx 2A$. If $A \oplus B \approx A \oplus C$, then $B \approx C$.*

Proof. According to 19.64 E, there exist decompositions $A = A_1 \oplus A_2$ and $B = B_1 \oplus B_2$ such that $A_1 \oplus B_1 \approx A$ and $A_2 \oplus B_2 \approx C$. By 19.64 A, either A_2 is submorphic to B_1 or B_1 is submorphic to A_2.

If A_2 is isomorphic to a proper submodule of B_1, then $A = A_1 \oplus A_2$ is isomorphic to a proper submodule of $A_1 \oplus B_1 \approx A$, which contradicts 19.64 D. The same contradiction arises if B_1 is isomorphic to a proper submodule of A_2, hence the only possibility is $A_2 \approx B_1$.

Therefore $B = B_1 \oplus B_2 \approx A_2 \oplus B_2 \approx C$. $\square$

19.65 Proposition. *For any infinite cardinal α, $H(\alpha)$ is an ideal of R.*

Proof. Case I. $\alpha = \aleph_0$. If $H(\alpha)$ is not a two-sided ideal, then there exist $x, y \in H(\alpha)$ and $z \in (Rx R + Ry R) \smallsetminus H(\alpha)$. According to 19.62 we may assume that $Rx R \leq Ry R$, hence $z \in Ry R$. Note that $z \neq 0$ and $y \neq 0$. Now $zR \approx 2(zR)$, and from 19.59 we see that zR is submorphic to $n(yR)$ for some positive integer n, whence 19.64 C yields $n(yR) \approx 2n(yR)$. However, $yR \approx 2(yR)$, so by inducting on 19.64 F we reach the contradiction $0 \approx n(yR)$.

Case II. $\alpha > \aleph_0$. Given $x, y \in H(\alpha)$ and $z \in Rx R + Ry R$, we must show that $z \in H(\alpha)$. According to Theorem 19.62 we may assume that $Rx R \leq Ry R$, hence $z \in Ry R$. If $y \in H(\aleph_0)$, then $z \in H(\aleph_0) \subset H(\alpha)$ by Case I, hence we need only consider the possibility $y \notin H(\aleph_0)$. Thus $y \neq 0$ and $yR \approx 2(yR)$. Inasmuch as zR is submorphic to $n(yR)$ for some positive integer n (19.59), we infer from $yR \approx 2(yR)$ that zR is also subisomorphic to yR. Since $yR \not\approx E[\alpha(yR)]$, 19.64 C says that either $z = 0$ or else $zR \not\approx E[\alpha(zR)]$, whence $z \in H(\alpha)$.

19.66 Proposition. *Any nonzero ideal H of R must be equal to $H(\alpha)$ for some infinite cardinal α.*

Proof. If δ is an infinite cardinal strictly larger than the cardinality of R, then no right ideal of R can contain an independent family of δ nonzero right ideals. Thus $xR \not\approx E[\delta(xR)]$ for all nonzero $x \in R$, whence $H(\delta) = R$. Since $H \leq H(\delta)$, there must be a smallest infinite cardinal α for which $H \leq H(\alpha)$, and we of course claim that $H = H(\alpha)$.

Suppose on the contrary that there exists a $y \in H(\alpha) \setminus H$. Choosing a nonzero $x \in H$, we see from 19.59 that yR cannot be submorphic to any finite direct sum of copies of xR. In particular, yR is not submorphic to xR, so by 19.64 A, xR must be submorphic to yR. Choosing a maximal independent family $\{A_i\}$ from those submodules of yR which are isomorphic to xR, we get $yR = E[\oplus A_i] \oplus B$ for some B. The maximality of $\{A_i\}$ ensures that xR cannot be submorphic to B, hence by 19.64 A, B must be submorphic to xR. Letting τ denote the cardinality of the set $\{A_i\}$, we see that $E[\tau(xR)]$ is submorphic to yR, while yR is submorphic to $E[(\tau+1)(xR)]$. Inasmuch as yR is not submorphic to any finite direct sum of copies of xR, τ must be infinite. Thus $\tau = \tau + 1$, and with the help of 19.64 B we conclude that $yR \approx E[\tau(xR)]$.

We now have an infinite cardinal τ, and we infer that $yR \approx E[\tau(yR)]$. Inasmuch as $yR \not\approx E[\alpha(yR)]$, we obtain $\tau < \alpha$, and thus $H \nleq H(\tau)$. Choosing $z \in H \setminus H(\tau)$, we have $zR \approx E[\tau(zR)]$. Inasmuch as $xR \not\approx yR \approx E[\tau(xR)]$, we see from Lemma 10 that zR cannot be submorphic to xR. Thus xR must be submorphic to zR (19.64 A), whence $E[\tau(xR)]$ is submorphic to $E[\tau(zR)]$. We infer from this that yR is submorphic to zR, from which it follows via 19.59 that $y \in H$, which is a contradiction.   □

Proof of Theorem 19.64. Given any nonempty collection of nonzero ideals of R, there must be a smallest infinite cardinal α for which $H(\alpha)$ belongs to the collection, and then $H(\alpha)$ is the smallest ideal in the collection. Any other nonempty collection of ideals must contain 0, which is then the smallest ideal in the collection.   □

How far can a prime, regular, right selfinjective ring R be from a *full linear ring*? R is isomorphic to a full linear ring if and only if R has a minimal right ideal. Moreover, in general R is not even isomorphic to a factor ring of a full linear ring because a theorem of Osofsky [66a] which says that a factor of a full linear ring by a nontrivial ideal cannot be right selfinjective. Another plausible hypothesis might be that R is isomorphic to the maximal right quotient ring of a factor of a full linear ring, but Goodearl has disproved this (see Goodearl [74]).

Kaplansky's Conjecture

Kaplansky's conjecture is now an easy consequence of Goodearl's theorem:

19.67 Corollary to 19.64. *A regular right selfinjective ring R is prime if and only if it is primitive.*

Proof. Inasmuch as all primitive rings are prime, we need only consider the case when R is prime. The collection $\mathscr{P}$ of primitive ideals of R is nonempty (since $\mathscr{P}$ contains all the maximal ideals), hence by 19.64, $\mathscr{P}$ has a smallest element P. The Jacobson radical of a regular ring is 0, hence $\bigcap \mathscr{P} = 0$ and thus $P = 0$. Therefore 0 is a primitive ideal of R.   □

19.68 **Corollary.** *Any prime right nonsingular ring R has a primitive maximal right quotient ring.*

Proof. $\hat{R}$ is a prime right selfinjective regular ring. □

Thus, any regular prime ring has a primitive (right or left) maximal quotient ring.

Prime Ideals in Selfinjective Rings

Goodearl's theorem also has implications for the linear ordering of ideals R/P for certain prime ideals in an arbitrary right selfinjective ring.

19.69 **Theorem.** *Let R be right selfinjective. If P is any prime ideal of R containing the right singular ideal* sing R ($=$ rad R), *for example, if P is any primitive ideal, or if* rad R *is nilpotent, then the ideals of R containing P are linearly ordered.*

Proof. By Utumi's theorem 19.28, rad $R =$ sing R, and $R/$rad R is right self-injective and regular. Moreover, the correspondence theorem for ideals (I, p. 70) yields a bijection between prime ideals P containing rad R and prime ideals $P/$rad R of $R/$rad R. Therefore, Goodearl's theorem 19.62 applies, and asserts that ideals R/P is linearly ordered for any prime $P \supseteq$ rad R. Since any right or left primitive ideal contains rad R, and any prime ideal contains any nilpotent ideal, the proof is complete. □

Exercises for Chapter 19

1. A ring R is said to be **subdirectly irreducible** in case R satisfies the two equivalent conditions: (a) R contains a least nonzero ideal which is contained in any nonzero ideal of R; (b) if (as is defined in Chapter 26) R is a subdirect product of a set $\{R_i\}_{i \in I}$ of rings, then $|I| = 1$, and $R \approx R_i$.

2. (Jacobson) Any simple ring is subdirectly irreducible.

3. (Jacobson) A subdirectly irreducible primitive ring is right and left primitive.

4. (Birkhoff) Any ring is a subdirect product of subdirectly irreducible rings.

5. (Jacobson) A commutative primitive ring is a field.

6. (Jacobson) A primitive ring R is right Artinian iff R is $\approx$ full linear ring on a finite dimensional vector space.

7. For a right selfinjective ring R the following conditions are equivalent: (a) left Noetherian, (b) right Noetherian, (c) left Artinian, (d) right Artinian. In this case every right and left ideal of R is an annulet, and R is left selfinjective.

8. A commutative domain R is hereditary iff for any nonzero ideal I it is true that R/I is selfinjective and Artinian. This condition also holds in any (not necessarily commutative) right and left Noetherian hereditary prime ring for any ideal $I \neq I^2$ (Chapter 25).

9. (Johnson [53]) Let R be a n.s. prime ring with a uniform right ideal U, and a uniform left ideal V. Then $K = U \cap V$ is a ring-1 right and left Ore domain. If $x_1, \ldots, x_n$ are K-linearly independent elements of U, and $y_1, \ldots, y_n$ arbitrary elements of U, then there exists elements $k \in K$, and $a \in R$ such that $x_i a = k y_i$, $i = 1, \ldots, n$ (Johnson transitivity theorem. See also Faith [67, Chapter 14] on Johnson rings, and problem 14, p. 129, *loc.cit.*).

10. For any ring R, the intersection of the annihilators of indecomposable injective right R-modules is zero. (What about the annihilators of the finitely generated indecomposable quasinjectives?)

11. (Fisher and Snider [74]) Verification of Kaplansky's conjecture for rings with a countable base of nonzero ideals $\{I_i\}_{i=1}^{\infty}$: If R is a prime regular ring, and if every nonzero ideal contains one of the I_j, then R is primitive.

12. (Formanek [72b] and Passman [73]) Let $S_\omega = \bigcup_{n < \omega} S_n$ be the infinite symmetric group. Then, for any field F, the group ring FS_ω is primitive. (Formanek proved zero radical.) [Hint: apply 11.]

13. (Formanek and Snider [72]) A union of a countable chain of semisimple Artinian rings is primitive iff prime. (Apply problem 11.)

14. (Auslander [57], Harada [56]) Let G be a group which is **locally finite** in the sense that every finite subset generates a finite subgroup. For a ring R, the group ring RG is von Neumann regular iff (a) R is regular and (b) R is uniquely divisible by the order of each element of G. Moreover, for any group G, the group ring RG is regular only if (b) holds and G is a torsion group.

15. (Connell [63]) A group G is said to be **prime** if G has no finite normal subgroups $\neq 1$. A group ring RG is a prime ring iff R is a prime ring and G is a prime group.

16. (Formanek and Snider [72]) Let G be a countable locally finite group, and let R be a prime regular ring satisfying the regularity conditon (b) of problem 14. Then, RG is a primitive ring iff G is prime.

17. (Formanek and Snider) Let F be any field. Then any group G is embeddable in a group H such that FH is primitive. (Take a sequence $\{G_i\}$ of groups and a sequence of modules $\{M_i\}$, defined inductively by

$$G_1 = G \qquad\qquad M_1 = FG_1$$
$$G_2 = \mathrm{Aut}_F M_1 \qquad M_2 = FG_2 + M_1$$
$$G_3 = \mathrm{Aut}_F M_2 \qquad M_3 = FG_3 + M_2$$

Then, $M = \bigcup_{i=1}^{\infty} M_i$ is a faithful, irreducible FH-module, where $H = \bigcup_{i=1}^{\infty} G_i$. (Hint: G_{i+1} acts transitively on the nonzero elements of M_i, so that H acts transitively on the nonzero elements of M. [Incidentally, M_i is a faithful irreducible FG_{i+1} module.])

18. (Formanek [73a]) If F is a field, and if G is the free product of nontrivial groups and if $G \neq \mathbb{Z}_2 * \mathbb{Z}_2$, then FG is primitive.

19. (Formanek [73a]) If R is a commutative integral domain, and if G is a free group of cardinality not less than that of R, then RG is primitive. (Thus,

RG may be primitive even when R is not [cf. problem 17, and also Exercise 19.23 B.6].)

20. (Connell [63]) If R is right selfinjective, and G is finite, then RG is right selfinjective. Moreover, for any group G, and ring R, the group ring RG is right selfinjective only if G is finite and R is right selfinjective. (Connell proved G is torsion, and finiteness was independently proved by Farkas, Jain, and Renault.

21. A group ring RG is right selfinjective and regular iff G is finite and R is right selfinjective regular.

22. A countable union of semisimple Artinian rings need not be selfinjective. (Apply problem 20.)

23. (Goodearl [73b]) A regular right selfinjective ring R is a product of prime rings iff every nonzero ideal contains a minimal nonzero ideal of R. (This also characterizes R as a product of rings which are either prime, primitive, or indecomposable, in view of 19.57 B and 19.67. Also see 19.37.)

24. (Trivial Generalization of 19.63) If in a ring R the ideals are linearly ordered, and idempotent, then they are all prime.

25. (Converse of 19.63) If all of the ideals of a ring R are prime, then they are linearly ordered.

26. (Trivial Generalization of 19.67) If in a ring R the ideals are well ordered, and if every ideal I is **semiprimitive** in the sense that modulo I the ring has zero Jacobson radical, then every ideal is primitive. (Semiprimitivity is taken up in Chapter 26.)

27. (Another one) If R satisfies the d.c.c. on ideals, and if every ideal is semiprimitive, then every prime ideal is primitive.

28. Generalize problem 24 to "primitive" ideals (if possible).

29. (McCarthy [73]) If R is a von Neumann regular ring, then the polynomial ring $R[x]$ is semihereditary. Also prove or disprove the converse. Cf. Gilmer [73] and Jøndrup [71].

30. (Stringall [71]) A **generalized Boolean ring** R is defined to be a ring for which there is a prime number p and identities $x^p = x$, $px = 0$, holding in R. For a fixed prime p, let C_p denote the subcategory of RINGS consisting of all such generalized Boolean rings. For any two primes p and q, there is a category equivalence $C_p \approx C_q$.

31. (Teply [70]) If R is a ring, and if for every right R-module M, the singular submodule splits off, then (R is called a "splitting ring" and) R is a right nonsingular ring of gl. dim ≤ 2.

32. (Goodearl [73]) If R is a splitting ring with zero right socle, then (R is right nonsingular by 31. and) R is isomorphic to a triangular matrix $\begin{pmatrix} A & 0 \\ B & C \end{pmatrix}$ where A is a semiprime ring, C is a left and right Artinian ring, and B is a (C, A)-bimodule. Conversely, the displayed triangular matrix ring is a splitting ring if it has zero right socle, A is a right splitting ring, B is an injective right A-module, all nonsingular right C-modules are projective, and $\operatorname{End} B_A$ is an essential extension of C in mod-C.

33. (Goodearl [72]) If every singular module of mod-R is injective, or equivalently, if every singular module is semisimple, then R is right hereditary.

34. (Faith [74]) For any ring R, a right injective module E is **fieldendo** ($=\operatorname{End} E_R$ is a field) iff E is indecomposable and very right ideal $A \neq R$ such that R/A embeds in E is "critical".

Notes for Chapter 19

Injective modules were commented on in the Notes for Chapter 3 (I, p. 184). Quasinjectives ($=$quasi-injectives) were introduced by Johnson and Wong [61], although, as indicated, Jacobson proved that they satisfied the double annihilator condition (19.10) in Jacobson [55]. For modules over semisimple algebras, this condition was proved by Hall [39], and extended to quasi-Frobenius algebras and rings [2] by Nakayama [39, 41]. Nagao and Nakayama [53] introduced M_0 and M_u as symbols denoting respectively projective and injective modules (see their note on p. 170 of this paper). Over right Artinian rings, they proved that any projective is isomorphic to a direct sum of right prindecs (see their remark on p. 167), a theorem which Bass [60] extended to rings with the d.c.c. on principal right ideals. (This is taken up in Chapter 22.) Moreover, for an algebra A of finite dimensions over a field k, injective left modules are direct sums of k-duals of right prindecs, and conversely.

Noncommutative (and nonsemisimple) von Neumann regular rings originated as the "coordinate rings" for the continuous geometries of von Neumann [36a, b]. Utumi [65] proved that these rings are selfinjective, but only recently was it discovered (by Goodearl [74]) that these rings are simple! (We discuss this in the following paragraph.)

To be specific, a continuous geometry is any complemented modular lattice which is **continuous** (both upper [that is, $a \wedge (\bigvee B) = \bigvee_{b \in B}(a \wedge b)$ for upper directed B] and lower). Von Neumann's coordinatization theorem states that there exists a continuous geometry L which is **irreducible** (in the sense that only 0 and 1 have unique relative complements) and which is isomorphic to the lattice of principal right ideals of a regular ring $\mathscr{R}(L)$. Recently, Goodearl [74] established the astonishing theorem:

$R = \mathscr{R}(L)$ is a (right and left selfinjective) simple ring.

The proof is a direct application of Goodearl's theorem 19.66 which determines that every ideal of a selfinjective regular ring R has the form $H(\alpha)$ for some infinite cardinal α. (R is right and left selfinjective by an application of the criterion of Utumi [65]: namely, R is a direct sum of n isomorphic left (right) ideals, where n is the order of L.) Since $\alpha \leq \beta\ H(\alpha) \subseteq H(\beta)$, then R will be simple if $H(\aleph_0) = R$, that is, if $1 \in H(\aleph_0)$. But, by another theorem of Utumi 19.41, R is Dedekind finite ($xy = 1 \Rightarrow yx = 1$), and therefore, injectivity implies that R is not isomorphic to any proper right ideal. In particular, $R \not\approx E[\aleph_0(R_R)]$, so $1 \in H(\aleph_0)$, and therefore R is simple.

In my *Lectures* [67, p. 130] I raised the question of the structure of non-Artinian simple selfinjective ring A. Is A necessarily left selfinjective? And if so, then does this still hold if A is the maximal right quotient ring of a simple

---

[2] Quasi-Frobenius rings are the subject of Chapter 25.

domain K? The latter question awaited solution until Cozzens, in answering a question of Jategaonkar, constructed an example of a simple principal left ideal domain R which is not a right Ore domain. (Cozzens [72] showed that it was not right Noetherian, but the proof shows it is not right Ore; it appears unknown whether the latter can hold without the former! Cf. Camillo and Cozzens [73].) If one takes the maximal right quotient ring A of R, then A is right selfinjective 19.35. Moreover, any nonzero element $x \in R$ has a left inverse in A. By injectivity of A, there exists $y \in A$ such that $y x R = R$, and $y x = 1$. If A were also left selfinjective, then 19.41 would imply that every element of R is a unit of A, so A would be a field, and R a right Ore domain.

The proof has two immediate corollaries: (1) any integral domain R, $A = \hat{R}$ is a simple regular ring which is both right and left injective iff R is a right Ore domain, a fact first pointed out to me by J.E. Roos, answering the first question; but it required the existence of a simple not Ore domain to answer the second.

(2) The second consequence of the proof, and the observation that von Neumann's coordinate ring is non-Artinian and twosided selfinjective, is that not every right selfinjective ring is the maximal quotient ring of some integral domain. (However, Handelman has pointed out (in a letter) that this still leaves that possibility for a Dedekind infinite right selfinjective regular ring. Also see Goodearl and Handelman [75].)

We have remarked on Goodearl's partial solution to Kaplansky's question on the implication prime $\Rightarrow$ primitive in regular rings (19.67), but there has been progress on the problem for regular group rings by Formanek, Passman, and Formanek and Snider. Instead of reporting on this, however, we have cited some of the fundamental tools, including the Auslander and Harada characterizations of regular group rings and Connell's characterization of prime group rings, in a sequence (Problems 11–19) including the simplifying lemma of Fisher and Snider [74]. (For related questions on semisimplicity of group algebras over fields of characteristic 0, consult Amitsur [59] or Passman [62, 71].)

Exercises 19.37-8 offer a number of results on the structure of the maximal right quotient ring Q of a ring R (usually assumed to be right nonsingular). For example, when Q is: a product of full linear rings (Chase and Faith [64]); also a left quotient ring (Utumi [63 b, c]); left flat over R (Sandomierski [67], Cateforis [69 a]), an epic image of R in RINGS (Silver [67]); and other results on when finitely generated nonsingular modules embed in freebees[3] (Goodearl [71]); or are projective (Catefois [69 b]); or when is Q projective (Viola-Prioli [73], Handelman-Lawrence [75], and Handelman [75]). Also, the structure of nonsingular modules over selfinjective rings are determined (Sandomierski [68] and Zelmanowitz [71]). Compare Boyle and Goodearl [76] who classifies nonsingular injectives up to isomorphism by their "rank".

The Chevalley-Jacobson density theorem has been placed in a categorical setting (and in a Morita context) by Amitsur [71], Fuller [74], and Zelmanowitz [73]. (The latter also studies regular modules in [72].) Also see Müller [73].

---

[3] In American slang, freebee = anything free = a free thing (that is, something that does not cost anything), hence a convenient abbreviation for a free object, or module.

Courter [69] characterized when every module is rationally closed (and also the dual notion of corationally closed). Storrer [71a, b] studied maximal rational extensions in general and over perfect rings.

Warfield [72b] introduced the concept of exchange ring (defined *sup.* 19.21), and Monk [72] characterized them, showing that they were more general than lift/rad rings which are regular modulo radical.

Any Boolean ring R is a commutative regular ring, and, as noted in (I, 3.2.9, p. 115), every Boolean ring R can be embedded in the Boolean ring of subsets Pow X of some set X (Stone [23]). Moreover, if A is any commutative ring, there is a Boolean ring Idem A consisting of all idempotent elements of A, with the operations of addition and multiplication

$$a \dotplus b = a + b - 2ab \quad \text{and} \quad a \cdot b = ab$$

The functor $A \mapsto \text{Idem} A$ (from COMM RINGS $\leadsto$ BOOLEAN RINGS) has a left adjoint which assigns to each Boolean ring B the ring $Z[B]$ defined by the generators $\{[b]\}_{b \in B}$ and the relations:

$$[1] = 1, \quad [0] = 0, \quad [a+b] = [a] + [b] - 2[a][b], \quad [a \cdot b] = [a][b].$$

Furthermore, the map $B \to \text{Idem} Z[B]$ sending $b \mapsto [b]$ is a ring monic. For details, sharpenings, and generalizations see Bergman [72]. For generalizations of the Stone representation theorem for Boolean rings to (bi)regular rings, consult Arens and Kaplansky [48], Kaplansky [50], and Jacobson [64], especially Chapter IX of the latter on the structure space of a ring (= the topology on the set of primitive ideals).

McCoy and Montgomery [37] show that any generalized Boolean ring for a prime p embeds in a direct sum of copies of $GF(p)$. We have noted the remarkable generalization of Stringall [71] in Chapter Exercise 30.

Unquestionably, still the best source book on the structure of primitive rings is *The Structure of Rings* (Jacobson [64]; see especially the examples, pp. 35–37, and pp. 251–255 of Appendix A). Lambek's *Lectures on Rings and Modules* is *non-pareil* on the subject of maximal quotient rings; and for quotient rings less than maximal, there is *Torsion Theories, Additive Semantics, and Rings of Quotients* (Lambek [71]). (We have been forewarned, if not prepared, by Gabriel's localizations and quotient categories in Chapters 15 and 16 (Volume I, esp. pp. 525–537), that practically *everything* is a "torsion theory"! But *semantics*?)

There is a torsion theory in the sense of (I, 16.8B, p. 527) which assigns to each module M its "torsion" submodule, namely, sing M so that the nonsingular modules are the torsion freebees. Also, there is a torsion theory which assigns to each M the "torsion" submodule consisting of all $x \in M$ which annihilate a right ideal I for which R is a rational extension. (Such a right ideal is said to be **dense** in R. See (I, p. 529).) The torsion theory thus generated is the largest for which R is torsion free. (See (I, 1.3, p. 535) in this connection.)

Much, but not all, of the basic theorems on quasiinjectives and their endomorphism rings, the Findlay-Lambek rational extensions, and the Johnson-Utumi maximal quotient rings are contained in my Lectures [67].

Right nonsingular rings, a subject originated by R. E. Johnson [51, 53, 61, 65a], generalize integral domains, simple rings, and regular rings. For commutative integral domains, the singular submodule of a module is just the torsion sub-module. Kaplansky [52] showed that all finitely generated modules split (= the torsion submodule splits off) iff R is semihereditary. Moreover, all torsion sub-modules of bounded order split iff the domain is Dedekind (Kaplansky [52], Chase [60]). Goodearl [72, 73a] has given a massive study for singular splitting in nonsingular rings, including when the singular submodule is projective or injective. We have included as an exercise a theorem of Teply on when every singular submodule splits off: the ring must have r.gl.dim ≤ 2 (Teply [70]), an upperbound which is attained (Fuelberth and Teply [72]). Goodearl [73a] has characterized these (so-called) splitting rings which have zero right socle (see Exercises 32–33).

A ring R is said to be a **right intrinsic** extension of a subring A provided that every nonzero right ideal of R has nonzero intersection with A; and **strong intrinsic** if every closed right ideal of A is the contraction of a right ideal of R. Faith and Utumi [64b] characterized the Johnson maximal right quotient ring $\hat{A}$ of a right nonsingular ring A by the strong right intrinsic property in the case that $\hat{A}$ has no strongly regular ideal. (The latter condition avoids the case that A is an integral domain, and $\hat{A}$ is a field.) Moreover, then any strongly intrinsic ring extension R of A embeds in $\hat{A}$. Assuming that $\hat{A}$ is also a left quotient ring, then the conditions weaken to intrinsicity. In certain other cases (e.g., when A is semiprime right Goldie, or a right nonsingular prime ring containing a uniform right ideal), then right intrinsicity suffices. (See Hutchinson [69] and O'Meara [73]; O'Meara [75] also determined right orders in infinite dimensional full linear rings.)

Harada [65] and Miyashita [65] proved also many theorems on QI modules, including 19.6(a) (proved by Faith-Utumi [64a]). Defining the closure cl(N) of a submodule N of a module M to be the set of all $x \in M$ such that ann x/N (or $x^{-1}N$) is an essential right ideal, then Harada proved that cl(cl(0)) is a direct summand of any quasinjective module M. (This is 19.30(a).) Harada applies this to decompose a QF ring into a product of a semisimple ring and a ring R such that cl(cl(0)) = R. (Compare the concept of Hall [40] and Brown-McCoy [50] of ring **bound to its radical**.)

Harada [65] also determines the structure of the indecomposable QI modules over a Dedekind domain (cf. also Fossum [71]). This determination follows from the fact that a quasinjective module over a right Noetherian ring is a direct sum of indecomposable quasinjective modules. (Compare the theorem of Matlis-Papp 20.6. See also 20.6A.) Next, Harada [72] determines the structure of Noetherian quasinjectives over commutative rings (i.a. they are Artinian). This partially solves for commutative rings the question which I raised in my Lectures [67, p. 127, Problem 2], namely: characterize which rings are endomorphism rings of quasinjective modules [of finite length]. (Another question raised there is still open: what is the structure of quasinjective modules of finite length relative to the d.a.c. stated in 19.10?)

Rings with quasinjective right ideals have been studied in by Jain, Mohamed, and Singh [69], and S.H. Mohamed [69, 70]. If the cyclic modules are all quas-injective, consult Ahsan [73]; and if the proper cyclic modules are injective, look at Boyle [74] or Faith [73]. If every submodule of an indecomposable injective module is quasinjective, then refer to Dickson and Fuller [69] for the solution for finite dimensional algebras. (These algebras all have finite module type (FFM) in the sense defined in Chapter 20, and, in fact, Harada [65, p. 356] shows that any generalized uniserial ring (= serial ring in our terminology, Chapter 25) has the stated property.) Also see Fuller [75].

A ring R is right **FP-injective** provided that every map $f: I \to R$ of a finitely generated right ideal is extendable to a map $f': R \to R$. Thus, these rings generalize both regular rings and right selfinjective rings. Jain [73a] has shown that R is right FP-injective iff every finitely presented left R-module is torsionless; and for right coherent rings this can be strengthened to reflexivity. It would be interesting to report on the vast number of papers on FP-injective rings, but this is patently impossible (cf. Jain [73b]).

Elizarov [69] has compiled a massive study of quotient rings with several hundred bibliographic entries.

Ivanov [70] determined the structure of a nonsingular ring with minimal onesided ideals, generalizing the work of Goldie [64], Colby and Rutter [68], and others.

Lambek [76] has worked with localizations using QI modules in place of injectives, and generalizes the density theorem.

References

Ahsan [73], Amitsur [56a, 59, 71], Arens and Kaplansky [48], Armendariz [73], Artin [50], Asano [49], Azumaya [66], Bass [60], Beachy [71], Bergman [72], Birkhoff [67], Boyle [74], Brown and McCoy [50], Camillo and Cozzens [73], Cateforis [68, 69, 70], Chase [60], Chase and Faith [65], Colby-Rutter [68], Connell [63], Courter [69], Cozzens [72], Crawley and Johnson [64], Dickson and Fuller [69], Faith [67a, 72b, 73], Faith and Utumi [64a, b, 65a], Faith and Walker [67], Findlay and Lambek [58], Fisher and Snider [74], Formanek [72b, 73a], Fossum [71], Formanek and Snider [72], Fuchs [69], Fuchs and Rangaswamy [70], Fuelberth and Teply [72], Fuller [69b, 74], Goldie [64], Goodearl [72, 73a, b, d, 74], Hall [39, 40], Handelman [75], Handelman-Lawrence [75], Harada [65, 72], Hutchinson [69], Ivanov [70], Jacobson [50, 64], Jain [73], Jain, Mohamed and Singh [69], Johnson [51, 53, 61, 65], Johnson and Wong [61], Kaplansky [50, 52, 69], Lambek [66, 71], Levy [63a], Malcev [36], McCarthy [73], Miyashita [65], Mohamed [69, 70], Monk [72], Müller [73], Nagao and Nakayama [53], Nakayama [39, 40, 41], O'Meara [73, 75], Osofsky [66a, b, 68a], Passman [62, 71, 73], Sandomierski [67, 68], Silver [67], Storrer [71a, b, 73], Stringall [71], Teply [70], Utumi [56, 61, 63b, c, 65, 67], Vamos [68], Viola-Prioli [73], von Neumann [36a, b, 60], Warfield [69a, 72b], Wong and Johnson [59], Zelmanowitz [67, 71, 72, 73].

Other References

Goodearl and Handelman [75], Jøndrup [71], Gilmer [73], Masaike [70b, 71a, b], Morita [70, 71a, b], Nastasescu and Popescu [70], Rangaswamy [73], Richman and Walker [72], Sandomierski [70], Vámos [68], Boyle-Goodearl [76], Faith [74], Goodearl [75a, b], Lambek [76], Rowen [74].

Chapter 20. Direct Sum Representations of Rings and Modules

If C is a class of right R-modules, a module M will be called **sigma** C, in case M is isomorphic to a direct sum of modules in C. We will write $\sum$-C for short. Also, if $\mathcal{M}$ is a class of right R-modules which is $\sum$-C then we say that R is right $\mathcal{M}\sum$-C. For example, if $\mathcal{M}$ is the class of (injective) right R-modules, and C is the class of finitely generated right R-modules, then the corresponding statement is that R is right **(injective)** $\sum$**-finitely generated.** Also, the statement R is right **(injective)** $\sum$**-cyclic** means that every (injective) right R-module is a direct sum of cyclic modules. (Note that one applies these designations to R as an object of RINGS rather than as an object of mod-R; strictly speaking, it is mod-R (not R) that is (injective) $\sum$-finitely generated when we say that R is.)

Let fin. gen. mod-R represent the subclass of finitely generated modules in mod-R. If C is a subclass of mod-R, then σ-C is the class $\sum$-$C \cap$ fin. gen. mod-R, namely the class of all finitely generated modules which are in $\sum$-C. Thus, R is right σ-**cyclic** if every finitely generated right R-module is a direct sum of cyclics.

Let α be any cardinal number. A module M is an α-gen in case M is generated by α or fewer elements. Then M is said to be α-**gened.** The ring R is said to be α-gened in case every indecomposable $M \in$ mod-R is α-gened. In this case, we also say that R is BG **(bounded generator).** If there is an integer $n > 0$ such that every finitely generated module is n-gened, then R is said to be FBG (finite BG). The notation n-gened with roman letter n will be reserved for finite cardinals n.

If C is a class of modules over R such that every indecomposable $M \in C$ is α-gened, then R is C-α-gened, for example, injective-α-gened. If the finitely generated modules of C are n-gened, then R is said to be FC (n-gened).

We say cyclic-gened instead of 1-gened. If for some integer $n > 0$ every (right) R-module is a direct sum of n-gened modules, then, R is said to be (right) $\sum$-n-gens. (Then, R is (right) n-gened.) If every finitely generated right R-module is a direct sum of n-gens, for a fixed integer n, then R is σ-n-gens. Every n-gened ring A is similar to a cyclic-gened ring 20.39.

A ring R is said to have (finite) **finite module type** provided that the number of isomorphism classes of indecomposable (finitely generated) right modules is finite, and then we say that R is right **(F)FM.** (Actually, only right FFM rings are considered here, and these rings are also called rings of **finite representation type** in the literature.) We have discussed, but do not prove, the Brauer-Thrall conjecture in the Introduction to Volume II, namely

(Brauer-Thrall Conjecture):   right FBG $\Rightarrow$ right FFM that is, right bounded module type implies right finite module type. (As stated, Roiter [68] verifies this for a ring R which is a finite dimensional algebra over a field, and Auslander [74] using different methods, extended the truth of the Brauer-Thrall conjecture to Artinian rings. Also see Chapter Notes.)

Summary of Results on Direct Sum Decompositions of Modules

We now list some theorems proved in this Chapter (including one proved in Chapter 24), and alert the reader to see Chapter 25 (Serial and Sigma Cyclic Rings) for additional theorems of this genre.

20.7   Theorem (Faith-Walker [67]). *A ring R is right injective BG if and only if R is right Noetherian.*

20.6 and **20.9   Theorem** (Matlis [58], Papp [59]). *A ring R is right injective $\sum$-indecomposable if and only if R is right Noetherian.*

20.17–8   Theorem (Faith-Walker [67]). *If R is right injective $\sum$-fin. gen., then R is right Artinian, and conversely for a commutative ring R.*

24.14   Theorem (Faith-Walker [67], and Faith [66b]). *A quasi-Frobenius ring R is right injective $\sum$-cyclic and conversely for commutative R[1].*

20.23   Theorem. *The following conditions, which are equivalent, imply that a ring R is right Artinian.*

(a) *There is a set $\{M_i\}_{i \in I}$ of right R-modules such that every right module is (isomorphic to) a direct sum of modules of the family $\{M_i\}_{i \in I}$.*

(b) *There exists a cardinal α such that every right module is a direct sum of modules each generated by a set of cardinal $\leq \alpha$.*

(c) *R is right $\sum$-α-gens.*

20.42   Theorem (Warfield [70]). *A commutative local FBG ring is a valuation ring (in the sense that the ideals are linearly ordered).*

20.49   Theorem (*Kaplansky* [52], Matlis [66], Gill [71], Lafon [70], Warfield [70]). *A local ring R is σ-cyclic iff R is an almost maximal valuation ring (in the sense of Kaplansky [52]).*

Summary of Results on Direct Decompositions of Rings

We also include in this chapter theorems on decompositions of rings into finite products of (semi)prime rings and Artinian rings, including:

20.30 Chatters's Theorem for Noetherian hereditary rings.

20.32 Levy's Theorem for semiprime right hereditary right Goldie rings.

20.36 Robson's Criterion for Noetherian rings.

20.37 The Krull-Asano-Goldie theorem for principal ideal rings.

---

[1] We remind the reader to look in Chapter 24 for this result!

A concept which has proved useful in widely varying applications is that of $\sum$-injectivity of a module M, namely the condition when M is injective and every direct sum of copies of M is injective. This can be characterized by the a.c.c. on right ideals of which are the annihilators of subsets of M; and this implies that $R/\mathrm{ann}_R M$ satisfies the a.c.c. on direct summands 20.3A. This theorem is used in characterizing QF rings (Chapter 24).

Another theorem with many applications is a theorem of Chase which states that if every product R^a of copies of R is a $\cap$-pure submodule of a direct sum of right modules of cardinals bounded by a cardinal independent of a, then R satisfies the d.c.c. on principal left ideals 20.21. Thus, Chase's theorem implies that $\mathbb{Z}^\omega$ is not free (20.22). Moreover, combined with 20.7, Chase's theorem implies 20.23 (also see 22.31 B).

Direct Sum Representations of Injective Modules

In any category, products of injective objects are injective; however, coproducts of injective objects are not, in general, injective.

20.1   Theorem (Cartan-Eilenberg-Bass). *The following conditions on a ring R are equivalent:*

(a) *R is right Noetherian.*

(b) *Any direct sum of injective right R-modules is injective.*

(c) *Any countable direct sum of injective right R-modules is injective.*

Proof. (a) $\Rightarrow$ (b). Let $f: A \to \sum_{i \in I} \oplus M_i = M$ be a map of a right ideal A into a direct sum M of injective modules. Since A is finitely generated, im $f \subseteq M'$, where M' is a sum of finitely many of the M_i. Since the sum is direct, M' is injective, so $f: I \to M'$, whence $f: I \to M$ is induced by a left homothetic. Then M is injective by Baer's criterion (I, 3.41, p. 157).

(c) $\Rightarrow$ (a). Let $I_1 \subseteq I_2 \subseteq \cdots \subseteq I_n \subseteq \cdots$ be a chain of right ideals, let

$$E_n = \text{inj hull } R/I_n \qquad\qquad (n \geq 1)$$

as R-modules, let $I = \bigcup_{n=1}^\infty I_n$, let $E = \coprod_{n=1}^\infty E_n$, let $f_n: I \to E_n$ be the map, where $f_n(x) = [x + I_n]$, and let $f: I \to E$ be the direct sum map $\coprod_{n=1}^\infty f_n$. Since E is injective, f is induced by a left homothetic by $m \in M$, so

$$\text{im } f \subseteq m R \subseteq \sum_{n=1}^t \oplus E_n$$

for some $t < \infty$. (This is because $m R$ is finitely generated.) Then $I_n = I_{t+1} \ \forall n > t$, so R is Noetherian. $\square$

$\sum$-Injective Modules

For an R-module M, with endomorphism ring $S = \text{End } M_R$, we let $\mathscr{A}_l(M, R)$ denote the set of S-submodules of M that are annihilators of subsets of R. Thus,

$X \in \mathscr{A}_l(M, R)$ if and only if

$$X = \operatorname{ann}_M \operatorname{ann}_R X = {}^{\perp}(X^{\perp})$$

where ${}^{\perp}Y$ and $X^{\perp}$ denote annihilation. Similarly, $\mathscr{A}_r(M, R)$ denotes the set of right ideals of R of the form $X^{\perp}$ for a subset X of M. Thus $I \in \mathscr{A}_r(M, R)$ if and only if $I = \operatorname{ann}_R \operatorname{ann}_M I$ (cf. 19.10 ff.). Note for any faithful module M, that any summand of R is an annihilator, that is, any right ideal generated by an idempotent $e \in R$ is the right annihilator of the subset $M(1 - e) = \{m \in M \mid m\, e = 0\}$.

Let $\{M_a\}_{a \in A}$ be a family of right R-modules, indexed by a set A, and if M_a is isomorphic to a fixed right module $M \; \forall\, a \in A$, then

$$M^A = \prod\nolimits_{a \in A} M_a \qquad\qquad \text{(direct product)}$$

and

$$M^{(A)} = \sum\nolimits_{a \in A} \oplus M_a \qquad\qquad \text{(direct sum).}$$

If A is countably infinite, then set $M^\omega = M^A$, $M^{(\omega)} = M^{(A)}$.

If M is injective, then M^A is injective for any index set A; M will be said to be $\sum$**-injective** if $M^{(A)}$ is injective for any index set A; M is **countably $\sum$-injective** if $M^{(\omega)}$ is injective.

20.2 A   Proposition (Faith [66a]). *If $M \in$ mod-R, then $\mathscr{A}_r(M, R)$ satisfies the a.c.c. if and only if to each right ideal I of R there corresponds a finitely generated subideal I_1 such that ${}^{\perp}I = {}^{\perp}I_1$.*

Proof. Assume a.c.c. for $\mathscr{A}_r(M, R)$, or equivalently, the d.c.c. for $\mathscr{A}_l(M, R)$, let I be a right ideal of R, and let I_1 be a finitely generated subideal that is minimal in $\{{}^{\perp}K\}$, where K ranges over all finitely generated subideals of I, and ${}^{\perp}K$ is taken in M. If $x \in I$, then $Q = I_1 + xR$ is a finitely generated subideal of I satisfying ${}^{\perp}Q \subseteq {}^{\perp}I_1$. By the choice of I_1, necessarily ${}^{\perp}Q = {}^{\perp}I_1$, so ${}^{\perp}I_1\, x = 0$. Since this is true $\forall\, x \in I$, then ${}^{\perp}I_1\, I = 0$; that is, ${}^{\perp}I_1 \subseteq {}^{\perp}I$. But $I_1 \subseteq I$ implies ${}^{\perp}I_1 \supseteq {}^{\perp}I$, so ${}^{\perp}I_1 = {}^{\perp}I$ as asserted.

Conversely, let $I_1 \subseteq I_2 \subseteq \cdots \subseteq I_n \subseteq \cdots$ be a chain of right ideals of R lying in $\mathscr{A}_r(M, R)$, let $X_i = {}^{\perp}I_i$, $i = 1, 2, \ldots$, be the corresponding elements of $\mathscr{A}_l(M, R)$, let $I = \bigcup_{i=1}^{\infty} I_n$, and let J_1 be the finitely generated subideal of I such that ${}^{\perp}I = {}^{\perp}J_1$. Since J_1 is finitely generated, there is an integer q such that $J_1 \subseteq I_k$, $k \geq q$; that is, ${}^{\perp}J_1 \supseteq X_k = {}^{\perp}I_k$, $k \geq q$. But

$$ {}^{\perp}J_1 = {}^{\perp}I = \bigcap\nolimits_{n=1}^{\infty} X_n $$

that is, $X_k = {}^{\perp}J_1$, $k \geq q$. Then, $I_k = X_k^{\perp} = I_q$, $k \geq q$, proving the proposition. $\square$

In the case $M = R$, $\mathscr{A}_l(M, R)$ [resp. $\mathscr{A}_r(M, R)$] is simply the lattice of left (resp. right) annulets of R, producing the following corollary.

20.2 B   Corollary. *A ring satisfies the a.c.c. on right annulets if and only if each right ideal I contains a finitely generated right ideal I_1 with the same left annihilator ${}^{\perp}I = {}^{\perp}I_1$.*

20.3 A   Proposition (Faith [66]). *The following conditions on an injective module $M \in$ mod-R are equivalent:*

(a) M is countably $\sum$-injective.

(b) R satisfies the a.c.c. on the right ideals in $\mathscr{A}_r(M, R)$.

(c) M is $\sum$-injective.

When this is so, then $R/\text{ann}_R M$ satisfies the a.c.c. on right ideals generated by idempotents (equivalently, by 22.28, $R/\text{ann}_R M$ contains no infinite sets of orthogonal idempotents).

Proof. (a) $\Rightarrow$ (b) (indirect proof). Let $I_1 \subset I_2 \subset \cdots \subset I_m \subset \cdots$ be a strictly ascending sequence of right ideals in $\mathscr{A}_r(M, R)$, let $I = \bigcup_{n=1}^{\infty} I_n$, and let x_n be an element of $^{\perp}I_n$ (taken in M) not in $^{\perp}I_{n+1}$ $n = 1, 2, \ldots$. If $r \in I$, then there exists q such that $r \in I_k \ \forall k \geq q$, and since $^{\perp}I_q \supset {}^{\perp}I_k, \ \forall k \geq q$, then $x_k r = 0 \ \forall k \geq q$. Therefore the element $r' = (x_1 r, \ldots, x_n r, \ldots)$ lies in $M^{(\omega)}$, even though $x = (x_1, \ldots, x_n, \ldots)$ lies in M^{ω}. Let f denote the map defined by $f(r) = r' \ \forall r \in I$. Assuming momentarily that $M^{(\omega)}$ is injective, there is given, by Baer's criterion 3.41, an element $y = (y_1, \ldots, y_m, 0, \ldots) \in M^{(\omega)}$ such that

$$f(r) = y\,r = (y_1\,r, \ldots, y_m\,r, 0, \ldots)$$

$$= (x_1\,r, \ldots, x_m\,r, \ldots) \hspace{3cm} \forall r \in I.$$

But this implies that $x_t r = 0, \ \forall t > m, \ \forall r \in I$; that is, $x_t \in {}^{\perp}I \subseteq {}^{\perp}I_{t+1}$, contrary to the choice of x_t. Thus, (a) $\Rightarrow$ (b).

(b) $\Rightarrow$ (c). Let I be a right ideal of R, and let $I_1 = r_1 R + \cdots + r_n R$ be the finitely generated subideal given by 20.2A such that $^{\perp}I = {}^{\perp}I_1$. Let $f: I \to M^{(A)}$ be any map. Since M^A is injective, there exists an element $p \in M^A$ such that $f(r) = p\,r \ \forall r \in I$. Since $f(r_i) = p\,r_i \in M^{(A)}, \ i = 1, \ldots, n$, there exists an element $p' \in M^{(A)}$ such that $p_a\,r_i = p'_a\,r_i \ \forall a \in A, \ i = 1, \ldots, n$, where g_a is the a coordinate of any $g \in M^A$. Since $r_1, \ldots, r_n$ generate I_1, this implies that $p\,r = p'\,r \ \forall r \in I_1$, whence $(p_a - p'_a) \in {}^{\perp}I_1 \ \forall a \in A$. Since $^{\perp}I_1 = {}^{\perp}I$, it follows that $p_a\,x = p'_a\,x \ \forall a \in A, \ \forall x \in I$; that is $p\,x = p'\,x \ \forall x \in I$. Thus, $f(x) = p'\,x \ \forall x \in I$, with $p' \in M^{(A)}$, so $M^{(A)}$ is injective by Baer's criterion 3.41. Thus, (b) $\Rightarrow$ (c), completing the equivalence of (a)–(c). Since M is $\sum$-injective and faithful as a right $R/\text{ann}_R M$ module, then (b) applied to $R/\text{ann}_R M$ implies the a.c.c. on right ideals generated by idempotents. $\square$

20.3 B **Proposition** (Goursaud and Valette [75]). *If R has a faithful $\sum$-injective right module, then R satisfies* (acc)$\oplus$.

Proof. Let $I = \bigoplus_{n=1}^{\infty} x_n R$ be an infinite direct sum of nonzero cyclic right ideals of R. Since M is faithful, then for each integer n there exists an element $y_n \in M$ such that $y_n\,x_n \neq 0$. Then, the morphism $f: I \to M^{(\mathbb{N})}$ such that

$$p_m(f(x_n)) = \delta_{mn}\,y_n\,x_n \hspace{3cm} (\text{Kronecker-}\delta)$$

can not be extended to a mapping of $R \to M^{(\mathbb{N})}$, contradicting $\sum$-injectivity of M. $\square$

If the injective hull of M is $\sum$-injective, we say that M has $\sum$-*injective hull*.

20.3 C **Corollary.** *A ring R is right Noetherian iff mod-R possesses a $\sum$-injective cogenerator.*

Proof. If R is Noetherian, every injective is $\sum$-injective. Conversely, if E is a cogenerator, then, by Exercise (f) of 20.4C, every right ideal of R is the annihilator of a subset of E. Then, by 20.3 A, E is $\sum$-injective only if R satisfies the a.c.c. on right ideals. □

20.3 D Corollary (Kurshan [70]). *A ring R is right Noetherian iff the direct sum of the injective hulls of any collection of simple modules is injective.*

Proof. For then mod-R has a $\sum$-injective cogenerator. □

20.3 E Corollary (Kurshan [70]). *If every semisimple right R-module is injective, then R is right Noetherian.*

These rings are, then, precisely the right Noetherian right V-rings.

Note: Kurshan's result is stronger than stated in 20.3 D in that he requires the condition only for countable collections of simple modules (Exercise).

QI Rings

The next remark is almost obvious.

20.4 A *Remark.* *If R is a ring, and M in mod-R is such that $M \oplus \hat{R}$ is QI, then M is injective.*

Proof. For then $M \oplus \hat{R}$ is a QI module containing R, hence is injective. □

This proves the equivalence of (a) and (b) in the next proposition.

20.4 B Proposition (Koehler [70]). *The two conditions on a ring A are equivalent:*

(a) *A is a right QI ring.*

(b) *The direct sum of any two QI right A-modules is again a QI module.*

When this is so, then A is right Noetherian.

Proof. Since any semisimple module is injective, then A is right Noetherian by 20.3 E. □

20.4 C *Exercise* (Faith [72 b]). Recall from Chapter 19: Finendo = finite over endomorphism ring. Let M be a right R-module, and $\bar{R} = R/\operatorname{ann}_R M$.

(a) Prove the equivalence of (1)–(4):

    (1) If $\bar{R}$ is semisimple, and a flat left R-module, then M is $\sum$-injective and finendo.

    (2) If M has $\sum$-injective hull $\hat{M}$, and if $\bar{R}$ is a regular ring, then $\bar{R}$ is semisimple.

    (3) If M has $\sum$-injective hull $\hat{M}$, and if $R/\operatorname{ann}_R \hat{M}$ is a regular ring, then $M = \hat{M}$ is finendo, and $\bar{R}$ is semisimple.

    (4) If M is any semisimple finendo right R-module, then $\bar{R} = R/\operatorname{ann}_R M$ is semisimple. Furthermore, M injective in mod-R iff $\bar{R}$ is a flat left R-module.

(b) For a right module M over a regular ring R, the following conditions are equivalent:

(1) M has $\sum$-injective hull in mod-R.

(2) M has $\sum$-injective hull in mod-$R/\text{ann}_R M$.

(3) $R/\text{ann}_R M$ is semisimple.

(4) M is a semisimple, injective finendo right R-module.

(c) If E is an indecomposable injective module over R, and if E has no fully invariant submodules $\neq 0$ ($= E$ is NFI), then $\text{End} E_R$ is a field ($= E$ is fieldendo) under the hypothesis that there is a right ideal of R maximal in $\{x^\perp | x \in E, x \neq 0\}$. Thus, any NFI $\sum$-injective indecomposable module is fieldendo.

(d) A ring R is right Goldie iff $\hat{R}$ is $\sum$-injective.

(e) If R has maximal right quotient ring S, and if S is semisimple, then S is $\sum$-injective.

(f) If M is a cogenerator of mod-R, then M is faithful, and every right ideal of R belongs to $\mathscr{A}_r(M, R)$ (Rosenberg and Zelinsky [61]).

*(g) (Dlab and Ringel [72b]) Any cogenerator M of mod-R which is finendo is balanced (I, p. 323). (Since M is faithful this implies $R = \text{Biend}\, M_R$ canonically.)

(h) (Morita [58]) Any injective cogenerator over a right Artinian ring is balanced (apply (g)).

20.5 Proposition. *If R is a right QI ring, then R is a right Noetherian ring, a finite product of simple rings, and every indecomposable right injective module is fieldendo.*

Proof. By 20.4B, R is right Noetherian. Also, since R is a right V ring, $\text{rad}\, R = 0$ (I, 7.32A, p. 356), so R is semiprime right Goldie (I, 9.9, p. 394) and a right order in a semisimple ring. Then, R is a finite product of simple rings by (I, 7.36A, p. 357) which are manifestly right QI. The last part is Exercise 20.4C(a). $\square$

Completely Decomposable Modules

A module M is **completely decomposable** if M is a direct sum of indecomposable submodules. (Notice the anomaly: Every indecomposable module is completely decomposable!)

20.6 Theorem (Matlis [58], Papp [59]). *If R is a Noetherian ring, then every injective module is completely decomposable.*

Proof. By 7.17, there exists a nonzero uniform submodule, and by Zorn's lemma there is a maximal independent set $X = \{U_i | i \in I\}$ of nonzero uniform submodules. For each $i \in I$, let M_i be an inj hull of U_i contained in M. Each module M_i is then indecomposable, and the set $\{M_i | i \in I\}$ also is independent: $\sum_{i \in I} M_i = \sum_{i \in I} \oplus M_i$. Since R is Noetherian, by Cartan-Eilenberg-Bass, $M' = \sum_{i \in I} \oplus M_i$ is injective, whence a direct summand of M: $M = M' \oplus N$. If $N \neq 0$, N would contain a uniform submodule W, and $W \cup X$ would be an independent set of uniform submodules violating the maximality of X. Hence $N = 0$, so $M = M' = \sum_{i \in I} \oplus M_i$ is completely decomposable. $\square$

Papp has shown that this property characterizes Noetherian rings (see 20.9).

20.6 A **Corollary** (Cailleau-Renault [70]). *A quasinjective module M is $\sum$-quasinjective iff $\hat{M}$ is $\sum$-injective (and iff R satisfies the a.c.c. on right ideals annihilated by subsets of M). This is so iff M is a direct sum of $\sum$-quasinjective indecomposable modules.*

Proof. Exercise (see 20.3 A and the proof of 20.6). $\square$

Characterizations of Noetherian Rings

20.7 **Theorem** (Faith-Walker [67]). *R is right Noetherian if and only if there exists a cardinal number c such that each injective right R-module is a direct sum of modules, each generated by c elements* (cf. 20.19).

Proof. If R is right Noetherian, then by 20.6, each injective module is a direct sum of indecomposable injective modules. Since an indecomposable injective module D is the injective hull $\hat{C}$ of any nonzero cyclic submodule C, it suffices to show that there exists a cardinal number c such that each such D is generated by c elements. Since the collection of all isomorphism classes of cyclic modules is a set, it follows that the collection of all isomorphism classes of indecomposable injective modules is a set $\{M_i | i \in I\}$. If $M \in M_i$ is generated by c_i elements, then $c = \sum_{i \in I} c_i$ (cardinal sum) is the desired cardinal.

Conversely, assume that such a cardinal number exists. R is right Noetherian if and only if each direct sum of injective modules is injective (20.1). By our assumption, it suffices to show that if M is a direct sum $\sum_{i \in I} \oplus M_i$ of injective modules M_i, each generated by c elements, then M is injective. For simplicity, let c be an infinite cardinal $\geq |R|$. We may assume that I is infinite.

Let B be a set with cardinality $> 2^{cd}$, where $d = |I|$. For each $i \in I$, let $N_i = \prod_{b \in B} M_{i,b}$, the direct product of $|B|$ copies of M_i, and let $P = \prod_{i \in I} N_i$. Since a direct product of injective modules is always injective, N_i is injective for all i, and P is injective. By hypothesis, we may write $P = \sum_{g \in G} \oplus Q_g$, where Q_g is generated by c elements. Well order I, and take one of the direct summands M_{1,b_1} of N_1. Since M_{1,b_1} is generated by c elements, since c is infinite, and since each element of M_{1,b_1} is contained in a direct sum of finitely many $\{Q_g | g \in G\}$, then M_{1,b_1} is contained in $P_1 = \sum_{g \in G_1} Q_g$, where G_1 is a subset of G consisting of c elements. Since each Q_g is generated by c elements, P_1 is generated by $c^2 = c$ elements. Consequently, $|P_1| \leq c |R| \leq c^2 = c$, so P_1 has at most 2^c subsets. Since $\{M_{2,b} \cap P_1 | b \in B\}$ is an independent collection of submodules of P_1, and since $|B| > 2^c$, then $M_{2,b_2} \cap P_1 = 0$, for some $b_2 \in B$. The projection φ of M_{2,b_2} into $\sum_{g \in G} Q_g$ is a monomorphism, $\varphi(M_{2,b_2}) \subseteq P_2 = \sum_{g \in G_2} Q_g$, where G_2 is a subset of G consisting of c elements, and $G_1 \cap G_2$ is empty.

For $\alpha \in I$, assume that there exist mutually disjoint subsets $\{G_\gamma\}_{\gamma < \alpha}$ of G such that $|G_\gamma| = c$, and such that $P_\gamma = \sum_{g \in G_\gamma} Q_g$ contains a copy M_{γ, b_γ} of M_γ. Let $H_\alpha = \bigcup_{\gamma < \alpha} G_\gamma$, and set $S_\alpha = \sum_{g \in H_\alpha} Q_g = \sum_{\gamma < \alpha} P_\gamma$. Since each P_γ is generated by c

elements, and since $|H_\alpha| \le c \cdot d$, it follows that $|S_\alpha| \le c \cdot d \cdot c = c \cdot d$, so that S_α has at most $2^{c \cdot d}$ subsets. Since $|B| > 2^{cd}$, by the preceding reasoning, there exists a $b_\alpha \in B$ such that $M_{\alpha, b_\alpha} \cap S_\alpha = 0$. Thus there exists a subset G_α disjoint from H_α such that G_α is generated by c elements, and such that $P_\alpha = \sum_{g \in G_\alpha} Q_g$ contains a copy M_{α, b_α} of M_α. By transfinite induction, G_α and P_α exist $\forall \alpha \in I$. Let $H = \bigcup_{\alpha \in I} G_\alpha$. Then

$$P = \sum_{\alpha \in I} P_\alpha \oplus \sum_{g \notin H} Q_g.$$

Since M_{α, b_α} is injective, and since P_α contains an isomorphic copy, M_{α, b_α} is isomorphic to a direct summand of P_α $\forall \alpha \in I$. Since $\sum_{\alpha \in I} P_\alpha$ is a direct summand of P, and since P is injective, it follows that $\sum_{\alpha \in I} M_{\alpha, b_\alpha}$, being isomorphic to a direct summand of P, is injective. Since $M \approx \sum_{\alpha \in I} M_{\alpha, b_\alpha}$, M is injective. $\square$

The condition stated in the theorem seems to have been first studied by Chase [60, Lemma 4.1]. There, in a lemma contributed by Bass, it is proved that any semiprimary ring R satisfying it is right Artinian.

20.8 Corollary. *The following conditions on a ring R are equivalent:*

(a) *R is right Noetherian.*

(b) *There exists a cardinal number d such that every injective right R-module is a direct sum of the injective hulls of modules (generated by a set) of cardinal $\le d$.*

(c) *There exists a subclass S of mod-R satisfying the following two conditions:*

(1) *S is a set.*

(2) *Every injective object in mod-R is a isomorphic to a direct sum of objects in S.*

Proof. The two forms of (b), with or without parentheses, are equivalent (provided that different cardinals d and d' are used!).

(a) $\Rightarrow$ (b) follows immediately from the theorem.

(b) $\Rightarrow$ (c). Let F denote the free module $\approx R^{(d)}$. Then each injective module is a direct sum of injective hulls of factor modules of F, that is, of modules of the form $\widehat{F/K}$. Since $\{\widehat{F/K} \mid K \subseteq F\}$ is a set, then $\{\widehat{F/K} \mid K \subseteq F\}$ is a set S satisfying (c).

(c) $\Rightarrow$ (a). If c denotes the cardinality of S, then c is a cardinal with the property of the theorem, so R is right Noetherian by the theorem. $\square$

20.9 Corollary (Theorem of Papp [59]). *If each injective right R-module is a direct sum of indecomposable modules, then R is right Noetherian.* $\square$

20.10 Exercises

20.10.1 (a) If R is a right order in a ring S, and if S satisfies the a.c.c. on R-submodules, then $S = R$.

(b) Let R be a right Noetherian ring, with maximal nilpotent ideal N such that inj hull$_R R/N$ is finitely generated (in mod-R). Then R is right Artinian. (Hint: First assume $N = 0$. Then R is a right order in a semisimple ring by Goldie's theorem. In the general case, show that $\bar{R} = R/N$ inherits the hypothesis; that is, inj hull$_R \bar{R}$ is finitely generated.)

(c) If R is right Noetherian, and if the injective hull of cyclic modules are finitely generated, then R is right Artinian. (The converse is false, cf. Rosenberg-Zelinsky [57].)

20.10.2   (a) If a uniform R-module is finitely generated and projective, it is isomorphic to a right ideal of R.

(b) Every indecomposable injective projective module is isomorphic to a summand of R.

Injective Hulls of Finitely Generated Modules

Let $\hat{M}$ denote the injective hull of a module M. In this section we encounter the following condition: If C is a cyclic or finitely generated module, then $\hat{C}$ is finitely generated. This condition for Artinian rings was   .died by Rosenberg and Zelinsky [57]. We show that any Noetherian ring satisfying this condition must be Artinian. Characteristically, a single cyclic module, namely, R modulo the maximal nilpotent ideal N, does the damage[2].

20.11   Lemma. *If R is a right Noetherian semiprime ring, and if $\hat{R} = \text{inj hull}_R(R)$ is finitely generated, then R is semisimple, and $R = \hat{R}$.*

Proof. By Goldie's theorem (9.12), R has a unique classical right quotient ring $Q = \{ab^{-1} | a, \text{ regular } b \in R\}$, which is a semisimple ring. If $q = ab^{-1} \in Q$, and if $q \neq 0$, then $a = qb$ is a nonzero element of $qR \cap R$, showing that Q is an essential extension of R, so we may assume that $R \subseteq Q \subseteq \hat{R}$. [Actually, in general, Q is also injective over R (proof?), so we could assume that $Q = \hat{R}$, but we do not need this.]

Since $\hat{R}$ is finitely generated, it is Noetherian. If b is a regular element of R, then $b^{-1} \in Q$, and $b^{-n} a = b^{-(n+1)}(ba) \ \forall a \in R$, showing that

$$R \subseteq b^{-1} R \subseteq \cdots \subseteq b^{-n} R \subseteq \ldots.$$

Since $\hat{R}$ is Noetherian, $b^{-n} R = b^{-(n+1)} R$ for some n, so $b^{-(n+1)} = b^{-n} a$ for some $a \in R$. But then $b^{-1} = a \in R$. Since this is true for all regular $b \in R$, it follows that $Q = R$, that is, R is semisimple. $\square$

20.12   Theorem. *If R is a right Noetherian ring, and N is its maximal nilpotent ideal, and if $\widehat{R/N} = \text{inj hull}_R(R/N)$ is finitely generated in mod-R, then R is right Artinian.*

Proof. Let Q be the injective hull of R/N in mod-R/N. Then Q is an R-module and as such is an essential extension of R/N. Thus we may assume the R-module inclusions $R/N \subset Q \subset \widehat{R/N}$. Since $\widehat{R/N}$ is finitely generated and R is Noetherian, Q is finitely generated as an R-module and hence as an R/N-module. By the lemma, R/N is semisimple. This implies that R is a semiprimary right Noetherian ring, so the theorem of Hopkins and Levitzki (18.12) yields R right Artinian. $\square$

---

[2] An English colleague of mine asks, "What damage?"

20.13   Corollary. *If R is right Noetherian, and if the injective hulls of cyclic (resp. finitely generated) modules in M_R are finitely generated, then R is right Artinian.* $\square$

20.14   Proposition. *Let a, b be cardinals with $a > b$. Suppose C is generated by b elements in mod-R and $\hat{C}$ is contained in a direct sum of modules, each of which is generated by fewer than a elements. Then*

(i) *if $a = \aleph_0$, then $\hat{C}$ is finitely generated;*

(ii) *if $b \geq \aleph_0$, then $\hat{C}$ is generated by a elements.*

Proof. Let $\{Q_i | i \in I\}$ be objects in mod-R, each generated by a elements such that $\hat{C}$ is contained in their direct sum. Since each generator of C is contained in a direct sum of finitely many of the Q_i, it follows that C is contained in $K = \sum_{i \in I'} \oplus Q_i$, where I' is a subset of I with the properties:

$$b < \aleph_0 \Rightarrow c = \operatorname{card} I' < \aleph_0,$$

$$b \geq \aleph_0 \Rightarrow c = \operatorname{card} I' = b.$$

Now let f denote the projection of $\sum_{i \in I} \oplus Q_i$ onto K. Since $\ker f \cap C = 0$, then $\ker f \cap \hat{C} = 0$, showing that f maps $\hat{C}$ monomorphically into K. If $a = \aleph_0$, then $b < \aleph_0$ and $c < \aleph_0$, so K is a direct sum of finitely many finitely generated modules, that is, K is finitely generated. But then so is any direct summand of K, in particular, $f(\hat{C})$, whence $\hat{C}$ is finitely generated.

If $b \geq \aleph_0$, then $c = b$, so K is generated by $b a = a$ elements, so $f(\hat{C})$ and $\hat{C}$ are each generated by a elements in this case. $\square$

20.15   Theorem. *If R is any ring, then an indecomposable injective and projective right R-module M is isomorphic to a summand of R, that is, there exists an idempotent $e \in R$ such that $M \approx e R$.*

Proof. Write $M = \hat{C}$, where C is any nonzero cyclic submodule of M. Since M is projective, $\hat{C}$ is contained in a direct sum of copies of R, and the proof of 20.14 shows that $\hat{C}$ is contained in a direct sum $R^{(n)} = R_1 \oplus \cdots \oplus R_n$ of n copies of R. Hence, there is a least integer k such that $R^{(k)} = R_1 \oplus \cdots \oplus R_k$ contains a copy B of $\hat{C}$. Since B is indecomposable and injective, any two nonzero submodules of B have nonzero intersection. Thus, if $k > 1$, then B cannot have nonzero intersection with each component of R_i of $R^{(k)}$. But if $B \cap R_k = 0$, for example, then the projection of $R^{(k)}$ on $R^{(k-1)} = R_1 \oplus \cdots \oplus R_{k-1}$ maps B monomorphically into $R^{(k-1)}$, which contradicts the definition of k. Thus, $k = 1$, so $B \subseteq R_1$, and B, being injective, is a summand of R_1. Thus, $\hat{C}$ is isomorphic to a summand of R. $\square$

20.16   Corollary. *A uniform R-module U embeds in a free R-module F iff U embeds in R.*

Proof. Any module is the direct limit of its finitely generated submodules 14.7. Let B be any finitely nonzero submodule of U. Then, B is contained in a finitely generated free submodule S of F, and the proof of 20.15 shows that B embeds in R. If $\{B_i\}_{i \in I}$ denotes the family of nonzero finitely generated submodules of U, and if $\{h_i : B_i \to R\}_{i \in I}$ is a set of monics, then there is an isomorphism $h : U \to K$, where K denotes the right ideal of R generated by the set of all images $\{\operatorname{im} h_i\}_{i \in I}$. (This follows since a direct limit of monics is a monic 14.6.6, and since K is the direct limit of the images $\{\operatorname{im} h_i\}_{i \in I}$, which are cofinal in the set of finitely generated submodules of K.) $\square$

Direct Sums of Finitely and Countably Generated Modules

A theorem of Kaplansky states that if a module M is a direct sum of countably generated modules, then each summand of M has the same property (Kaplansky [58a]). We use this to prove the following theorem (Faith-Walker [67]) which generalizes the theorem of Cohen-Kaplansky [51] and Chase [60].

20.17 Proposition. *If each module in* mod-R *is contained in a direct sum of finitely generated modules, then R is right Artinian (cf. 20.23).*

Proof. If M is injective, then M is a summand of each over-module, so M is a direct sum of countably generated modules by Kaplansky's theorem. Then theorem 20.7 implies that R is right Noetherian. Now let C be any cyclic module. Then $\hat{C}$ is contained in a direct sum of finitely generated modules, so $\hat{C}$ is finitely generated by 20.14. Then R is right Artinian by 20.13. $\square$

20.18 Corollary. *Let R be a commutative ring. Then R is Artinian if and only if each injective R-module is a direct sum of finitely generated modules.*

Proof. One way follows from the preceding theorem. Conversely, let R be a commutative Artinian ring. Then, since R is Noetherian, an injective module M is a direct sum of indecomposable modules. By a theorem of Morita (see end-of-chapter Exercises), an indecomposable injective module over R is finitely generated. $\square$

If R is a ring with the property of the theorem, then each finitely generated module C is contained in a direct sum of finitely generated modules and then 20.14 implies that $\hat{C}$ is finitely generated. Expressed otherwise, if C is a module of finite length, then $\hat{C}$ has finite length. Rosenberg and Zelinsky [57] have shown that, in general, right Artinian rings do not enjoy this latter property, and consequently cannot have the former property.

C. Walker [66] has generalized Kaplansky's theorem as follows: If M is a module that is a direct sum of modules each generated by c elements, where c is an infinite cardinal, then each direct summand of M is a direct sum of modules, each generated by c elements. Using this theorem, we can generalize 20.7 as follows:

20.19 Theorem. *A ring R is right Noetherian if and only if there exists a cardinal number c such that each right R-module is contained in a direct sum of modules generated by c elements.*

Proof. If R is right Noetherian, 20.7, and the fact that every module is contained in an injective module, gives the desired c. Conversely, suppose such a c exists. Then the generalization of Kaplansky's theorem, together with 20.7, yields R right Noetherian. $\square$

Rings with d.c.c. on Principal Right Ideals

These rings include, in addition to right Artinian rings, all semiprimary rings by the theorem of Chase 18.14. Such rings are called left perfect rings, for reasons fully discussed in Chapter 22. In this paragraph, another theorem of Chase is proved stating that a ring R is left perfect if there exists a cardinal a such that every direct product of copies of R is a direct sum of modules generated by a elements. Together with 20.7, this implies that any ring with the latter condition holding for all modules is necessarily right Artinian 20.23.

An ordered class C is an **increasing filter** provided that C is a directed set. C is also said to be *directed from above*. Then opposite or dual class C^* is a **decreasing filter,** and *directed from below*. A set of principal right ideals is said to be a decreasing filter provided that the order by inclusion induces a decreasing filter. Thus

$$a, b \in R \ \& \ aR, bR \in F \ \Rightarrow \ \exists c \in R \ \& \ cR \subset aR \cap bR.$$

20.20   Chase's Lemma. *Let $A = \prod_{i \in \omega}{}^{(i)}A$, and $A_n = \prod_{i \le n}{}^{(i)}A$, where $\{{}^{(i)}A\}_{i \in \omega}$ is a countable sequence of objects of R-mod. Let $C = \sum_{\alpha \in I} \oplus C_\alpha$ be a direct sum of a family $\{C_\alpha\}_{\alpha \in I}$ of left R-modules, let $f: A \to C$ be a morphism of R-mod, and let $f_\alpha: A \to C_\alpha$ be the compose of f and the canonical projection $C \to C_\alpha$, for every $\alpha \in I$. Then, if F is any decreasing filter of principal right ideals, there exists $aR \in F$ and an integer $n > 0$ such that*

$$f_\alpha(a A_n) \subseteq \bigcap_{bR \in F} b C_\alpha$$

for all but a finite number of $\alpha \in I$.

Proof (Chase [61]). Assume that the statement is false, and inductively construct sequences $\{x_n\} \subseteq A$, $\{a_n R\} \subseteq \mathscr{F}$, and $\{\alpha_n\} \subseteq I$ such that the following conditions hold:

(i) $a_n R \supseteq a_{n+1} R$.

(ii) $x_n \in a_n A_n$.

(iii) $f_{\alpha_n}(x_n) \not\equiv 0 \pmod{a_{n+1} C_{\alpha_n}}$.

(iv) $f_{\alpha_n}(x_k) = 0$ for $k < n$.

We proceed as follows. Select any $a_1 R$ in $\mathscr{F}$. Then there exists $\alpha_1 \in I$ such that $f_{\alpha_1}(a_1 A_1) \not\subseteq \bigcap_{bR \in \mathscr{F}} b C_{\alpha_1}$, and hence we may select bR in $\mathscr{F}$ such that $f_{\alpha_1}(a_1 A_1) \not\subseteq b C_{\alpha_1}$. Since $\mathscr{F}$ is a filter of principal right ideals, there exists $a_2 \in a_1 R \cap bR$ such that $a_2 R \in \mathscr{F}$, in which case $f_{\alpha_1}(a_1 A_1) \not\subseteq a_2 C_{\alpha_1}$. Hence there exists $x_1 \in a_1 A_1$ such that $f_{\alpha_1}(x_1) \not\equiv 0 \pmod{a_2 C_{\alpha_1}}$. Then conditions (i)–(iv) above are satisfied for $n = 1$.

Proceed by induction on n; assume that the sequences $\{x_k\}$ and $\{\alpha_k\}$ have been constructed for $k < n$ and the sequence $\{a_k R\}$ has been constructed for $k \le n$ such that conditions (i)–(iv) are satisfied. Now, there exist $\beta_1, \ldots, \beta_r \in I$ such that, if $\alpha \ne \beta_1, \ldots, \beta_r$, then $f_\alpha(x_k) = 0$ for all $k < n$. We may then select $\alpha_b \ne \beta_1, \ldots, \beta_r$ such that $f_{\alpha_n}(a_n A_n) \not\subseteq \bigcap_{bR \in \mathscr{F}} b C_{\alpha_n}$; for, if we could not do so, then the theorem would be true. Hence there exists $bR \in \mathscr{F}$ such that $f_{\alpha_n}(a_n A_n) \not\subseteq b C_{\alpha_n}$. Since $\mathscr{F}$

is a filter of principal right ideals, there exists $a_{n+1} \in a_n R \cap bR$ such that $a_{n+1} R$ is in $\mathscr{F}$, in which case $f_{\alpha_n}(a_n A_n) \not\subseteq a_{n+1} C_{\alpha_n}$. Thus we may select $x_n \in a_n A_n$ such that $f_{\alpha_n}(x_n) \not\equiv 0 \pmod{a_{n+1} C_{\alpha_n}}$. It is then clear that the sequences $\{x_k\}$ and $\{\alpha_k\}$ for $k \leq n$ and $\{a_k R\}$ for $k \leq n+1$ satisfy conditions (i)–(iv), and hence the construction of all three sequences is complete.

Now write $x_k = (x_k^{(i)})$, where $x_k^{(i)} \in {}^{(i)}A$. Since $x_k \in a_k A_k$, $x_k^{(i)} = 0$ for $k > i$, and $x^{(i)} = \sum_{k=1}^{\infty} x_k^{(i)}$ is a well-defined element of ${}^{(i)}A$. Also, since $a_n R \supseteq a_{n+1} R \supseteq \cdots$, it follows that there exists $y_n^{(i)} \in {}^{(i)}A$ such that $x^{(i)} = x_1^{(i)} + \cdots + x_n^{(i)} + a_{n+1} y_n^{(i)}$. Therefore, setting $x = (x^{(i)})$ and $y_n = (y_n^{(i)})$, we see that $x = x_1 + \cdots + x_n + a_{n+1} y_n$ for all $n \geq 1$.

It follows immediately from inspection of conditions (iii) and (iv) above that $\alpha_i \neq \alpha_j$ if $i \neq j$. Hence there exists n such that $f_{\alpha_n}(x) = 0$. Writing $x = x_1 + \cdots + x_n + a_{n+1} y_n$ as above, we may then apply f_{α_n} and use condition (iv) to conclude that $f_{\alpha_n}(x_n) = -a_{n+1} f_{\alpha_n}(y_n) \equiv 0 \pmod{a_{n+1} C_{\alpha_n}}$, contradicting condition (iii). $\square$

As always, the symbol $|X|$ denotes the cardinality of the set X.

A submodule A' of a left R-module A is $\cap$-pure if $A' \cap aA = aA'$ $\forall a \in R$. (Note: Every summand of A is such a submodule. Cf. the concept of purity discussed in, and preceding, 19.21.)

20.21 Corollary (Chase). *Assume that there is an infinite set J such that the product $A = R^J$ is a $\cap$-pure submodule of a direct sum $\sum_{\gamma \in G} \oplus C_\gamma$ of left R-modules such that $|C_\gamma| \leq |J|$. Then R satisfies the* d.c.c. *on principal right ideals. (And by 22.29, R satisfies the* d.c.c. *on finitely generated right ideals.)*

Proof. Since J is an infinite set, it is easy to see that $A \approx \prod_{i=1}^{\infty} {}^{(i)}A$, where ${}^{(i)}A \approx A$, and so without further ado we shall identify A with $\sum_{i=1}^{\infty} {}^{(i)}A$. Let $f: A \to C$ be the inclusion mapping, and $f_\beta: A \to C_\beta$ be the composition of f with the projection of C onto C_β. Finally, set $A_n = \prod_{i=n+1}^{\infty} {}^{(i)}A$.

Suppose that the statement is false. Then there exists a strictly descending infinite chain $a_1 R \supsetneq a_2 R \supsetneq \cdots$ of principal right ideals of R. These ideals obviously constitute a filter of principal right ideals of R, and so we may apply 20.20 to conclude that there exists $n \geq 1$ and $\beta_1, \ldots, \beta_r$ such that $f_\beta(a_n A_n) \subseteq a_{n+1} C_\beta$ for $\beta \neq \beta_1, \ldots, \beta_r$.

Now let $C' = C_{\beta_1} \oplus \cdots \oplus C_{\beta_r}$; then the projection of C onto C' induces a $\mathbb{Z}$-homomorphism $g: a_n C/a_{n+1} C \to a_n C'/a_{n+1} C'$. Also, the restriction of f to A_n induces a $\mathbb{Z}$-homomorphism $h: a_n A_n/a_{n+1} A_n \to a_n C/a_{n+1} C$. A_n is a direct summand of A, which is a pure submodule of C, and so A_n is likewise a pure submodule of C. Hence h is a monomorphism. We may then apply the conclusion of the preceding paragraph to obtain that the composition gh is a monomorphism. In particular, $|a_n A_n/a_{n+1} A_n| \leq |a_n C'/a_{n+1} C'| \leq |C'|$.

Observe that $|C'| \leq |J|$, since J is infinite and $|C_\beta| \leq |J|$ for all β. However, since $a_n R \neq a_{n+1} R$, $a_n R/a_{n+1} R$ contains at least two elements; therefore $|a_n A_n/a_{n+1} A_n| = |a_n A/a_{n+1} A| \geq 2^{|J|} > |J|$. We have thus reached a contradiction, and the corollary is proved. $\square$

20.22 Corollary (Baer-Chase). *A product $\mathbb{Z}^\omega$ is not a pure subgroup of a direct sum of countably generated Abelian groups. In particular, $\mathbb{Z}^\omega$ is not a free Abelian group.* $\square$

The theorem of E. Specker states that any countably generated subgroup of $\mathbb{Z}^\omega$ is free, and in the group of all bounded functions $\omega \to \mathbb{Z}$ any subgroup of card $\leq \aleph$ is free (Specker [50]). See Dubois [66] and Nobeling [68], for certain generalizations. Bergman [70] gives an alternative proof. Kaplansky [69b, p. 83] comments on several aspects of Specker's theorem.

The next theorem was proved independently by Faith [71c], Griffith [70], and Vámos [71].

20.23   Theorem. *If R is a ring, and if there is a cardinal number α such that every right R-module is a direct sum of modules generated by α (or fewer) elements, then R is right Artinian.*

Proof. By the corollary above, R must have the d.c.c. on principal left ideals, and this implies that $J = \mathrm{rad}\,R$ is a nil ideal, that is, every element $x \in J$ is nilpotent. (Cf. the proof of 22.29.) Since R is Noetherian by 20.7, then every nil ideal of R is nilpotent by a theorem of Levitzki 9.17. The d.c.c. on principal left ideals of R implies that on R/J which implies the d.c.c. on R/J. (Cf. proof of 22.29.) Then 18.12 applies, and R is right Artinian. $\square$

20.24   Corollary. *Let R be a right Noetherian ring which is not right Artinian. Then there exists a cardinal α such that every right R-module is contained in a direct sum of modules generated by α elements, but there does not exist a cardinal β such that every right R-module is isomorphic to a direct sum of modules generated by β elements.*

Proof. 20.7 and 20.23. $\square$

A ring R is said to be a **right AD ring** if every right module has an AD. (See *sup.* 18.15.)

20.25   Corollary. *A right AD ring R is a semilocal right Noetherian lift/rad ring. Moreover, if the isomorphism class of indecomposable right modules is a set, then R is $\sum$-α-gens, for some cardinal α, and right Artinian.*

Proof. R is lift/rad semilocal 18.26, right Noetherian 20.7, and right Artinian 20.23. $\square$

It may be that *any* right AD ring is right Artinian. (Note, however, that over a local ring R with maximal ideal m, complete in the m-adic topology, every finitely generated module has an AD 27.22. This includes the ring of formal power series over a field.)

Direct Decompositions of Noetherian Rings

The Wedderburn-Artin theorem can be formulated as follows:
   Every ring R of global dimension 0 is a finite product of Artinian prime rings.
   Of course, the prime factors are all simple rings, and in fact, total matrix rings over division rings ($=$fields). (Furthermore the converse also holds true.) References for these statements are in Volume I, notably: (8.12, p. 370), (8.8, p. 369), (8.20, p. 375).

Since ring R of global dimension $= 0$ is therefore left and right Noetherian, to generalize this theorem to rings of global dimension $= 1$, namely, to hereditary rings, one ought to invoke this hypothesis. A. W. Chatters [72] has done this.

Chatters' Theorem (20.30). *A Noetherian hereditary ring R is a finite ring product of Artinian rings, and of prime rings.*

Thus, the structure of hereditary rings is reducible to two categories; namely, the category of Artinian rings, and the category of prime rings.

We start the proof of Chatters' theorem with a lemma on when a cyclic module embeds in a free module.

20.26 Lemma. *If I is a right ideal, then R/I embeds in a free module F (resp. in a product $F = R^\alpha$ of copies of R) iff $\exists$ a finite subset (resp. a subset) $X \subseteq R$ such that $X^\perp = I$, where $X^\perp = \{r \in R \,|\, x\,r = 0 \ \forall\, x \in X\}$. In this case, R/I embeds in a finitely generated free module.*

Proof. Let $f \colon R/I \to R^\alpha$ be an embedding in a product R^α. Then, $I = X^\perp$, where X is the set of projections in R of the element $f([1 + X])$ in R^α. Conversely, given $I = X^\perp$, with $X \subseteq R$, then there is an embedding $f \colon R/I \to R^\alpha$ sending $[a + I] \mapsto (\ldots, x_i\, a, \ldots) \in R^\alpha$, $x_i \in X$. Furthermore, if X is finite, then $\operatorname{im} f$ is contained in a finitely generated free submodule of R^α. $\square$

This lemma will be used again in Chapter 24.

20.27 Lemma. *If R is a left Noetherian right semihereditary ring, then each annihilator right ideal I of R is projective, and $I = e R$ for an idempotent $e \in R$.*

Proof. The fact that I is a right annihilator, say $I = A^\perp$, for some left ideal A of R, implies by the lemma that R/I embeds in R^n. Then, the fact that R is semihereditary implies that R/I is projective. Thus, $R \to R/I \to 0$ splits, and therefore $I = e R$ for some $e = e^2 \in R$. $\square$

A module M is said to have the **restricted minimum condition** if M/N is Artinian for every essential submodule N. Chatters's theorem 2 depends on a lemma of his stating that any left Noetherian left hereditary ring satisfies a kind of restricted right minimum condition, namely R satisfies the d.c.c. on finitely generated right ideals containing any finitely generated essential right ideal. The proof his lemma shows more generally the validity of the following lemma.

20.28 Lemma. *If R is left Noetherian and right nonsingular, and J is a finitely generated projective essential right ideal, then R satisfies the d.c.c. on finitely generated projective right ideals containing J.*

Proof. Since R is right nonsingular, the maximal right quotient ring Q exists, and is injective as a right R-module 19.35. Thus, for any essential right ideal I, we may identify $I^* = \operatorname{Hom}_R(I, R)$ with the set

$$I^* = \{q \in Q \,|\, q\,I \subseteq R\}.$$

(Clearly any element $f \colon I \to R$ is induced by an element $q_f \in Q$, and q_f is unique, since the singular submodule of $Q = 0$.) If, moreover, I is finitely generated and projective, then I^* is finitely generated (and projective) in R-mod, hence Noetherian and $I^{**} = \{k \in Q \mid I^* k \subseteq R\}$ coincides with I. To see the latter assertion, by projectivity and finite generation of I there are elements $\{x_i\}_{i=1}^n$ of I, $\{q_i\}_{i=1}^n$ of I^* such that

$$y = \sum_{i=1}^n x_i q_i(y) \qquad\qquad \forall\, y \in I.$$

Therefore, $t = (1 - \sum_{i=1}^n x_i q_i)$ annihilates I. Since I is an essential right ideal, then t lies in the right singular ideal of R, hence $t = 0$, so $1 = \sum_{i=1}^n x_i q_i$. But then $I^* k \subseteq R$ implies

$$k = \sum_{i=1}^n x_i(q_i k) \in I$$

as asserted. Now, for any chain

$$I_1 \supset I_2 \supset \cdots \supset I_n \supset \cdots$$

of finitely generated projective right ideals $\{I_n\}_{n=1}^\infty$ containing a given finitely generated projective essential right ideal J,

$$J^* \supset \cdots \supset I_n^* \supset \cdots \supset I_2^* \supset I_1^*.$$

Since J^* is finitely generated over the left Noetherian ring R, then J^* is Noetherian which implies $I_n^* = I_{n+1}^*$, whence $I_n = I_{n+1}$, for some n. $\square$

20.29   Corollary (Webber [70], Chatters [71]). *If R is a hereditary Noetherian ring, then R satisfies the right and left restricted minimum conditions. Furthermore, any finitely generated R-module M, satisfies the restricted minimum condition.*

Proof. The first assertion follows directly from the lemma, whereas the second is an exercise. $\square$

Chatters Theorem

20.30   Decomposition Theorem (Chatters [72]). *If R is a right and left Noetherian hereditary ring, then R is a finite ring product $\prod_{i=1}^n R_i$, where R_i is either Artinian, or a prime ring.*

Proof. Write R as a ring product $R \approx \prod_{i=1}^n R_i$ of finitely many ring-indecomposable rings. (This is possible since R is Noetherian.) Hence, we may suppose that R is indecomposable, and not prime. Then R has prime ideals, $P \neq 0, R$. If P is essential as a right or left ideal, then R/P is Artinian by 20.29, and then R is Artinian by Lemma 18.34 B. Thus, we may assume that P is not essential, and so $P \cap I = 0$ for some onesided ideal I, say I is a right ideal $\neq 0$. Then, $IP \subseteq I \cap P = 0$, so $A = {}^\perp P \neq 0$. Let $B = A$. Then $AB = 0 \subseteq P$, and since $A \nsubseteq P$, then $B \subseteq P$. But $B \supseteq P$, so $B = P$ is a right annihilator ideal. Then Lemma 20.27 shows that $P = eR$, for some $e = e^2 \in R$. Dually, $P = Rf$ for $f = f^2 \in R$, and then, $e = f$ is a central idempotent. Thus, $R = P \oplus A$ is a ring product, contradicting indecomposability of R. This contradiction proves the theorem. $\square$

20.31 *Example.* The theorem cannot hold for onesided hereditary rings, since, as the author points out, the ring R of upper triangular

$$\begin{pmatrix} \mathbb{Z} & \mathbb{Q} \\ 0 & \mathbb{Q} \end{pmatrix}$$

is right hereditary right Noetherian but not left hereditary (essentially because $\mathbb{Q}$ is not projective), and hence not left Noetherian. Furthermore, R is neither prime nor Artinian.

Although 20.31 shows that Chatters's theorem fails for Noetherian right hereditary rings, we next prove a theorem of Levy which yields the desired decomposition for semiprime rings under a hypothesis considerably weaker than right Noetherian right hereditary. Recall from Notes for Chapter 17 that a ring R is **right PP** if every principal right ideal is projective. (Note that any integral domain is trivially PP.)

20.32 Theorem (Levy [63]). *A right Goldie semiprime right PP ring R is a finite product of prime right Goldie right PP rings. Moreover, this product is unique, and the prime factor rings are the minimal annihilator ideals of R. In particular, any right semihereditary right Goldie semiprime ring has such a decomposition.*

Proof. Exercise (cf. Exercise 19.37(a)). □

20.33 Corollary. *If the quotient ring $Q(R)$ of a commutative PP ring is a finite product of fields, then R is a finite product of domains, and conversely.* □

20.34 Theorem (Small [66a]). *A (right) hereditary right Noetherian ring R has a (right) quotient ring $Q(R)$.*

Proof. This follows from Chatters' theorem assuming the conditions on two sides, since then R is a finite product of prime and Artinian rings. Now a prime direct factor A is also right Noetherian, hence $Q(A)$ exists by the Goldie and Lesieur-Croisot theorem (I, 9.10, p. 396). And, of course, $Q(A) = A$ for any Artinian ring. Thus, $Q(R)$ is the product of the quotient rings of the direct factors.

The onesided case is assigned to be an exercise. □

Robson's Theorem

There are three theorems about certain types of (right and left) Noetherian rings which have the same conclusion—namely, that each ring decomposes as the direct sum of a semiprime ring and an Artinian ring—but whose proofs are quite distinct. One, of Krull [24], Asano [38, 49], Goldie [62], concerns principal right ideal rings; another, of Chatters [72], concerns hereditary rings; and the third, of Warfield [75], concerns serial rings [3]. The next theorem is Robson's criterion for such a decomposition which is readily checked for the cited examples.

The Wedderburn radical of a ring R is the maximal nilpotent ideal, assuming existence. (It always exists in a right or left Noetherian ring, of course.)

For a ring R with radical N, we let $\mathscr{C}(N)$ denote the set $\{c \in R \mid c + N$ is regular in $R/N\}$.

20.35   Theorem (Robson [74]). *If R is a Noetherian ring with Wedderburn radical N, then R decomposes as the direct sum of a semiprime ring and an Artinian ring precisely when $cN = Nc = N$ for all $c \in \mathscr{C}(N)$.*

Proof. One direction is obvious. So suppose $cN = Nc = N$ for all $c \in \mathscr{C}(N)$. This hypothesis on R is inherited by the factor rings R/N^m. It implies that each $c \in \mathscr{C}(N)$ is regular; for the map $N \to cN$, being a surjective endomorphism of the Noetherian module N, is an automorphism. It also, in an obvious fashion, makes N^m/N^{m+1} into a left and right module over Q, the quotient ring of R/N (which exists by the Goldie-Lesieur-Croisot theorem (I, 9.9, p. 394)). Let $Q = Q_1 \oplus Q_1'$, where Q_1 is a simple Artinian component of Q. Suppose that $Q_1 N^m/N^{m+1} \neq 0$ for some m. Then, being a left Q_1-module, $Q_1 N^m/N^{m+1}$ is a direct sum of minimal left ideals of Q_1; but, as a left R/N-module, it is finitely generated. These facts combine to show that Q_1 is a finitely generated left R/N-module, via the embedding of R/N in Q. It follows easily that $Q_1 \subseteq R/N$, and

$$R/N = Q_1 \oplus (Q_1' \cap R/N).$$

(There exists $b \in \mathscr{C}(N)$ such that $\bar{b} Q_1 \subseteq \bar{R} = R/N$. Since $\bar{b} Q_1 = Q_1$, then $Q_1 \subseteq R/N$, and the claim follows.)

Hence, we reduce to the case when $R/N = S \oplus T$ with S semisimple Artinian and T a semiprime ring such that

$$TN^m/N^{m+1} = N^m/N^{m+1} T = 0$$

for all m. Let $e = e^2$ be a lifting to R of the identity element of T. Then $eN^m \subseteq N^{m+1}$ and $N^m e \subseteq N^{m+1}$ for all m and so $eN = Ne = 0$. Hence

$$R = eRe \oplus (1-e)R(1-e).$$

Of course $eRe = T$, and $(1-e)R(1-e)$, being a Noetherian ring which, modulo its prime radical, is isomorphic to S, is Artinian (18.12). $\quad\square$

If one assumes, for a Noetherian ring R, merely that $cN = N$ for all $c \in \mathscr{C}(N)$, the proof above gives an isomorphism

$$R \approx \begin{pmatrix} (1-e)R(1-e) & (1-e)Re \\ 0 & eRe \end{pmatrix}$$

where $(1-e)R(1-e)$ is Artinian and eRe is semiprime. An example due to Small [66c]

$$R = \begin{pmatrix} \mathbb{Z}/(p) & \mathbb{Z}/(p) \\ 0 & \mathbb{Z} \end{pmatrix}$$

illustrates this, and shows R need not decompose. As a special case, we have:

20.36   Theorem. *If R is a Noetherian ring whose Wedderburn radical N is prime, and if $cN = N$ for all $c \in \mathscr{C}(N)$, then R is prime or Artinian.* $\quad\square$

Chatters's Theorem Revisited

We next observe that it is now possible to give a partially different proof of Chatters' theorem, if we employ Robson's theorem 10.35. We now check the hypotheses of Theorem 20.35:

Now $c \in \mathcal{C}(N)$ is regular in R (for its right and left annihilators are contained in N yet have idempotent generators) and also $R/^{\perp}N$ is Artinian, by 20.29 and 9.4.3. The latter fact shows that $c^n R + {}^{\perp}N = c^{n+1} R + {}^{\perp}N$ for some n, and hence $c^n N = c^{n+1} N$. The former enables us to deduce $c N = N$. $\square$

20.37 (Krull [24], Asano [38, 49], Goldie [62]). *A left Noetherian principal right ideal ring R is a finite product of prime rings and primary Artinian rings.*

Proof. Let N be the maximal nilpotent ideal. Then, $\bar{R} = R/N$ is a semiprime principal right ideal ring. Now, each right ideal I of $\bar{R}$ is a summand of an essential right ideal $K = I + J$, where J is a relative complement right ideal of I. Moreover, any essential right ideal in a right Goldie semiprime ring contains a regular element (proof of (I, 9.9, p. 394)). Now $K = x\bar{R}$ is a principal right ideal, and if $y \in K$ is regular, then x is regular, so that $K \approx \bar{R}$ is projective. This proves that I is projective, that is, $\bar{R}$ is right hereditary. Then, Levy's theorem 20.32 permits $\bar{R}$ to decompose into a finite product of prime rings, and by a key lemma of Robson [67, Theorem 2.2] on lifting decompositions, this decomposition can be lifted: R is a finite product of rings each with maximal nilpotent ideal which is a prime ideal. Assuming that R is thus, 20.36 is applicable. Let $c \in \mathcal{C}(N)$. Then $cR + N = dR$ for some d. Since $d \in \mathcal{C}(N)$ and $dR \supseteq N$ it is clear that $dN = N$. But then $c N = N \pmod{N^2}$ and so, by induction $c N = N$.

Therefore, by 20.36, the indecomposable ring summands of R are either prime or Artinian primary. $\square$

Thus, we have an alternative proof of 18.38.3.

The proof has the following corollary implicit in the work of Goldie (compare (I, 10.22, p. 413)).

20.38 Corollary. *Any semiprime principal right ideal ring is right hereditary.* $\square$

Rings Similar to σ-Cyclic Rings

The subject of σ-cyclic rings is examined in detail in Chapter 25; here we content ourselves with just one rather obvious theorem which shows that n-gened rings are similar to σ-cyclic rings.

20.39 Theorem. *Let $n > 0$ an integer.*

(a) *Then a ring A is n-gened if and only if the matrix ring A_n is cyclic-gened.*

(b) *A is σ-n-gens if and only if the matrix ring A_n is σ-cyclic.*

(c) *Every n-gened ring A is similar to a cyclic-gened ring, and conversely, every ring similar to a cyclic-gened ring is n-gened for some n. This remains true when "σ-cyclic" replaces "cyclic-gened" and "σ-n-gen" replaces "n-gened".*

Proof. Exercise. $\square$

---

[3] See Chapter 25, esp. 25.3.5.

Commutative FBG and Sigma Cyclic Rings

In this section are a number of theorems on FBG and σ-cyclic commutative rings of Kaplansky [49, 52], Warfield [69b, 70], Gill [70], Lafon [70], and others. Some of these theorems require a knowledge of some of the basic techniques of commutative algebra, but these are mostly elementary things such as localization at a prime ideal and can be administered *ad hoc*.

A commutative ring R is a **valuation ring** (VR) if it satisfies one of the following three equivalent conditions:

   (i) for any two elements a and b, either a divides b or b divides a;

   (ii) the ideals of R are linearly ordered by inclusion;

   (iii) R is a local ring and every finitely generated ideal is principal.

An **arithmetical ring** is a variant term for a VR appearing in the literature.

A submodule A of an R-module B is **relatively divisible** or **pure** if for all $r \in R$, there holds $rA = A \cap rB$. (Compare $\cap$-pure in 19.21(h).)

A module P is **pure-projective** if for every short exact sequence

$$0 \to A \to B \to C \to 0$$

with A relatively divisible in B the induced sequence

$$0 \to \mathrm{Hom}_R(P, A) \to \mathrm{Hom}_R(P, B) \to \mathrm{Hom}_R(P, C) \to 0$$

is exact.

A module M is **cyclic presented** (CP) provided that there is an exact sequence $R \to R \to M \to 0$.

20.40 *Exercise* (Kaplansky [69b], Warfield [69b]). 1. An R-module M over a commutative ring R is pure-projective iff M is a direct summand of a direct sum of CP modules.

2. Over a valuation ring, any finitely presented cyclic module is cyclic presented. (Hint: Schanuel's lemma 11.28 (I, p. 436).)

3. Define **pure-injective** dually to pure-projective, and show that a module Q is pure-injective iff Q is a direct summand of any module containing it as a pure submodule. Thus, conclude that pure-injectivity is a (weakened) form of algebraic compactness (AC), defined *sup.* 19.21, inasmuch as any "Cohn pure" submodule is relatively divisible.

20.41 **Theorem.** *If M is a finitely presented module over a valuation ring then M is a direct sum of CP modules*[4].

Proof (Warfield [70]). Let $\mathfrak{m}$ be the maximal ideal of R and $y_1, \ldots, y_n$ a basis for the $R/\mathfrak{m}$-vector space $M/\mathfrak{m}M$. If $x_1, \ldots, x_n$ are elements of M with $x_i + \mathfrak{m}M = y_i$, then the x_i generate M. One of the basis elements $y_1, \ldots, y_n$ (say y_1) has the property that any element x in M with $x + \mathfrak{m}M = y_1$ has order ideal equal to the annihilator ideal of M. (If this were not the case, we could choose a set of

---

[4] This theorem is implicit in Kaplansky [49] and the proof involves matrices, however, Kaplansky attributes it to W. Krull (Letter of June 1974).

generators $x_1, ..., x_n$ as above such that the order ideals of the x_i were all greater than the annihilator ideal of M. Since the ideals of R are linearly ordered, and the annihilator ideal of M is the intersection of the order ideals of the x_i, this is impossible.) If y_1 is chosen in this way and x_1 is chosen so that $x_1 + \mathfrak{m}M = y_1$, then the submodule (x_1) generated by x_1 is relatively divisible. (To see this, suppose $r x_1 = s z$, where $z \in M$ and s does not divide r. Then $s = r t$ for some $t \in \mathfrak{m}$. If, then, $x_1^* = x_1 - t z$, then $x_1^* + \mathfrak{m}M = y_1$ and $r x_1^* = 0$, so (by the condition on y_1) $r x_1 = 0$. Hence $r x_1$ is divisible by s, which shows that (x_1) is a relatively divisible submodule.)

We now look at the sequence $0 \to (x_1) \to M \to M/(x_1) \to 0$. By induction, $M/(x_1)$ is a direct sum of CP modules, and is therefore pure-projective. Hence $M \approx (x_1) \oplus M/(x_1)$, from which it follows that (x_1) is also finitely presented and therefore a CP module, so the theorem is proved. $\square$

20.42 Theorem (Warfield [70]). *Let R be a local ring which is not a valuation ring. Then for any $n > 0$, there are finitely presented modules which are indecomposable and which cannot be generated by fewer than n elements.*

Proof. Let a and b be elements of R, neither dividing the other. By taking a suitable quotient ring, we may assume $(a) \cap (b) = 0$, and $\mathfrak{m}(a) = \mathfrak{m}(b) = 0$ (where $\mathfrak{m}$ is the maximal ideal). Let F be a free module with generators $x_1, ..., x_n$, let K be the submodule generated by $a x_1 - b x_2, ..., a x_{n-1} - b x_n$, and let $N = F/K$. Clearly $N/\mathfrak{m}N \approx F/\mathfrak{m}F$, so N cannot be generated by fewer than n elements. We will show that N is indecomposable. If I is the ideal $a\mathfrak{m} + b\mathfrak{m}$ then we let $M = N/IN$, and show that M is indecomposable. This will imply the indecomposability of M by Nakayama's lemma (18.4).

Let y_i and z_i be the images of x_i in M and $M/\mathfrak{m}M$ respectively. Let $S = aM + bM$. We observe that $aF + bF$ is an $R/\mathfrak{m}$-vector space with the elements $a x_1, ..., a x_n, b x_1, ..., b x_n$ as a basis. K is a subspace of $aF + bF$ of dimension $n - 1$, and S is the corresponding quotient space, with dimension $n + 1$. A sample basis for S is $b y_1, a y_1, ..., a y_n$, from which it follows that aM is a subspace of codimension 1, as is bM. There are natural homomorphisms α, β taking $M/\mathfrak{m}M$ into S given by multiplication by a and b respectively, (so $\alpha(z_i) = a y_i$) and these are injective (by dimension count). Now suppose $M = A \oplus B$. Clearly $aM = aA \oplus aB$, and $S = (aA + bA) \oplus (aB + bB)$. Since aM has codimension one in S, one of these summands of S is in aM, so we may suppose $bB \subseteq aB$. Choose $w = c_1 z_1 + \cdots + c_n z_n$ in $B/\mathfrak{m}B$ so as to minimize the index of the first nonzero coefficient. (Note that $B/\mathfrak{m}B \neq 0$ by Nakayama's lemma 18.4(c).) Note that $c_1 = 0$, since otherwise $\beta(w) \notin aM$ (since a basis of S is $b y_1, a y_1, ..., a y_n$ and the last n elements generate aM) and we know, in fact, that $\beta(w) \in aB$. If $w = c_2 z_2 + \cdots + c_n z_n$, then $\beta(w) = \alpha(c_2 z_1 + \cdots + c_n z_{n-1})$, so $c_2 z_1 + \cdots + c_n z_{n-1}$ is in $B/\mathfrak{m}B$, contradicting the choice of w (since α is injective). This contradiction shows that M is indecomposable. $\square$

20.43 Corollary (Warfield [70]). (a) *If R is a local ring and every finitely presented module is a summand of a direct sum of cyclic modules, then R is a valuation ring.*

(b) *Any commutative local FBG ring is a valuation ring.*

Proof. Any finitely generated direct summand M of a direct sum of cyclic modules is a direct summand of finitely many cyclic modules, and then by the Unique Decomposition theorem 18.18, M is itself a direct sum of cyclic modules, so Theorem 20.42 applies. (b) is obvious from 20.42. Actually, in (b) more can be said; see 20.49. □

The next exercise and theorem requires the concept of a local ring $R_{\mathfrak{m}}$ at a maximal or prime ideal $\mathfrak{m}$, and also that of a local property.

20.44   *Exercise.* Over a commutative ring R, for finitely presented module M, both projectivity and pure projectivity are local properties, that is, M has the property iff $M_{\mathfrak{m}} = M \otimes_R R_{\mathfrak{m}}$ has the property for all maximal ideals $\mathfrak{m}$.

20.45   Theorem. *A commutative ring R has the property that every finitely presented module is a summand of a direct sum of cyclic modules if and only if $R_{\mathfrak{m}}$ is a valuation ring for each maximal ideal $\mathfrak{m}$ in R.*

Proof. If R is such a ring so is $R_{\mathfrak{m}}$, since any finitely presented $R_{\mathfrak{m}}$-module is of the form $M_{\mathfrak{m}}$ for some finitely presented R-module M. Hence the condition is necessary by the above corollary. Conversely, by 20.44 a finitely generated R-module M is pure-projective if and only if M is finitely presented and $M_{\mathfrak{m}}$ is pure-projective for each maximal ideal $\mathfrak{m}$, from which the result follows. □

Almost Maximal Valuation Rings

A valuation ring is said to be **maximal** if every system of pairwise solvable congruences of the form

$$x \equiv x_\alpha(I_\alpha) \qquad\qquad (\alpha \in A,\ x_\alpha \in R,\ I_\alpha \text{ an ideal of } R),$$

has a simultaneous solution in R. We say R is **almost maximal** if the above congruences have a simultaneous solution whenever $\bigcap_{\alpha \in A} I_\alpha \neq 0$. These definitions are straightforward generalizations of those of maximal and almost maximal valuation *domains*, given by Kaplansky [52], where the terminology "maximal" is justified[5].

We define a *uniserial* module to be one with a totally ordered lattice of submodules. (Cf. Serial rings in Chapter 25.)

In this section we give equivalent conditions on a ring for it to be an almost maximal valuation ring. For a local ring R, with maximal ideal M, the following are equivalent: (i) $E(R/M)$ is uniserial, where, for any module X, $E(X)$ is the injective hull; (ii) R is an almost maximal VR; (iii) every indecomposable injective R-module is uniserial; (iv) every finitely generated R-module is a direct sum of cyclic R-modules ($= \sigma$-cyclic).

We write $\subset$ to mean "is strictly contained in".

Let $f: R \to S$, be a ring homomorphism, and let I be an ideal of R, and J an ideal of S. Then we denote by J^c the *contraction* of J, and by I^e the *extension* of I. Thus,

$$J^c = f^{-1}(J \cap f(R)) \quad \text{and} \quad I^e = f(I)S;$$

---

[5] Also see Chapter Exercise 21.

so we have $J^{ce} \supseteq J$ and $I^{ec} \supseteq I$. A basic result on commutative rings is that relative to any multiplicative set S (for example, the complement $S = R - P$ of any prime ideal P), the extension map from the ideals of R to those of the ring R_S of partial fractions (for example, when $S = R - P$, this is just the local ring R_P) is a surjection. (See 16.9 and 16.10 (I, pp. 529–31).)

For any nonempty subsets x and y of R, let $x : y = \{a \in R | ya \subseteq x\}$. If x is an ideal, then $x : y = \mathrm{ann}_R \, y/x$.

20.46 Proposition (Gill [70]). *Let R be a valuation ring, which is not an integral domain. Then R is almost maximal if and only if R is maximal.*

Proof. Clearly R maximal implies R almost maximal. Assume that R is almost maximal. Let P be the intersection of all prime ideals of R. Since R is a valuation ring, P is a prime ideal, and $P \neq 0$, as 0 is not a prime ideal. P consists of all the nilpotent elements of R. Consider the following system of pairwise solvable congruences:

(1) $\qquad x \equiv x_\alpha (I_\alpha)$ $\hspace{6cm}$ ($\alpha \in A$).

We can assume that $\bigcap_{\alpha \in A} I_\alpha = 0$, for, if $\bigcap_{\alpha \in A} I_\alpha \neq 0$, (1) has a solution, as R is almost maximal. Since $P \neq 0$, there exists $t (\neq 0) \in P$, with $t^2 = 0$. As $\bigcap_{\alpha \in A} I_\alpha = 0$, there exists $\alpha_0 \in A$ such that $I_{\alpha_0} \subset Rt \subseteq P$. Let $A' = \{\alpha \in A | I_\alpha \subseteq I_{\alpha_0}\}$. Put $J_\alpha = I_\alpha : t$. Then $\bigcap_{\alpha \in A} J_\alpha \neq 0$, since $t \in \bigcap_{\alpha \in A} J_\alpha$. From the pairwise solvability of (1), $x_\alpha - x_{\alpha_0} \in I_{\alpha_0}$ for all $\alpha \in A'$. So $x_\alpha - x_{\alpha_0} = z_\alpha t$, where $z_\alpha \in R$, $\alpha \in A'$. Now consider the following system of congruences:

(2) $\qquad z \equiv z_\alpha (J_\alpha)$ $\hspace{6cm}$ ($\alpha \in A'$).

Consider $z_\alpha - z_\beta$ $(\alpha, \beta \in A')$, and suppose that $I_\alpha \subseteq I_\beta$. Then $x_\alpha - x_\beta \in I_\beta$, and $t(z_\alpha - z_\beta) \in I_\beta$. Thus $z_\alpha - z_\beta \in J_\beta$, and so (2) is pairwise solvable. Since $\bigcap_{\alpha \in A'} J_\alpha \neq 0$, (2) has a simultaneous solution, say z_0. It is now easily seen that $x_{\alpha_0} + z_0 t$ is a solution of (1), and so R is maximal. $\square$

20.47 Lemma. *Let R be a valuation ring, and let P be a prime ideal of R. Then R is maximal (almost maximal) $\Rightarrow$ R_P is maximal (almost maximal).*

Proof. We know that R_P is a local ring with maximal ideal PR_P. Consider the natural homomorphism $\Phi \colon R \to R_P$, given by $\Phi(r) = r/1$. Let I be an ideal of R_P. Then $I^{ce} = I$, and since the ideals of R are totally ordered, so are the ideals of R_P.

Assume that R is maximal (almost maximal). Suppose that we have the following system of pairwise solvable congruences:

(1) $\qquad x \equiv x_\alpha (I_\alpha)$ $\hspace{6cm}$ ($\alpha \in A$),

(where $x_\alpha \in R_P$, and I_α is an ideal of R_P). We can assume that $I_\alpha \subseteq PR_P \,\, \forall \alpha \in A$, since the congruence $x \equiv x_\alpha (R_P)$, is satisfied for all $x \in R_P$.

Suppose $x_\alpha \in \Phi(R)$, $\forall \alpha \in A$. Let $\Phi(z_\alpha) = x_\alpha$, where $z_\alpha \in R$. Consider the following system of congruences:

(2) $\qquad z \equiv z_\alpha (I_\alpha^c)$ $\hspace{6cm}$ ($\alpha \in A$).

(Note that if $\bigcap_{\alpha \in A} I_\alpha \neq 0$, then $\bigcap_{\alpha \in A} I_\alpha^c \neq 0$.) Since (1) is pairwise solvable, clearly (2) is also pairwise solvable. As R is maximal (almost maximal), (2) has a solution, say z_0, in R. Clearly $\Phi(z_0)$ is a solution of (1), and so R_P is maximal (almost maximal).

Thus we can assume that there exists $\alpha_0 \in A$, such that $x_{\alpha_0} \notin \Phi(R)$. Note that $PR_P \subseteq \Phi(R)$. Consider any congruence of (1), say $x \equiv x_\beta(I_\beta)$. Pairwise solvability gives

$$x_{\alpha_0} - x_\beta \in I_\beta \cup I_{\alpha_0} \subseteq PR_P \subseteq \Phi(R).$$

Suppose that $x_\beta \in \Phi(R)$. Then $x_{\alpha_0} - x_\beta + x_\beta \in \Phi(R)$ and so $x_{\alpha_0} \in \Phi(R)$, which is not so. Thus $x_\beta \notin \Phi(R) \; \forall \beta \in A$. Therefore (1) becomes

$$(1') \qquad x \equiv x_\alpha(I_\alpha) \qquad\qquad (\alpha \in A)$$

where $x_\alpha \notin \Phi(R)$, $\forall \alpha \in A$. Now $x_\alpha \notin \Phi(R)$ implies $x_\alpha = \Phi(r_\alpha)^{-1}$, where $r_\alpha \in R - P$. Hence $\Phi(r_\alpha) = x_\alpha^{-1}$. Consider the following system of congruences:

$$(3) \qquad z \equiv r_\alpha(I_\alpha^c) \qquad\qquad (\alpha \in A).$$

(3) is pairwise solvable for, if $I_\alpha \subseteq I_\beta$, I_β^c. Now

$$x_\alpha - x_\beta = \Phi(r_\alpha)^{-1} - \Phi(r_\beta)^{-1} = \Phi(r_\alpha)^{-1} \, \Phi(r_\beta)^{-1} \, \Phi(r_\beta - r_\alpha).$$

Thus

$$\Phi(r_\alpha)^{-1} \, \Phi(r_\beta)^{-1} \, \Phi(r_\beta - r_\alpha) \in I_\beta, \qquad \Phi(r_\beta - r_\alpha) \in I_\beta, \qquad r_\beta - r_\alpha \in I_\beta^c.$$

Since R is maximal (almost maximal), (3) has a solution, say z_0, in R. Now $z_0 - r_\alpha \in I_\alpha^c \subseteq P$, $\forall \alpha \in A$. Hence $z_0 \in R - P$. Clearly $\Phi(z_0)^{-1}$ becomes a solution of (1), and so R_P is maximal (almost maximal). $\quad\square$

20.48 Lemma. *Let R be a valuation ring. Then the following are equivalent.*

(1) ann ann $Rb = Rb$, *for all $b \in R$.*

(2) *Every element of R is either a unit a unit or a zero divisor of R.*

Proof. Assume (1), and suppose m is not a zero divisor. Then ann $Rm = 0$, and so ann ann $Rm = R$. Thus, by the hypothesis, $Rm = R$, and so m is a unit.

Assume (2), and let M be the maximal ideal of R. Clearly $Rb \subseteq$ ann ann Rb for all $b \in R$. Suppose there exists $c \in R$, such that $Rc \subset$ ann ann Rc. Then there exists $y(\neq 0) \in R$, such that $y \in$ ann ann $Rc \setminus Rc$. Now Rc Ry, so $c = my$, for some $m \in M$. Now

$$Ry \subseteq \text{ann ann } Rc = \text{ann ann } Rmy.$$

Taking annihilators

$$\text{ann } Ry \supseteq \text{ann } Rmy.$$

Thus if $rmy = 0$, then $ry = 0$. Hence $Ry \cap \text{ann}(m) = 0$. But $Ry \neq 0$, and as the ideals of R are totally ordered, $\text{ann}(m) = 0$. This is a contradiction, since m is a zero divisor. Thus ann ann $Rb = Rb$, for all $b \in R$. $\quad\square$

Kaplansky [52], Matlis [66], Gill [70], Lafon [70], and Warfield [70] all contributed to the next theorem.

20.49   Theorem. *Let R be a local ring, with maximal ideal M. Then the following are equivalent.*

   (i) *The unique simple module R/M has uniserial injective hull $E(R/M)$.*

   (ii) *R is an almost maximal valuation ring.*

   (iii) *Every indecomposable injective R-module is uniserial.*

   (iv) *Every finitely generated R-module is a direct sum of cyclic R-modules.*

Proof. (Gill [70]). (i) $\Rightarrow$ (ii). Suppose that $E(R/M)$ is uniserial. We know that $E = E(R/M)$ is an injective cogenerator (I, 3.31, p. 148). Consider the mapping by annihilation from the ideals of R into the submodules of E, given by $I \mapsto 0:_E I$. It is known that this is an order-reversing, injective mapping, and $0:(0:_E I) = I$. (Exercise, or see 23.13.) Thus since the submodules of E are totally ordered, so are the ideals of R, and so R is a valuation ring.

Consider the following system of pairwise solvable congruences:

(1) $x \equiv x_\alpha(I_\alpha)$ $(\alpha \in A, \ \bigcap_{\alpha \in A} I_\alpha \neq 0)$.

We can consider $I_\alpha \subseteq M$, $\forall \alpha \in A$, since the congruence $x \equiv x_\alpha(R)$ is satisfied $\forall x \in R$. Put $E_\alpha = 0:_E I_\alpha = \{x \in E \mid x I_\alpha = 0\}$. As above we note that if $E_\alpha \subseteq E_\beta$, then $I_\beta \subseteq I_\alpha$. Also $I_\alpha = 0: E_\alpha = \{r \in R \mid E_\alpha r = 0\}$. Since $\bigcap_{\alpha \in A} I_\alpha \neq 0$, $E' = \bigcup_{\alpha \in A} E_\alpha \subset E$. Define $f': E' \to E$, by $f'(e) = x_\alpha e$, if $e \in E_\alpha$. To show that f' is well defined, we must show that, if $e \in E_\alpha$, and $e \in E_\beta$, then $x_\alpha e = x_\beta e$. Suppose that $E_\alpha \subseteq E_\beta$. Then $I_\beta \subseteq I_\alpha$. The pairwise solvability of (1) gives $x_\alpha - x_\beta \in I_\alpha$. Hence, since $e \in 0:_E I_\alpha$, $(x_\alpha - x_\beta)e = 0$, and so $x_\alpha e = x_\beta e$.

Now since $E' \subset E$, there exists $e_0 \in E$, such that $E' \subset R e_0 \subseteq E$. As E is injective, f' can be extended to $f: E \to E$, as in the diagram. We claim that $f(e_0) = r e_0$ for

$$0 \longrightarrow E' \xrightarrow{\ \text{in}\ } E$$

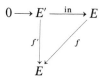

some $r \in R$. Suppose that this is not true. Then since E is uniserial, $e_0 = s f(e_0)$ for some $s \in R$. Now let $e \in E_\alpha$ for some $\alpha \in A$. Clearly $e = t e_0$ for some $t \in R$. Then

$$s f(e) = s f(t e_0) = t s f(e_0) = t e_0 = e.$$

Thus

$$e - s f(e) = 0, \quad \text{and so} \quad e - s x_\alpha e = 0.$$

It follows that $1 - s x_\alpha \in I_\alpha \subseteq M$, hence $s x_\alpha \notin M$, and so $s \notin M$. Thus s is a unit, and so $f(e_0) = s^{-1} e_0$. Hence we have shown that $f(e_0) = r e_0$ for some $r \in R$.

Let $e \in E'$. Then $e = t e_0$ for some $t \in R$, and

$$f'(e) = f'(t e_0) = f(t e_0) = t r e_0 = r e.$$

Thus $f'(e) = r e$, for all $e \in E'$. We claim that r is a solution of (1). Consider $r - x_\alpha$, for some $\alpha \in A$. Let $e \in E_\alpha$. Then

$$(r - x_\alpha)e = r e - x_\alpha e = f(e) - f(e) = 0.$$

Thus $r - x_\alpha \in I_\alpha$, $\forall \alpha \in A$. Hence r satisfies (1), and so R is an almost maximal valuation ring.

(ii) $\Rightarrow$ (iii). The proof of this part requires the following:

20.50 *Exercise* (Klatt and Levy [69]). Suppose R is an almost maximal valuation ring. Then R is selfinjective iff

(1) ann ann $Rb = Rb$, for all $b \in R$, and

(2) R is maximal.

It is evident that the set of zero divisors of a valuation ring is a prime ideal, say P. If R is not a domain, then R_P is maximal (from 20.46 and 20.47). This is clearly also true if R is a domain. From 20.48, ann ann $R_P b = R_P b$, for all $b \in R_P$. Thus R_P is selfinjective. This implies that R_P is injective as an R-module. But R_P is clearly an essential extension of R. Thus $E_R(R) = R_P$, and so $E_R(R)$ is uniserial. Now let I be a proper ideal of R. Since R is an almost maximal valuation ring, $\bar{R}$ is a maximal valuation ring, and so, using the same argument as above, we get that $E_R(\bar{R})$ is uniserial, where $\bar{R} = R/I$.

Let E be any indecomposable injective R-module. Let Rx and Ry be submodules of E, where $x, y \in E$. Let $0 : x = \{r \in R \mid xr = 0\}$. To show that E is uniserial, we must show that $Rx \subseteq Ry$ or that $Ry \subseteq Rx$. Now either $0 : x \subseteq 0 : y$ or $0 : y \subseteq 0 : x$. We can assume, without loss of generality, that $0 : x \subseteq 0 : y$. Put $I = 0 : x$. Then Rx, $Ry \subseteq E' = 0 :_E I$. Also E' is $\bar{R}$-*injective*, and $E' = E_R(\bar{R})$, which is uniserial. Thus $Rx \subseteq Ry$, or $Ry \subseteq Rx$, as $\bar{R}$-modules, and so also as R-modules. Thus E is uniserial.

(iii) $\Rightarrow$ (iv). Suppose every indecomposable injective R-module is uniserial. $E(R/M)$ is an indecomposable injective R-module, and so is uniserial. Thus, by (i) $\Rightarrow$ (ii), R is a valuation ring.

Let A be a finitely generated R-module, and let $a_1, \ldots, a_n$, be a set of generators of A. Then $0 : A = \bigcap_{i=1}^n (0 : a_i)$, and since R is a valuation ring, $0 : A = 0 : a_s$, for some s $(1 \leq s \leq n)$. Clearly $0 : a_s \subseteq 0 : x$, for all $x \in A$. Put $I = 0 : a_s$. Then $R a_s \approx R/(0 : a_s) = R/I$. Consider the diagram

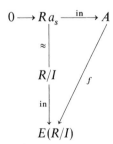

Denote by α the combined homomorphism $R a_s \xrightarrow{\approx} R/I \xrightarrow{\text{in}} E(R/I)$. Clearly α is a monomorphism. Since $E(R/I)$ is injective, there exists a homomorphism $f : A \to E(R/I)$, which makes the diagram commutative. Consider $f(A) \subseteq E(R/I)$. Since A is finitely generated, $f(A)$ is finitely generated. Since I is coirreducible, $E(R/I)$ is an indecomposable injective R-module. Thus $E(R/I)$ is uniserial, and so every finitely generated submodule of $E(R/I)$ is cyclic. Hence $f(A)$ is cyclic, say

$f(A) = R\,y$, where $y \in E(R/I)$. We now have

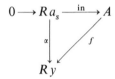

Since f is an epimorphism from A onto $R\,y$, there exists $z \in A$, such that $f(z) = y$. Now if $rz = 0$, $f(rz) = 0$. Thus $r f(z) = 0$ and $r y = 0$. So $0 : z \subseteq 0 : y$. Also $\alpha(a_s) = t\,y$, for some $t \in R$. If $r y = 0$,

$$\alpha(r\,a_s) = r\alpha(a_s) = r t\,y = 0.$$

But α is a monomorphism and so $r a_s = 0$. Thus $0 : y \subseteq 0 : a_s$. We have then

$$0 : z \subseteq 0 : y \subseteq 0 : a_s,$$

and so from the choice of a_s, $0 : z = 0 : y = 0 : a_s$. Thus $R z \xrightarrow{\text{in}} A \xrightarrow{f} R y$ is an isomorphism, and $A = R z \oplus B$, where B is an R-module.

The result now follows from Nakayama's lemma 18.4(c) and induction on $\dim_{R/M} A/MA$.

(iv) $\Rightarrow$ (i). Suppose that every finitely generated R-module is a direct sum of cyclic R-modules. Since $E = E(R/M)$ is an indecomposable injective R-module, every finitely generated R-submodule of E is cyclic. Now let A, B be submodules of E such that $A \nsubseteq B$, and $B \nsubseteq A$. To prove that E is uniserial we must show that this assumption leads to a contradiction. There exist $x, y \in E$ such that $x \in A$, $x \notin B$, $y \notin A$, $y \in B$. From above there exists $z \in E$, such that $R x + R y = R z$. Clearly $z \neq 0$. Thus there are elements $a, b, r, s, \in R$, such that

$$z = ax + by, \quad x = rz, \quad y = sz.$$

So

$$z = arz + bsz, \quad \text{and} \quad (1 - (ar + bs))z = 0.$$

Since R is a local ring, the set of non-units of R is M. Thus, for any $t \in R$, either t or $1 - t$ is a unit. Hence either $ar + bs$ or $(1 - (ar + bs))$ is a unit. But $z \neq 0$, so $ar + bs$ must be a unit. Thus either ar or bs is a unit. We can assume, without loss of generality, that ar is a unit. Thus r is a unit, so x generates the same module as z. Thus $y \in A$, and we have a contradiction. Hence E is uniserial. $\quad\square$

Exercises of Chapter 20

1. Prove that if the lengths of the finitely generated indecomposable right R-modules over a right Artinian ring R are bounded [or equivalently, if there exists an integer n such that every such module is generated by n (or fewer) elements] then every indecomposable injective right R-module is Noetherian (and generated by $\leq n$ elements). Show that mod-R then has an injective co-generator U of finite length. (This implies a U-duality in the sense of Morita. See 23.25.7.)

2. Prove the Morita theorem stating that an Artinian commutative ring R has a finitely generated injective cogenerator. [This is equivalent to the assertion that the injective hull of R (resp. socle R) is finitely generated, and it suffices to assume that R is a local ring.]

*3. (Cf. Chase [61, p. 851]) Let A be a product of finitely generated modules over a commutative principal ideal domain R. If K is a submodule of A such that A/K is a torsion group, then A is the direct sum of a divisible (hence injective) module D, and a module B of bounded order.

4. (a) Let e be an idempotent in a ring R, let M be a right eRe-module. If

$$R^n \to M \otimes_{eRe} eR \to 0 \qquad \qquad \text{(in mod-}R\text{)}$$

is exact, then

$$Re^n \to M \to 0 \qquad \qquad \text{(in mod-}eRe\text{)}$$

is exact. — (b) If R is FBG, then so is eRe, for any idempotent e.

*5. A ring R with the property that every cyclic right module is either semisimple or isomorphic to R is a simple right V-domain or a semisimple ring.

6. The ring $R = k[y, D]$ of differential polynomials over a universal field k has the properties: (a) Every cyclic module $\neq R$ is a semisimple injective module of finite length; (b) Every quasi is injective; (c) There are but two equivalence classes of indecomposable right modules consisting of the unique simple module V, and the right quotient field K. (Cf. Chapter 19 Exercise 32, and also Goodearl's characterization of rings which are restricted right semisimple in the sense that R/I is semisimple for every essential right ideal (Goodearl [72, p. 47]).)

7. Let R be a left semihereditary Goldie semiprime ring. Then, R satisfies the d.c.c. on f.g. projective right ideals containing any fixed nonzero finitely generated essential projective right ideal iff R is left Noetherian (cf. 20.28).

8. (Converse of the Webber-Chatters theorem Goldie semiprime ring) Let R be a left semihereditary Goldie semiprime ring which satisfies the restricted right minimum condition. Then R is left Noetherian, and hence left hereditary.

9. (Camillo-Cozzens [73]) The following conditions are equivalent for a left Ore principal right ideal domain. (1) R is a left PID; (2) R is left Noetherian; (3) R has the a.c.c. on principal left ideals; (4) R satisfies the restricted right minimum condition. (It is not known if (1) ever fails.)

10. In any ring R, any direct sum of injective right R-modules is an epic image of a direct sum of copies of the injective hull $\hat{R}$ of R.

11. An R-module M is **completely injective** if every factor module is injective. Then, M is $\sum$-**completely injective** if every direct sum of copies of M is completely injective. A ring R is right Noetherian right hereditary iff every right R-module M is $\sum$-completely injective iff $\hat{R}$ is $\sum$-completely injective.

12. In a right Noetherian ring, every direct sum of completely injective right R-modules is completely injective. Conclude from # 11 that a right Noetherian ring R is right hereditary iff $\hat{R}$ is completely injective.

13. Let I be a right ideal in a (nonsemisimple) ring R which is maximal in the set of those right ideals K such that the injective hull $\widehat{R/K}$ is not semisimple. (I always exists in a nonsemisimple right Noetherian ring.) Then I is coirreducible. Moreover, if $E = \widehat{R/I}$ has no fully invariant submodules (as is the case when R is right QI), then E/K is semisimple for every submodule $K \neq 0$.

14. (Faith [76]) Boyle's Conjecture: If R is right QI, then R is right hereditary. Prove this under the assumption of the restricted right socle condition: if $I \neq R$ is an essential right ideal, then R/I has socle $\neq 0$.

15. Observe that the restricted right socle condition is necessary for the truth of Boyle's conjecture assuming R is both right and left QI.

16. A left and right principal ideal ring is σ-cyclic. (A left Noetherian principal right ideal semiprime ring is right σ-cyclic.)

17. (Kaplansky [49, 12.3, p. 486]) A commutative ring R is a PIR iff R is Noetherian and maximal ideals are principal.

18. (I.S. Cohen [50]) A commutative ring R is Noetherian iff every prime ideal is finitely generated. Conclude from Problem 17 that a ring R is a PIR iff every prime ideal is principal.

For the purpose of the next "exercise", a principal ideal ring R is said to be special if its radical is nilpotent.

*19. (Warfield [70]) The following conditions on a commutative Noetherian ring are equivalent.

(i) For every maximal ideal $\mathfrak{m}$, $\mathfrak{m}/\mathfrak{m}^2$ has $R/\mathfrak{m}$-dimension one.

(ii) R is a (finite) product of Dedekind domains and special PIR's.

(iii) For every maximal ideal $\mathfrak{m}$, $R_\mathfrak{m}$ is a discrete valuation ring or a special PIR.

(iv) Every finitely generated R-module is a summand of a direct sum of cyclic modules.

*20. If R is a commutative Noetherian ring then either for every positive integer n, there are finitely generated indecomposable R-modules which cannot be generated by fewer than n elements, or R satisfies the conditions of Exercise 19, in which case any finitely generated indecomposable R-module can be generated by two elements.

*21. (Kaplansky [52]) A principal ideal local domain R ($=$ a discrete valuation ring) is always almost maximal. Moreover, R is maximal iff R is complete in the p-adic topology, where p is the unique prime.

22. (Matlis [59]) A commutative valuation domain R with quotient field K is almost maximal iff K/R is injective.

Notes for Chapter 20

Kaplansky [49, p. 483, Theorem 11.1] proved every module generated by two elements over a commutative local domain R is σ-cyclic iff "every pseudo-convergent set of nonzero breadth in R has a limit in R". In Kaplansky [52, p. 336], it is explained that the latter condition is equivalent to R being an almost maximal valuation domain, and Theorem 14 of that paper states that then every finitely generated module is a direct sum of cyclic modules. The converse for domains then follows, as stated, from Kaplansky [49]; Matlis [66] gives a new proof, and Gill [70] extends the theorem to the general case of rings with zero divisors (see 20.49). Implicit in Kaplansky's work is Theorem 20.41 stating that every finitely presented module over a local ring R is σ-cyclic iff R is a valuation ring; and explicit in Lafon [70] and Warfield [70]. (See the theorems of Warfield expressed in Chapter exercises 19 and 20.)

Roux [72] partially generalized Theorem 20.41 to the noncommutative case (correcting Lafon [71 a]): if R is a local ring which has linearly ordered lattice of right (left) ideals, that is, if in the terminology of Chapter 25 R is a serial local ring, then every finitely presented module is a direct sum of cyclic modules. (A partial onesided theorem also is obtained.) This result is included in Warfield's generalization presented in Chapter 25, namely, a ring R is serial iff R is a semiperfect ring and every finitely presented right and left module is a direct sum of principal cyclic modules (Theorems 25.2.6 and 25.3.4).

Matlis [59] characterized almost maximal valuation domains among local domains by the property that Q/R is an injective module, $Q = Q(R)$ is the quotient field. Gill [70] (see 20.49) reports that this does not hold for non integral domains, but does not supply a proof of this. Of course, in this case Q is not necessarily injective. When Q is injective, then Q/R will be injective only under special hypothesis, e.g. when R is hereditary. Kaplansky [69 b, Notes, pp. 73 ff.] comments on this, and a number of other topics taken up in this chapter, including pure-projectivity and its dual pure-injectivity ($=$ algebraic compactness).

If for some cardinal α, a ring R is right $\sum$-α-gens, then R is right Artinian by 20.23. Warfield [72] has shown that if R is commutative, then R is a principal ideal ring, and hence $\sum$-cyclic. The problem of characterizing the noncommutative case remains open. Does $\sum$-α-gens imply $\sum$-indecomposables? A related question: does (right) $\sum$-indecomposables $\Rightarrow$ (right) Artinian and/or right FFM? Same question for right AD rings: if every right R-module is a direct sum of modules with local endomorphism rings, then is R right Artinian and right FFM? (See 20.25, however.)

We already commented on Roiter's verification of the Brauer-Thrall conjecture, in the Introduction to Volume II, and the fact that Auslander and Tachikawa have extended Roiter's theorem, and refined it, giving proofs markedly different from Roiter's. Moreover, Gabriel [72] and Dlab-Ringel [73] have been able to explicitly describe the indecomposable modules, and generalize theorems of Yoshii [56]. (Cf. Curtis-Jans [65] and Kawada [62–4].) Jans [57]— Tachikawa [60]—Colby [66] show that a right FFM right Artinian ring has finite lattice of ideals (see 25.4.5). (In fact, infinite lattice of ideals implies that the modules are "strongly unbounded" type.) Actually, then there are infinitely many indecomposable quasiprojective modules; similarly for quasiinjectives over finite dimensional algebras (Wu-Jans [67]).

The theorems of Chase appearing in this chapter originated in Chase's characterizations of when is every product of projective right R-modules projective. (It is equivalent to the requirement that every product R^α is projective.) The solution (22.31 B) uses Chase's characterization (I, 11.34, p. 439) of when R^a is flat, and Bass's characterization of when every flat right module is projective. The former are the left coherent rings, and the latter are the right perfect rings. Thus, Chase's solution: the left coherent right perfect rings. This generalized Krull's theorem. (See Kaplansky [69b, p. 82].)

Similarly, the theorems of Faith and Faith-Walker are used in Chapter 24 to characterize when is every injective right module projective (then every injective would have to be $\sum$-injective), and the dual problem of when is every projective right module injective. (Then every injective is $\sum$-fin. gens.) The answer in either case: R must be Quasi-frobenius, and conversely. (Thus, most of the theorems in this chapter will be used elsewhere in the text.)

Chatters theorem 20.30 is on the subject of Noetherian rings of finite global dimension ($=1$). The theorem of Auslander-Buchsbaum [57, 59] for a commutative Noetherian ring R of finite global dimension states that R is a finite product of unique factorization domains. (The emphasis in our context is on the domains, that is, prime rings.) For a proof, consult Osofsky [71a, p. 52].

Long after this chapter was written, I read Goodearl's Memoir [72], and noted the connection between QI rings, and some of his results for what he calls SI rings (singular submodule is injective). As we have noted in Exercise 6 (and also Exercise 32 of Chapter 19), these rings are restricted semisimple, and are right hereditary. Moreover, modulo socle, they are Noetherian and semisimple. In fact, after splitting off such a direct factor, a right SI ring is similar to a product of right SI domains. The latter are evidently right Ore domains, and every proper cyclic right module is injective. Rings with the latter property have been called right PCI rings, and have been studied by Boyle [71, 74] and Faith [73]. A right PCI ring is either semisimple, or a semihereditary right Ore domain; it is an open question whether a right PCI domain is right hereditary or equivalently, right Noetherian.

Right QI rings are of course right V rings (defined in Chapter 7, and also discussed in Chapter 19). V rings have been extensively studied, and I will be able to give only a limited bibliography in the references for this chapter. Here I mention that Cozzens and Johnson [72] (showing that Noetherian V rings need not be hereditary), Osofsky [71c] (showing that Noetherian V rings may have infinitely many simple modules), Cozzens [73] (extending Osofsky's examples to arbitrary characteristic, and simplying some proofs), and Boyle and Goodearl [75] (showing for the first time that Noetherian V rings are not necessarily QI). Michler and Villamayor [73] (for whom the appellation V ring was intended to honor (see Faith [67a, p. 130, Problem 17])) show that any right V ring of "Krull dimension at most 1" is right Noetherian and right hereditary (cf. 20.5). [This is related to Exercise 10.] Moreover, for a finite group G the group ring RG is a right V ring iff R is a right V ring and $|G|$ is a unit.

On the subject of (acc)$^\perp$, R.E. Johnson [69] has an interesting result: there exists an integral domain D such that the $n \times n$ matrix ring D_n for $n > 1$ fails to satisfy (acc)$^\perp$. Such a domain (called a Malcev domain by Johnson) can not be embedded in a field, since any subring of an $n \times n$ matrix ring over a field satisfies (acc)$^\perp$; and in fact, Johnson's example (like those of Bokut' [67], Bowtell [67], and Klein [67]) is such that the multiplicative semigroup D^* of nonzero elements is not embeddable in a semigroup.

Suppose that D is a domain with the property: (N) any nilpotent $n \times n$ matrix over D has index of nilpotency $\leq n$. Then the multiplicative semigroup D^* is embeddable in a group (Klein [69]). (In particular, (l.c.) any semifir has this property.) It is not known if D can be embedded in a field. However, D is an IBN ring and, in fact, is Dedekind finite (Klein [69, p. 151]).

Also see Notes for Chapter 21 and 25.

References

Asano [38, 49], Auslander [74], Auslander-Buchsbaum [57, 59], Beck [72], Bergman [72], Bokut' [67], Bowtell [67], Boyle [71, 74], Boyle-Goodearl [75], Bush [63], Cailleau-Renault [70], Camillo-Cozzens [73], Cartan-Eilenberg [56], Chase [60, 61, 62], Chatters [71, 72], Cohen [50], Cohen-Kaplansky [51], Colby [66], Cozzens [70, 73], Cozzens-Johnson [72], Curtis-Jans [65], Dlab-Ringel [72, 73], Dubois [66], Faith [66, 67a, 71c, 72b, 73, 75a, 76], Faith-Walker [67], Fuller [69], Gabriel [72], Gill [70], Goldie [62], Goodearl [72], Jans [57], Johnson [69], Kaplansky [49, 52, 69b], Kawada [62–64], Klatt and Levy [69], Klein [67, 68, 72], Krull [24], Kurshan [70], Lafon [70], Levy [63], Mal'cev [39], Levine [71], Matlis [58, 59, 66], Michler and Villamayor [73], Morita [58], Nöbeling [68], Osofsky [71a, c], Papp [59], Robson [67, 74], Roiter [68], Rosenberg-Zelinsky [57], Roux [72], Small [66a], Specker [50], Tachikawa [59, 60, 61, 73], Vámos [71], Warfiled [69b, 70, 72, 75], Webber [70], Yoshii [56].

Other References

Anderson and Fuller [72], Azumaya [50], Bass [60], Colby and Rutter [71], Crawley-Johnson [64], Faith and Walker [67], Fuchs [69], Fuller [73], Harada [71], Harada and Sai [70], Kaplansky [58a], Matlis [58], Monk [72], Shores [71], Shores and Wiegand [74], Swan [60, 68], C.L. Walker [66], Warfield [69b, c, 72a, b], Yamagata [73], Cozzens-Faith [75].

Chapter 21. Azumaya Diagrams

The UD theorem 18.18 is extended in this chapter to infinite Azumaya diagrams (AD's) in an $AB5$ category 21.6. Other results are (1) the X-lemma 21.1; (2) the cancellation theorem 21.2; (3) I-adically complete rings 21.7–8; (4) summands of AD's of injective modules are again AD's 21.15 (cf. 21.14).

21.1   The x-lemma. *If the following diagram has exact diagonals and if the composite hf is an isomorphism, then gj is an isomorphism:*

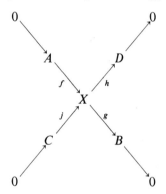

Proof. Apply the embedding theorem and check it for modules. □

21.2   Cancellation Theorem. *If C is an Abelian category, and if A is an object with $\text{End}_C A$ a local ring, then*

$$A \oplus B \approx A' \oplus B', \quad \text{and} \quad A \approx A' \;\Rightarrow\; B \approx B'.$$

Proof. Use the isomorphism of $A \oplus B$ with $A' \oplus B'$ to obtain maps $A \to A' \to A$ and $A \to B' \to A$ with sum $= 1_A$. Hence one of the two is an equivalence. If $A \to A' \to A$ is an equivalence, then let $X = A \oplus B$. Applying the X-lemma to

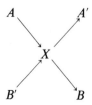

shows that $B' \to B$ is an equivalence.

Suppose $A \xrightarrow{f} B' \xrightarrow{g} A$ is an equivalence. Then $C' = \operatorname{im} f$ is a summand of B', where the splitting of the inclusion is the map $f (g f) \bar{g}^{1}$. Let $B' = C' \oplus C''$ with $C' \approx A$ and, consider the diagram

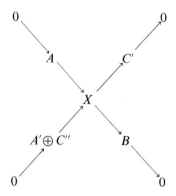

This yields $A' \oplus C'' \approx B$. Since C' is isomorphic to A', then $B' \approx B$. $\square$

21.3   Corollary. *If C is an Abelian category, and if objects A and A' have finite AD's, then*

$$A \oplus B \approx A' \oplus B' \quad and \quad A \approx A' \Rightarrow B \approx B'$$

for any objects B and B'. $\square$

21.4 *Exercise.* (a) Employ the cancellation theorem in another proof of the UD theorem 18.18 (cf. Swan [68, 2.8, p. 79]).

(b) If R is any semilocal ring, then an equivalence $R \oplus B \approx R \oplus B'$ in mod-R (R-mod) implies $B \approx B'$.

21.5   Lemma. *Let $\{A_i\}_{i \in I}$ be indecomposable subobjects with local endomorphism rings of an object X of on Abelian category C, and assume either that I is finite, or that C is AB5 (=has exact direct limits.) If $X = \sum_{i \in I} A_i$, and if B is any nonzero subobject such that $B \to X$ splits, $X = B \oplus B'$, then there is an $i \in I$ such that $A_i \to B \to A_i$ is an equivalence.*

Proof. There is a finite subset S such that the map $B \to \sum_{i \in I} A_i / \sum_{i \in S} A_i$ is not a monomorphism. This is obvious if I is finite. Let I be infinite and suppose there is no such set. Therefore the map $B \to \varinjlim \sum_{i \in I} A_i / \sum_{i \in S} A_i$ is a monic, where $\varinjlim$ is taken over finite subsets of I. But the limit is 0. This contradicts the assumption if B is nonzero.

The proof of the lemma proceeds by induction on the number of elements of S. Choose any $j \in S$. If $A_j \to B \to A_j$ is an equivalence, the lemma is proved. If not, then $A_j \to B' \to A_j$ is an equivalence, since $\operatorname{End}_C A_j$ is local. Then $B' = C' \oplus C''$ where C' is the image of A_j. Hence $A_j \to C'$ is an equivalence. Apply the X-lemma to the following diagram:

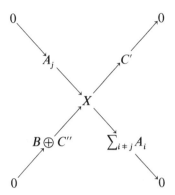

in order to obtain $B \oplus C''$ isomorphic to $\sum_{i \neq j} A_i$. Then $B \to \sum_{i \neq j} A_i / \sum_{i \in S - \{j\}} A_i$ is not monic. If $|S| = 1$, this is a contradiction. Otherwise, by induction, there is an $i \in S - \{j\}$ such that $A_i \to B \to A_i$ is an equivalence. $\square$

21.6 Unique Decomposition Theorem. *Let C be an Abelian category, let $A = \sum_{i \in I} \oplus A_i$, and $B = \sum_{j \in J} \oplus B_j$ be Azumaya diagrams for two objects A and B. Assume that either I is finite, or that C is AB5. Then, $A \approx B$ if and only if there is a bijection $p: I \to J$ and for each $i \in I$ there is an equivalence $g_i: A_i \approx B_{p(i)}$.*

Proof (Azumaya [50]; cf. Swan [68]). It suffices to prove the theorem under the assumption that $A = B$. If $n = |I|$ is finite, then the result follows by induction since, by 21.3, $A_1 \approx B_j$ for some j, and then by the cancellation lemma, $A/A_1 \approx A/B_j$ (cf. 18.18).

Now assume that I is infinite, and let M be one of the indecomposable summands, let a be the number of $i \in I$ such that $A_i \approx M$, and let b be the number of $j \in J$ such that $B_j \approx M$. Claim: $a = b$. If a is finite, this is clear by the argument for finite I. Otherwise, for each $i \in I$ the set S_i consisting of all $j \in J$ such that $A_i \to B_j \to A_i$ is an equivalence is a finite set, since there is a finite set S containing S_i for which $A_i \to \sum_{j \in J} B_j / \sum_{i \in S} B_i$ is not a monic. Since B_j is indecomposable, then $B_j \approx A_i \ \forall j \in S_i$. Since there is at least one A_i such that $A_i \to B_j \to A_i$ is an equivalence, then

$$\bigcup_{\{i \in I \mid A_i \approx M\}} S_i = \{j \in J \mid B_j \approx M\}$$

Thus, $b \leq a$. By symmetry, $a \leq b$, and hence $a = b$. The rest of the proposition is a consequence. $\square$

21.7 A Definition. Let I be an ideal in a ring R. Then R is I-**adically complete**, or briefly, I-**adic complete**, if the canonically map $R \to \varprojlim_n R/I^n$ is an equivalence.

Equivalently, R is I-adic complete if, in the topology on R determined by letting the ideals $\{I^n\}_{n=1}^{\infty}$ be a base of open neighborhoods of 0, every Cauchy sequence converges to a unique limit.

A necessary condition for R to be I-adic complete is residual nilpotence of I, and a sufficient condition is nilpotence.

The easiest example (other than when I is nilpotent) is the ring of formal power series $k\langle x \rangle$ over a field k. (Every element of $k\langle x \rangle$ is a limit of a sequence of polynomials and conversely.) Actually, these rings characterize "equicharacteristic" discrete valuation rings with a unique prime p, which are (p)-adic complete. (Teichmüller's theorem. See Zariski and Samuel [60, p. 307, Corollary] for I.S. Cohen's generalization for complete regular local rings.)

21.7B   Proposition. *If R is I-adic complete, then units and idempotents of R/I lift; in particular, I must be contained in the Jacobson radical of R.*

Proof. Consider the sequence of projections

$$R \to \cdots \to R/I^{n+1} \to R/I^n \to \cdots \to R/I \to 0.$$

Let $x \in R$ such that $\bar{x}_1$, the image of x in R/I, is a unit. Let $\bar{x}_n$ be the image in R/I^n. By 18.4 and 18.6, $\bar{x}_n$ is a unit in R/I^n. Let $\to y_n \to y_{n-1} \to \cdots \to y_1$ be the sequence of inverses and let $y = \lim y_i$. By completeness $y \in R$, and yx and xy have image 1 in every R/I^n. Therefore $xy = yx = 1$ and x is a unit. By 18.4, $I \subseteq \operatorname{rad} R$.

If $e_1 \in R/I$ is idempotent, then, since I^{n-1}/I^n is nilpotent, by 18.21, there exists idempotents $\{e_n\}_{n=1}^{\infty}$ such that $e_{n-1} \in R/I^{n-1}$ lifts to an idempotent e_n of R/I^n. Let $e = \lim e_n$. Then $e^2 - e = 0$ since $e_n^2 - e_n = 0$ for all n. Thus e is an idempotent with image e_1 in R/I.   □

21.8   Corollary. *Let C be an abelian category, and let A be an object for which there exists an ideal I of $R = \operatorname{End}_C A$ such that R is I-adic complete, and R/I left or right Artinian. Then A has a finite Azumaya diagram.*

Proof. Since R is I-adic complete, then $I = \operatorname{rad} R$, and R is lift/rad semilocal by 21.7. The corollary then follows from 18.26.   □

A complete local ring R is a commutative Noetherian local ring with maximal ideal J, and J-adic complete.

21.9   *Exercise* (Swan [60, p. 566]). If R is a complete local ring, then every finitely generated R-module has a finite AD (cf. 27.22).

If M is an R-module, and if α is a cardinal number, then $\dim M = \alpha$ will signify that M has an Azumaya diagram $\coprod_{i \in I} M_i$, where $|I| = \alpha$. The term dimension is much used in mathematics; we have encountered it thrice in this book: (1) vector space dimension; (2) Goldie dimension; (3) homological (injective or projective) dimension. The present use generalizes case (1), whereas inj. dim, proj. dim, or gl. dim modifies (3). There is some overlap with (2) (see the exercises to follow).

In the next corollary the statement $\dim P \leq n$ implies that P has an Azumaya diagram $\prod_{j=1}^m P_j$, and that $m \leq n$. Furthermore, if $\prod_{i=1}^n M_i$ is any Azumaya diagram for P, then the theorem implies that $n = m$ and that each $M_i \approx \text{some } P_j$.

21.10   Corollary. *Let M be an R-module of finite dimension n, and let P be a summand of M. Then, P has an AD, and $\dim P \leq n$. Equality holds if and only if $P = M$.*

Proof. Exercise.

21.11   Corollary. *If a module M satisfies* (acc) $\oplus$, *then M is a finite direct sum of indecomposable modules. If, in addition M has finite length, then M has finite dimension.*

Proof. The first statement is an exercise 7.18. If length $M < \infty$, then $M \approx \coprod_{i=1}^{n} M_i$, where M_i ($i = 1, \ldots, n$) is an indecomposable module of finite length. Since the endoring of M_i is a local ring 18.11, then M has finite dimension. $\square$

Summands of Azumaya Diagrams

If $M = \coprod_{i \in I} M_i$ is an Azumaya diagram, is every summand of M an Azumaya module? In general, the answer is unknown; however, the theorem of Matlis-Papp 20.6 gives an affirmative answer when M is an injective module over a Noetherian ring, and 21.10 when $\dim M < \infty$. In this paragraph these results to an arbitrary injective module over any ring, and also for the case where the indecomposable summands are countably generated injective (Faith-Walker [67]).

21.12   Lemma. *Let Q be a direct sum of indecomposable injective modules* $\{Q_i | i \in I\}$, *and let S be a direct summand of Q, $Q = S \oplus T$. Then:*

(a) *If M is a finitely generated submodule of S, then S contains an injective hull $\hat{M}$ of M, and $\dim \hat{M} < \infty$.*

(b) *If S_1 is any summand of S, $S = S_1 \oplus N$, and if $y \in S$, then there exists a summand T_1 of N such that $\dim T_1 < \infty$ and $y \in S_1 \oplus T_1$.*

Proof. (a) Let $x_1, \ldots, x_n$ generate M. Each x_i is contained in the sum P of finitely many of the Q_i's, whence $M \subseteq P$. Since P is injective, it contains an injective hull P_1 of M. Since $\dim P < \infty$, and P_1 is a summand, then $\dim P_1 < \infty$. The projection of P_1 into S along T fixes M, so it is a monomorphism since M is essential in P_1. The image of P_1 is clearly an injective hull $\hat{M}$ of M, so $\dim \hat{M} < \infty$.

(b) Write $y = s + t$, with $s \in S_1$, $t \in N$. Applying (a) to the summand N of Q, and the module $tR \subseteq N$, we obtain a summand T_1 of N, containing tR that has finite dimension, and $S_1 \oplus T_1$ contains y as required. $\square$

21.13   Proposition. *Let Q be a direct sum of indecomposable injective modules, and let S be a summand.*

(a) *If S is countably generated, then S is completely decomposable.*

(b) *If S is the injective hull of a finitely generated submodule, then $\dim S < \infty$.*

(c) *If S contains a finitely generated submodule M such that S/M is countably generated, then S is completely decomposable.*

Proof. (a) Let $x_1, x_2, \ldots, x_n, \ldots$ be a countable generating set of S. By the lemma, S contains a chain $S_1 \subseteq \cdots \subseteq S_n \subseteq \cdots$ of summands such that $x_1, \ldots, x_n \in S_n$, and such that $\dim S_n < \infty$ $\forall n$. Clearly $S = \bigcup_{n=0}^{\infty} S_n$, where $S_0 = 0$. Since S_n is a summand of S_{n+1}, $P_{n+1} = S_{n+1}/S_n$ has finite dimension $\leq \dim S_{n+1}$. Now $S_{n+1} \approx \coprod_{i=0}^{n+1} P_i$, and $S \approx \coprod_{i=0}^{\infty} P_i$ is completely decomposable.

(b) follows immediately from the lemma, and (c) is then a consequence of (a) and (b). $\square$

21.14 Corollary. *Let Q be a direct sum of (any number of) countably generated indecomposable injective modules. Then any direct summand S of Q is completely decomposable.*

Proof. By Kaplansky [58a], S itself is a direct sum of countably generated modules. Hence, it suffices to prove the corollary for the case when S is countably generated. But this case follows from (a) of the last proposition. □

The last corollary has been generalized by Warfield [69b, 7.1, p. 275].

21.15 Theorem. *Let Q be a completely decomposable injective right R-module, and write $Q = \sum_{i \in I} \oplus Q_i$, where $\{Q_i | i \in I\}$ is a set of indecomposable (injective) modules.*

(a) *Let S be any completely decomposable submodule which is a direct sum $\sum_{j \in J} \oplus S_j$ of indecomposable injective submodules $\{S_j | j \in J\}$. Then S is injective, and $|J| \leq |I|$.*

(b) *Any direct summand P of Q is completely decomposable.*

Proof. For any subset A of I, let $Q_A = \sum_{a \in A} Q_a$, and let π_A be the projection of Q on Q_A having kernel Q_{I-A}. A *homogeneous component* of Q is defined to be a submodule Q_A, where each A is the set of all those $b \in I$ for which Q_b is isomorphic to some fixed Q_a. Then $Q = \sum_A \oplus Q_A$, and if S is the submodule in the statement of (a), then also S is a direct sum of its homogeneous components: $S = \sum \oplus S_K$.

(a) If $k \in K$, then S_k, being injective, is a summand of Q, and, being indecomposable, is isomorphic to Q_a for some $a \in I$. We first show:

(i) S_K *is isomorphic to a submodule of Q_A, the homogeneous component of Q determined by a.* Suppose for the moment that $S_K \cap \ker \pi_A \neq 0$. If y is a nonzero element in this intersection, then the submodule T that it generates is contained in $S_{K'}$, where K' is a finite subset of K, and also $T \subseteq Q_B$, where B is a finite subset of $I - A$. (Recall that $\ker \pi_A = Q_{I-A}$.) Since $S_{K'}$ (resp. Q_B) is injective, it contains an injective hull E (resp. F) of T. Since $S_{K'}$ is finitely, completely decomposable and is homogeneous, so is E. Since E and F are isomorphic, this means that E contains submodule G, which is isomorphic to S_k, and hence to Q_a, which is also isomorphic to Q_b for some $b \in I - A$. But, by the definition of A, this is impossible, since $A = \{c \in I | Q_c \approx Q_a\}$. This contradiction shows that $\ker \pi_A \cap S_K = 0$, so π_A maps S_K monomorphically into Q_A. Hereafter, φ_K denotes a fixed monomorphism $S_K \to Q_A$.

We next show: (ii) S_K *is injective.* If $|A|$ is infinite, then Q_a is countably $\sum$-injective, hence by 20.2, Q_a is $\sum$-injective. This implies that S_K is injective. Next suppose $|A| = n$ is finite. If $|K| > n$, then S_K contains as a summand a submodule T, which is a direct sum of $n+1$ copies of Q_a. Furthermore, T is then injective, so $\varphi_K(T)$ is a summand of Q_A, which is a direct sum of $n+1$ copies of Q_a, violating the Krull-Schmidt-Remak theorem. Therefore, $|K| < \infty$, so S_K is injective in this case, too. This proves (ii), and also (iii).

(iii) *If W is a submodule of Q, and if W is a direct sum of isomorphic indecomposable injective submodules, then W is injective.* [(iii) is the homogeneous case of (a), that is, the case where $W = S = S_K$.]

In order to simplify notation, let $X = Q_A$ and $U = \varphi_K(S_K) = \operatorname{im} \varphi_K$. Since $U \approx S_K$, U is injective, so $X = U \oplus V$ for some submodule V. By Zorn's lemma, there exists a maximal independent set P of indecomposable injective submodules of V. The sum W of the submodules in P is direct. Since $X = Q_A$ is homogeneous, so is W, and W is therefore injective by (iii). Since W is therefore a summand of V, in order to avoid contradicting maximality of W, necessarily $W = V$. Together the completely decomposable modules U and V yield a decomposition of $X = U \oplus V$ into a direct sum of indecomposable modules, and the unique decomposition theorem then implies that the number of indecomposable summands of $U = \varphi_K(S_K)$ is less than $|A|$; that is, $|K| \leq |A|$. Thus, S_K is isomorphic to a summand of Q_A, and therefore $S = \sum_K \oplus S_K$ is isomorphic to a summand of Q. This proves that S is injective. Furthermore, since J (resp. I) is the disjoint union of the various K (resp. A), then $|K| \leq |A|$ implies $|J| \leq |I|$. This proves (a), and (b) is proved in exactly the same way that we proved V is completely decomposable. $\square$

The argument in the proof that showed that V is completely decomposable implies that $|K| \leq |A|$, a fact that has the following direct proof: Each S_K is contained in a finite direct sum of the $\{Q_a\}$, and this fact gives a mapping of K into the set of all finite subsets of A, whence $|K| \leq |A|$ when $|A|$ is infinite. [The proof of (ii) settled the inequality when $|A| = n < \infty$.]

Exercises of Chapter 21

1. If C is an Abelian category, and if objects, A, B, and B' have AD's, then $A \oplus B \approx A \oplus B'$ if and only if $B \approx B'$. Any equivalence $B \approx B'$ is extendable to an equivalence of $A \oplus B$ and $A \oplus B'$.

2. If A and B have Azumaya diagrams in an $AB5$ category, then there is an equivalence $A^{(I)} \approx B^{(J)}$ if and only if $|I| = |J|$ and $A \approx B$. (This also holds for finite sets I and J, for any abelian category.)

3. A ring R is right Noetherian iff R satisfies (acc)$\oplus$ and the a.c.c. on essential right ideals. Give an example of a commutative non Noetherian ring with the a.c.c. on essential right ideals.

4. A ring R is right fully Goldie if every factor ring is right Goldie. A commutative ring R is Noetherian iff fully Goldie (Camillo [75]).

5. (Yamagata [73]) The following conditions on a ring R are equivalent:

(I) R satisfies the a.c.c. on coirreducible right ideals (that is, on right ideals I such that R/I is uniform).

(II) R satisfies the a.c.c. on essential, coirreducible right ideals.

(III) Every completely decomposable module has a unique decomposition (as a direct sum of indecomposable injectives; unique in the sense of 21.6).

6. (Warfield [72b]) If R is an exchange ring, then every projective right (left) module is isomorphic to a direct sum of indecomposable right (left) ideals generated by idempotents.

7. (Warfield) A module M is said to have the n-exchange property for a finite integer n provided that whenever there is a direct decomposition

$$N = M \oplus K = \oplus_{i=1}^{m} A_i$$

with $m \leq n$, there is a subset $m' \subseteq m$ such that

$$N = M \oplus (\oplus_{i \in m'} A_i).$$

Then, M has the finite exchange property if this holds for all finite cardinals n. A right R-module M has the finite exchange property iff $\text{End}_R M$ has the finite exchange property on both sides. Moreover, R has the finite exchange property iff R has the exchange property.

8. (Monk [72]) A ring R has the finite exchange property iff given $a \in R$, there exist b and c in R such that

$$bab = b$$

and

$$c(1-a)(1-ba) = 1 - ba.$$

Notes for Chapter 21

The universal question, raised in various categories, is whether or not a summand of unique decomposition has a unique decomposition. This question is answered affirmatively for finite decompositions by the Krull-Schmidt theorem, and in a number of other cases discussed in this chapter. The general case, was raised by Matlis [58], solved by him for injective modules over Noetherian rings, and 20.15 dispenses with the Noetherian hypothesis. (Also see the dcs condition defined below.)

The Krull-Schmidt theorem has been discussed by Reiner [61, 62, 70]. It fails to hold for direct sums of indecomposable modules over integral group rings and rings of algebraic integers. For example, for any two nonzero ideals I and J in the ring R of algebraic integers of an algebraic number field (defined in (I, 12.19.2, p. 457)), or more generally in any Dedekind domain R (defined in (I, p. 457)), there holds

$$I \oplus J \approx R \oplus K$$

where K is isomorphic to the product IJ of the two ideals. (See Kaplansky [52, Theorem 2], where this theorem is attributed to Steinitz.) Since R is an integral domain, the summands are uniform, hence indecomposable, but $I \approx R$ iff I is a principal ideal. Similarly for projective modules of the group ring RG of any finite group G such that no prime divisor of $|G|$ is a unit of R (Swan [60]). (The integral group ring $\mathbb{Z}G$ is such an example.)

The exchange property has been defined in Chapter 19, *sup.* 19.21, and results of Crawley-Jonnson [64], Fuchs [69], and Warfield [69a] commented on in Exercise 19.21. Also Warfield [69a] applies results of Crawley and Jonnson in obtaining some generalizations of the theorems of Faith and Walker given in this chapter. In particular Proposition 21.14 holds for Q any direct sum of countably generated injectives: any two direct sum decompositions have isomorphic refinements.

Over rings with AD, this is, semilocal lift/rad rings, the projectives behave honorably: they are direct sums of principal indecomposable right ideals (=prindecs) (Theorem of Bass [60], proved in Chapter 22, for perfect rings, and in Müller [70a] and Warfield [72b] for semiperfect rings).

A direct decomposition of a module $M = \bigoplus_{i \in I} M_i$ is said to **complement direct summands** (=cds) provided that for each direct summand P of M there is a subset J of the index set I such that $M = P \oplus Q$, with $Q = \bigoplus_{j \in J} M_j$. Then M is said to have cds. For this to happen, each M_i must be indecomposable $\forall i \in I$. Along the lines of the proof of 20.15, one shows for any completely decomposable injective module that this direct decomposition complements direct summands. (See, for example, Anderson and Fuller [72, p. 251, Proposition 7] and Warfield [69b].) By 20.6, every injective right module satisfies cds iff the ring is right Noetherian; and every projective right module has cds iff R is right perfect (Anderson and Fuller *l.c.*). [Thus, for a local ring R, although every projective P is free, and hence every direct summand of a projective is a direct sum of indecomposable projectives, nevertheless, P does not in general have cds.]

For any right R-module M, and $S = \text{End}_R M$, the functor

$$\text{Hom}_R(M, \ ): \text{mod-}R \rightsquigarrow \text{mod-}S$$

induces on the category $\sum_M$ of direct summands of M an equivalence with the category of projective right S-modules, so the Fuller-Anderson theorem shows that every module in $\sum_M$ has cds iff S is right perfect. (This observation was made by Tachikawa and Fuller, in a note in Fuller [73, p. 178], together with the remark that every module over an Artinian FFM ring has cds, a result generalizing the one stated by Fuller [73].

As stated in Exercise 5, Yamagata [73] characterizes when a completely decomposable module has a unique decomposition; and the same condition also characterizes when every right module that is completely decomposable (as a direct sum of indecomposable injectives) satisfies dcs. (This happens iff R satisfies the a.c.c. on coirreducible right ideals, a condition considerably weaker than right Noetherian.)

As pointed out by Camillo [75] (and others) many theorems for Noetherian rings go over for fully Goldie rings, and, in view of Camillo's theorem stated in Problem 4, the question arises: are fully Goldie fully bounded rings Noetherian?

Also see Notes for Chapter 20.

References

Anderson and Fuller [72], Azumaya [50], Bass [60], Camillo [75], Crawley-Jonnson [64], Faith and Walker [67], Fuchs [69], Fuller [73], Harada [71], Harada and Sai [70], Kaplansky [58a], Matlis [58], Monk [72], Swan [60, 68], C.L. Walker [66], Warfield [69b, c, 72a, b], Yamagata [73], Zariski-Samuel [60].

Chapter 22. Projective Covers and Perfect Rings

A morphism $f: A \to B$ of R-modules is said to be **minimal** provided that $\ker f$ is a superfluous submodule of A. For example, for a right ideal I, the canonical map $R \to R/I$ is superfluous if and only if $I \subseteq \operatorname{rad} R$ 18.3. A module A is a **projective cover** (proj. cov.) of B provided that A is projective and there exists a minimal epimorphism $A \to B$. This notion is dual to that of injective hull, and yet, although each R-module has an injective hull, projective covers of modules may fail to exist. For example, as is shown in this chapter, a necessary condition that every right R-module have a projective cover is that $R/\operatorname{rad} R$ be semisimple, and $\operatorname{rad} R$ be a nil ideal.

A ring R is said to be right **perfect** (resp. **semiperfect**) provided that every right R-module (resp. every cyclic right R-module) has a projective cover. Left (semi)perfect rings are symmetrically defined.

An equivalence $T: \operatorname{mod}\text{-}R \rightsquigarrow \operatorname{mod}\text{-}S$ of abelian categories induces lattice isomorphism between the lattice of subobjects of an object $X \in \operatorname{mod}\text{-}R$ and that of $TX \in \operatorname{mod}\text{-}S$. It follows that superfluity of a submodule, and hence projective cover, are categorical properties. Since finite generation of a module in $\operatorname{mod}\text{-}R$ is also categorical, semiperfect and perfect are Morita invariants. Consider the simplest case, say where R is a local ring. Any proper right ideal I of R is contained in $\operatorname{rad} R$, and is superfluous 18.4, so the canonical map $R \to R/I$ is a projective cover. This shows that any local ring R is semiperfect, and so is the matrix ring R_n, for any n.

The following generalization of nilpotency of an ideal proves fundamental in the study of projective covers: a subset I of a ring R is said to be **left vanishing** provided that, for each infinite sequence $\{a_n\}$ of elements of I, there exists an integer k such that $a_k a_{k-1} \ldots a_1 = 0$. The concept right vanishing is defined by symmetry[1]. Thus, a right (left) vanishing ideal is a nil ideal. Moreover, the concept is related to that of an ideal being essentially nilpotent since an ideal N contains a right vanishing right ideal J which is right essential in N iff the same statement holds with J a nilpotent right ideal (Shock [71c]; compare 19.13 C).

A theorem 22.29 of Bass [60] states that the following conditions on a ring R are equivalent:

   I. *R is right perfect.*

   II. *$R/\operatorname{rad} R$ is semisimple and $\operatorname{rad} R$ is left vanishing.*

---

[1] Left vanishing is what is called right **T-nilpotent** (T for transfinite) by Bass [60]. The left signifies that the sequence $\{a_n \ldots a_1\}$ grows to the left, and the vanishing indicates the *eventual* vanishing of the terms, that is, left vanishing is short for eventually left vanishing.

III. *Every nonzero left R-module has nonzero socle, and R has a bounded number of orthogonal idempotents.*

IV. *R satisfies the* d.c.c. *on principal left ideals.*

22.0 *Exercise.* (a) Show that I, II, and III are Morita invariants.

(b) Prove the equivalence of I–IV for a local ring.

(c) Prove that any finite product of (semi)perfect rings is (semi)perfect. Show that an infinite product of rings is never semiperfect.

Besides the theorems of Bass just discussed, the main theorems of this chapter are: (1) Björk's theorem 22.30 stating that the d.c.c. on cyclic submodules of any module implies that on the finitely generated submodules; (2) a theorem of Chase characterizing when products of projectives are projective 22.31, and (3) Loewy series and the Brauer theory of blocks for left perfect rings 22.33 ff.

Björk's theorem shows that IV implies the d.c.c. on finitely generated left ideals.

Radical of Projective Modules

22.1 A **Proposition.** *If P is a nonzero projective R-module, then*

(1) $$\operatorname{rad} P = P \cdot \operatorname{rad} R \neq P.$$

Proof. First

(2) $$\operatorname{rad}\left(\sum_{i \in I} \oplus A_i\right) = \sum_{i \in I} \oplus \operatorname{rad} A_i$$

for any module $A = \sum_{i \in I} \oplus A_i$. For if $A \supseteq B$, then $\operatorname{rad} A \supseteq \operatorname{rad} B$, so the left side of (2) contains the right. If

$$a = \sum a_i \in \sum_{i \in I} \oplus A_i$$

and if $a_i \notin \operatorname{rad} A_i$ for some i, then (a_i) is not superfluous in A_i, and one sees immediately that a is not superfluous in A. Thus (2) holds, proving that $\operatorname{rad} F = F \cdot \operatorname{rad} R$ when F is free. In the general case, $P \oplus Q = F$ for some free module F, and

$$\operatorname{rad} F = \operatorname{rad} P \oplus \operatorname{rad} Q = F \cdot \operatorname{rad} R = P \cdot \operatorname{rad} R \oplus Q \cdot \operatorname{rad} R,$$

whence $\operatorname{rad} P = P \cdot \operatorname{rad} R$. It remains to show that $\operatorname{rad} P \neq P$. Let $\{x_i | i \in I\}$ be a free basis of F, and assume $0 \subset P \subset F \cdot \operatorname{rad} R$. Let x be any nonzero element of P, and assume that $\{x_i | i \in I\}$ has been chosen such that in an expression

$$x = \sum x_i r_i \qquad\qquad (r_i \in R)$$

the number of nonzero coefficients r_i is the smallest possible integer n, and so that $r_i \neq 0$, $i = 1, \ldots, n$. Now $F = P \oplus Q$, so we can write

$$x_i = p_i + q_i \qquad\qquad (p_i \in P,\ q_i \in Q,\ i = 1, \ldots, n).$$

Also

$$p_i = \sum_{j=1}^{n} x_j s_{ij} \qquad\qquad (s_{ij} \in R,\ i, j = 1, \ldots, n).$$

Since $P \subseteq F \cdot \operatorname{rad} R$, necessarily $s_{ij} \in \operatorname{rad} R$ $(i, j = 1, \ldots, n)$. Since

$$x = \sum_{i=1}^{n} x_i r_i = \sum_{i=1}^{n} p_i r_i + \sum_{i=1}^{n} q_i r_i \in P$$

necessarily $\sum_{i=1}^{n} q_i r_i = 0$. Thus,

$$x = \sum_{i=1}^{n} p_i r_i = \sum_{j, i=1}^{n} x_j s_{ij} r_i = \sum_{i=1}^{n} x_i r_i$$

so $r_1 = \sum_{i=1}^{n} s_{i1} r_i$, or $(1 - s_{11}) r_1 = \sum_{i=2}^{n} s_{i1} r_i$. Since $s_{11} \in \operatorname{rad} R$, $1 - s_{11}$ is a unit of R. If $s = (1 - s_{11})^{-1}$, then

$$r_1 = \sum_{i=2}^{n} s \, s_{i1} r_i$$

and

(3) $$x = \sum_{i=2}^{n} (x_i + x_1 \, s \, s_{i1}) r_i.$$

But $\{x_1, x_2 + x_1 \, s \, s_{21}, \ldots, x_n + x_1 \, s \, s_{n1}\}$ are linearly independent over R, and together with $\{x_i \mid i > n\}$ form a basis of F such that the expression (3) for x in terms of this basis has only $n - 1$ nonzero "coefficients" $r_2, \ldots, r_n$. This contradiction proves that $P \neq P \cdot \operatorname{rad} R$. $\square$

This proof by Hinohara [62] is modeled after an argument of Kaplansky (see also Bass [60]).

The next is a theorem of Eilenberg [56]–Kaplansky [58].

22.1 B   Corollary. *If R is a local ring, then any finitely generated projective R-module $P \neq 0$ is a free R-module. (In fact any projective R-module is free.)*

Proof. We first show that *any projective R-module is a generator of the form $R \oplus A$ for some $A \in M_R$.* Let $T = \operatorname{trace} P_R$, and $J = \operatorname{rad} R$. Since $P = PT$ by 12.1, the proposition implies that $T \nsubseteq J$, so there exist $p \in P$ and $f \in P^* = \operatorname{Hom}_R(P, R)$ such that $t = f(p) \notin J$. Since R/J is a simple R-module, then $R = tR + J$. Since J is a superfluous submodule of R by (b) of 18.3, necessarily $R = tR$. Thus, $f: P \to R$ is surjective; that is, $P \approx R \oplus A$ as claimed.

Now assume that, in addition, P is finitely generated. Then P/PJ is a vector space over the field R/J of finite dimensions n. Thus, P/PJ has at most n summands (isomorphic to R/J). Let k be an integer ≥ 1 and B an R-module such that $P \approx R^k \oplus B$. Then $P/PJ \approx (R/J)^k \oplus A_k/A_k J$, and consequently $k \leq n$. Let k be the maximal integer such that $P \approx R^k \oplus B$ for some module B. Since B is projective, the preceding result shows that $B \approx R \oplus C$ for some module C, if $B \neq 0$. But then $P \approx R^{k+1} \oplus C$, a contradiction that shows that $B = 0$, and that $P \approx R^k$ is free. (Actually $k = n$.)

The general case is a theorem of Kaplansky [58]. $\square$

Projective B-Objects

For the next several results, recall the definition 18.3 of a *B-object* of mod-R, namely an object M such that every proper submodule is contained in a maximal submodule.

22.2 Proposition. *If P is a projective R-module, and* $S = \text{End } P_R$, *then*

$$\text{rad } S = \{f \in S \mid P \bigcirc \text{im} f\}.$$

Proof. Dual to that of 19.27(a). $\square$

22.3 Corollary. *If P is a projective B-object of* mod-R, *then the radical of the endomorphism ring of P is*

$$\text{rad}(\text{Hom}_R(P, P)) = \text{Hom}_R(P, P \cdot \text{rad } R).$$

(This holds if P is finitely generated.) $\square$

22.4 Proposition. *Any row exact*

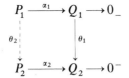

diagram, where P_1 *and* P_2 *are projective R-modules can be completed by a map* $\theta_2 : P_1 \to P_2$.

(a) *If* $P_2 \bigcirc \ker \alpha_2$, *then* θ_2 *is epic when* θ_1 *is.*

(b) *If both* $P_2 \bigcirc \ker \alpha_2$ *and* $P_1 \bigcirc \ker \alpha_1$, *then* θ_2 *is an isomorphism when* θ_1 *is.*

Proof. Let $\varphi_1 = \theta_1 \alpha_1$. Projectivity of P_1 then yields $\theta_2 : P_1 \to P_2$ such that $\alpha_2 \theta_2 = \varphi_1$, completing the diagram. If θ_1 is epic, so is φ_1. Thus, if $y \in P_2$, then $\alpha_2(y) = \varphi_1(x) = \theta_1 \alpha_1(x)$ for some $x \in P_1$. Since

$$\theta_1 \alpha_1(x) = \alpha_2 \theta_2(x) = \alpha_2(y)$$

then $\theta_2(x) - y \in \ker \alpha_2$. Since therefore $P_2 = \text{im } \theta_2 + \ker \alpha_2$, then $P_2 \bigcirc \ker \alpha_2$ implies θ_2 is epic, proving (a). Since P_2 is projective, $P_1 = A \oplus \ker \theta_2$, where $A \approx P_2$. Assuming θ_1 is monic, we have $\ker \theta_2 \subseteq \ker \alpha_1$. Now $P_1 \bigcirc \ker \alpha_1$ implies $P_1 \bigcirc \ker \theta_2$. Since $\ker \theta_2$ is a summand, necessarily $\ker \theta_2 = 0$. $\square$

22.5 Corollary. *Let* P_i *be a projective module which is a B-object of* mod-R *(e.g., let* P_i *be finitely generated projective)* $i = 1, 2$. *If* K_i *is a submodule of* rad P_i, $i = 1, 2$, *then*

(a) $P_1/K_1 \approx P_2/K_2 \Rightarrow P_1 \approx P_2$.

(b) $P_1/\text{rad } P_1 \approx P_2/\text{rad } P_2 \Leftrightarrow P_1 \approx P_2$.

The corollary follows immediately from 22.4, 22.1, and the fact 18.3 that rad P_i is superfluous. $\square$

22.6 Proposition. *Let P be a projective B-object of* mod-R, *and let* $S = \text{End } P_R$. *Then:*

(a) $S/\text{rad } S \approx \text{End}(P/P \cdot \text{rad } R)_R$.

(b) *If* $P/(P \cdot \text{rad } R)$ *is semisimple, then* $P/(P \cdot \text{rad } R)$ *is simple if and only if S is a local ring.*

Proof. Consider the diagram

$$P \xrightarrow{\quad \alpha \quad} P/PJ \longrightarrow 0$$

$$\left\downarrow \theta \qquad \left\downarrow \theta' \right. \qquad\qquad (J = \operatorname{rad} R)$$

$$P \xrightarrow{\quad \alpha \quad} P/PJ \longrightarrow 0$$

where $\alpha: P \xrightarrow{\text{nat}} P/PJ$. If $\theta \in S$, then since $\theta(PJ) = (\theta P)J \subseteq PJ$, θ induces a map $\theta': P/PJ \to P/PJ$. It is clear that the correspondence $\theta \to \theta'$ is a ring homomorphism

$$h: S \to \operatorname{End}(P/PJ)_R.$$

Then 22.4 implies that h is a ring epimorphism. By the commutativity of the diagram, $\theta \in \ker h$ if and only if im $\theta \subseteq PJ$. Hence by 22.3, $\ker h = \operatorname{rad} S$, proving (a).

(b) By Utumi's theorem 19.27, semisimplicity of P/PJ implies that $\operatorname{End}(P/PJ)_R$ is a regular ring. The following statements are easily seen to be equivalent: (i) P/PJ is simple; (ii) $\operatorname{End}(P/PJ)_R$ contains no idempotents; (iii) $\operatorname{End}(P/PJ)_R$ is a field. Thus (a) implies that S is a local ring. $\quad\square$

B-Rings and Left Vanishing Ideals

Next we investigate a class of rings, called *right B-rings*, so designated in honor of Bass [60] who initiated the study. These are rings with the property that every nonzero right module has a maximal submodule, or equivalently, every right module is a *B*-object 18.3.

22.7 A *Exercise.* (a) Any ring with a Noetherian cogenerator for mod-R is a right *B*-ring (cf. 7.41).

(b) Any semiprimary ring is a right and left *B*-ring.

(c) A module M with a projective cover P is a *B*-object. Hence any right perfect ring is a right *B*-ring.

22.7 B **Proposition** (Bass [60]). *If R is a right B-ring, then* rad R *is left vanishing.*

Proof (Rosenberg and Zelinsky [61]). Let $x_1, \ldots, x_n, \ldots$ be a countable basis of a free module P, let $a_1, \ldots, a_n, \ldots$ be an infinite sequence of elements of rad R, and let f be the element of $S = \operatorname{End} P_R$ mapping $x_i \mapsto x_{i+1} a_i$, $i = 1, 2, \ldots$. By 22.3, $f \in \operatorname{rad} S$, hence $(1-f)$ is a unit in S. Let $y = (1-f)^{-1} x_1$, and write

$$y = \sum_{i=1}^{\infty} x_i b_i$$

with $b_i \in R$, $b_n = 0$, $n \geq k$. Then

$$x_1 = (1-f)\, y = \left(\sum x_i b_i\right) - \sum x_{i+1} a_i b_i$$

$$= x_1 b_1 + x_2 (b_2 - a_1 b_1) + \sum_{n>2} x_n (b_n - a_{n-1} b_{n-1}).$$

Since $\{x_n \mid n \geq 1\}$ is a free basis, then $b_1 = 1$ and $b_n = a_{n-1} \ldots a_2 a_1$, $n \geq 2$. Thus $b_k = a_{k-1} \ldots a_2 a_1 = 0$. $\quad\square$

22.8   Proposition. *A right perfect ring is a right B-ring (and has left vanishing radical).*

Proof. Let M be a module such that rad $M = M$, and let $P \xrightarrow{f} M$ be a projective cover of M. If $M \neq 0$, then $P \neq 0$, so P has a maximal submodule P'. Let $f' = f|P' =$ restriction. The mapping $x + P' \mapsto f(x) + \text{im } f'$ is a homomorphism $P/P' \to M/\text{im } f'$. Since P/P' is simple, rad $M = M$ implies im $f' = M$; that is, $P = P' + \ker f$. Since $P \bigcirc \ker f$, then $P' = P$, a contradiction proving that rad $M = M$ implies $M = 0$.   □

22.9   Proposition. *If the principal left subideals of* rad R *satisfy the* d.c.c., *then* rad R *is left vanishing.*

Proof. Consider the sequences $\{a_n | n = 1, 2, \ldots\}$, $\{b_n | n = 1, 2, \ldots\}$, where, $\forall n$, $a_n \in$ rad R and $b_n = a_n \ldots a_2 a_1$. The chain of principal left subideals

$$R b_1 \supseteq R b_2 \supseteq \cdots \supseteq R b_n \supseteq \cdots$$

terminates, say $R b_k = R b_{k+1}$. Let $b = b_k$. Then $b_{k+1} = a_{k+1} b$. Since $b \in R b_{k+1}$, there exists x such that $b = x a_{k+1} b$. Since $u = x a_{k+1}$ lies in rad R, $1 - u$ is a unit of R. The identity $(1 - u) b = 0$ therefore implies $b = 0$.   □

Left Socular Rings

We next introduce terminology for a concept of Bass [60].

22.10 A   Proposition[2]. *A ring R is said to be* **left socular** *provided that the equivalent conditions hold:*

(a) *Every nonzero left module has nonzero left socle;*

(b) *Every cyclic left module R/I has a simple submodule when $I \neq R$.*

When this is so, then rad R *is left vanishing.*

Proof. Inductively define a sequence $\{S_\alpha | \alpha \in \Lambda\}$ of right ideals of R, where Λ is a well ordered set of ordinal numbers. Let 0 denote the least element of Λ, and set $S_0 = 0$. If $\alpha \in \Lambda$, then $S_{\alpha+1}/S_\alpha =$ left socle R/S_α; if $\beta \in \Lambda$ is a limit number, set $S_\beta = \bigcup_{\alpha < \beta} S_\alpha$. Since R is left socular, Λ can be chosen such that $R = S_\alpha$ for some $\alpha \in \Lambda$. For each $a \in R$, let $h(a)$ be the first element of Λ such that $a \in S_{h(a)}$. Note that $h(a)$ is not a limit number for any $a \in R$ (if $a \in S_\beta$, where β is a limit number, then $a \in S_\alpha$ for some $\alpha < \beta$). Let $h(a) = \beta + 1$, and let $J = $rad R. Since $J \cdot (S_{\beta+1}/S_\beta) = 0$, that is, $J \cdot S_{\beta+1} \subseteq S_\beta$, it follows that $h(ba) < h(a) \forall b \in J$, unless $\alpha = 0$, that is, unless $a = 0$. If $\{a_n\}$ is an infinite sequence of elements of J, and if $b_n = a_n \ldots a_1$, then

$$\cdots < h(b_n) < h(b_{n-1}) < \cdots < h(a_1)$$

is an infinite strictly decreasing set of numbers in Λ, unless $b_n = a_n \ldots a_1 = 0$ for some n. Since Λ is well ordered, the former case is impossible, so rad R is left vanishing.   □

---

[2] Since right perfect rings turn out to be left socular, I made up a rhyme connecting "socular" and "jocular", but a wiser colleague persuaded me to suppress it.

22.10 B   Summary. *The radical of a ring R is left vanishing whenever any one of the following conditions holds:*

(a) *R is a right B-ring 22.7.*

(b) *R is right perfect 22.8.*

(c) *R satisfies the* d.c.c. *on principal left ideals (or on the principal left subideals of* rad *R*) 22.9.

(d) *R is left socular 22.10 A.*

Uniqueness of the Projective Cover

The lemma shows that projective covers are unique when they exist.

22.11   Lemma. *Assume that* $P \xrightarrow{g} A$ *is a projective cover of a right R-module A, and let* $Q \xrightarrow{f} A \to 0$ *be any exact sequence, where Q is projective. Then:*

(a) *There exists a (splitting) epic* $h: Q \to P$, $Q = (\ker h) \oplus P'$, *where* $P' \approx P$, $\ker f \supseteq \ker h$, *and* $P' \bigcirc (\ker f \cap P')$.

(b) *If* $Q \xrightarrow{f} A$ *is a projective cover, then* $h: Q \approx P$.

Proof. Since Q is projective, the following diagram can be completed:

Since g is an epimorphism, $P = \operatorname{im} h + \ker g$. Then $P \bigcirc \ker g$ implies that h is epic. Since P is projective, h splits: $Q = (\ker h) \oplus P'$. Thus h induces an isomorphism $h': P' \to P$. If $x \in \ker f \cap P'$, then $g h(x) = f(x) = 0$, so $h': \ker f \cap P' \to \ker g$, proving that $h'(\ker f \cap P')$ is superfluous in $P = h'(P')$. Since h' is an isomorphism, this implies $P' \bigcirc (\ker f \cap P')$. This proves (a). In case (b), $Q \bigcirc \ker f$, so $Q = \ker f + P'$ implies $Q = P'$, whence $\ker h = 0$. $\square$

22.12 A   Lemma. (a) *If I is a right ideal of R, then* $R \xrightarrow{\text{canon}} R/I$ *is a projective cover if and only if* $I \subseteq$ rad R.

(b) *If R/I has a projective cover, and if* $I \not\subseteq$ rad R, *then I contains a nonzero summand of R; that is, I contains an idempotent of R.*

Proof. (a) is a consequence of the fact that rad R is a superfluous right ideal containing every superfluous right ideal of R (18.4).

(b) Let $P \to R/I \to 0$ be a projective cover, and by the lemma, write $R = P' \oplus H$, $H \subseteq I$. Since $I \not\subseteq$ rad R, then R is not a projective cover of R/I, R is not isomorphic to P', and $H \ne 0$ provides the desired summand of R contained in I. $\square$

22.12 B   Lemma. *Let I be an ideal of R.*

(a) *If* $P \xrightarrow{f} A \to 0$ *is a projective cover of a right R-module A, and if* $AI = 0$, *then* $P/PI \xrightarrow{\bar{f}} A \to 0$ *is a projective cover of the right R/I-module A, where* $\bar{f}$ *induces* f.

(b) *If R is (semi)perfect, so is R/I.*

Proof. (a) That P/PI is a projective $R|I$-module is trivially verified. Since $AI = 0$, $\ker f \supseteq PI$. This shows that the correspondence $x + PI \mapsto f(x)$, defined $\forall x \in P$, is a map $\bar{f} : P/PI \to A$. If S is a submodule of P, containing PI such that $P/PI = S/PI + K/PI$, where $K = \ker f$, then $P = S + K$. Then $P \supsetneq \ker f$ implies $P = S$ and $P/PI = S/SI$. This shows that K/PI is a superfluous submodule of P/PI. Since $\ker \bar{f} = K/PI$, P/PI is a projective cover of A (as an R/I-module). Obviously, (a) $\Rightarrow$ (b). $\square$

22.13   Proposition. *A semiperfect ring is semilocal.*

Proof. We may suppose that $\operatorname{rad} R = 0$, since $R/\operatorname{rad} R$ is semiperfect when R is, and then we must prove that R is semisimple. Let I be any right ideal, and let $f : P \to R/I$ denote a projective cover of R/I. Then, since $\ker f$ is superfluous, we must have $\ker f \subseteq \operatorname{rad} P = P \cdot \operatorname{rad} R$ (cf. 22.1 A and 18.3). Since $\operatorname{rad} R = 0$, then $\ker f = 0$, and therefore $R/I \approx P$ is projective. Then the canonical map $R \to R/I$ splits, so R is semisimple by 8.12. $\square$

22.14   Proposition. *Idempotents lift modulo any nil ideal.*

Proof. This is 18.21. $\square$

An ideal I of R is **locally nilpotent** provided that any ideal generated by a finite subset of I is nilpotent. If I is locally nilpotent, then, using the fact given by 22.14 that idempotents of R can be lifted mod I, direct computation reveals that in the matrix ring R_n, idempotents can be lifted modulo I_n, a fact used in the next proof. (The theorem is not used elsewhere.)

22.15   Proposition (Eilenberg [56], Swan [60]). *If I is an ideal of a ring R such that either* (a) *I is locally nilpotent, or* (b), *R is I-adically complete, then the mapping* $[P] \mapsto [P/PI]$ *is a bijection between the sets of equivalence classes of projective R-modules and projective R/I-modules.*

Proof. (a) In 22.12 we noted that P/PI is a projective over $S = R/I$ if P is projective over R. Conversely, if Q is a finitely generated projective S-module, then $Q \oplus U = S^n$ for some n, and some S-module U. Then, identifying S_n with $\operatorname{End}_S S^n$, there exists an idempotent $\bar{e} \in S_n$ such that $Q = \bar{e} S^n$. Now $S_n = (R/I)_n \approx R_n/I_n$. Identifying R_n/I_n with S_n, the idempotent $\bar{e}$ can be lifted to an idempotent $e \in R_n$. Then $P = eR^n$ is a projective R-module, and

$$P/PI \approx eR^n/eR^n I = eR^n/eI^n \approx \bar{e}(R/I)^n \approx \bar{e} S^n = Q.$$

$[(R^n/I^n) \approx (R/I)^n$ under $(a_1, \ldots, a_n) + I^n \to (a_1 + I, \ldots, a_n + I).]$

Thus the mapping $\varphi : P \to P/PI$ is surjective. Finally, let $P/PI \approx Q/QI$, where P, Q are finitely generated projective. Since $I \subseteq \operatorname{rad} R$, 22.5 implies $P \approx Q$, so φ is a bijection.

(b) By the proof of (a), it remains only to show that idempotents of R_n/I_n lift. To do this, it suffices by 21.7 to show that R_n is I_n-adically complete. The proof of this, in turn, depends on the fact that $(IK)_n = I_n K_n$ for ideals I and K of R, which follows from the correspondence theorem 4.6 (cf. 4.31). Thus, $I^m{}_n = I_n{}^m$, for any m and n, and

$$R_n \xrightarrow{\text{canon}} \varprojlim R_n/I^m{}_n \approx \varprojlim (R/I^m)_n$$

is an isomorphism since R is I-adically complete. Hence, R_n is I_n-adically complete, as required. □

22.16 *Exercise.* Let Proj. $\mathcal{M}_R$ denote the full subcategory of $\mathcal{M}_R$ whose objects are the projective modules of $\mathcal{M}_R$. Show that φ: Proj. $\mathcal{M}_R \to$ Proj. $\mathcal{M}_{R/I}$ of 22.15 is a category isomorphism, where if $f: P \to Q$, then $\varphi(f): P/PI \to Q/QI$ is the induced map $(x + PI) \mapsto (f(x) + QI)$.

22.17 **Lemma.** *Let I be an ideal of R contained in* rad R, *and assume that idempotents of R/I can be lifted. Let π be the map $R \xrightarrow{\text{canon}} R/I$. If g is an idempotent of R, then any idempotent of R/I orthogonal to $\pi(g)$ lifts to an idempotent in R orthogonal to g.*

Proof. If $y \in R/I$ is an idempotent orthogonal to $\pi(g)$, then by the lifting assumption, there exists an idempotent $f \in R$ such that $\pi(f) = y$. Since fg, gf lie in I, they are elements of rad R, and consequently $1 - fg$ has an inverse in R. Set

$$f' = (1 - fg)^{-1} f (1 - fg).$$

Since f' is conjugate to f, f' is an idempotent. Also $f'g = 0$, since $f(1 - fg)g = 0$. Now $(1 - fg)f' = f - fg$, so $f' - fgf' = f - fg$, and $f' - f = (fg)f' - (fg)$. Since $fg \in I$, this shows that $f' - f \in I$.

Now put $e = f' - g f' = (1 - g) f'$. Then $ge = eg = 0$, and

$$e \equiv (1 - g) f \equiv f \equiv y \qquad \text{(modulo } I\text{).}$$

Furthermore, $e^2 = (1 - g) f' (1 - g) f' = (1 - g) f' = e$. □

22.18 **Proposition.** *If I is an ideal of R contained in* rad R, *and if idempotents of R/I can be lifted, then any finite or countable set $\{u_i\}$ of orthogonal nonzero idempotents of R/I can be lifted to an orthogonal set of idempotents of R.*

Proof (Lambek [66]). Suppose we already have lifted orthogonal idempotents $u_1, \ldots, u_k$ in R/I to orthogonal idempotents $e_1, e_2, \ldots, e_k$. Then $g = e_1 + \cdots + e_k$ is an idempotent such that $u_{k+1} g$ and $g u_{k+1} \in I$. Hence, by the lemma, u_{k+1} lifts to an idempotent e_{k+1} orthogonal to g, and then e_{k+1} is orthogonal to $\{e_1, \ldots, e_k\}$. This completes the induction, and proves the theorem. □

22.19 **Proposition.** *If R is semiperfect, then R is a lift/rad semilocal ring, and direct decompositions of $R/$rad R can be lifted to direct decompositions of R.*

Proof. R is semilocal by 22.13. Let π be the canonical map $R \to R/J$, $J = $ rad R, and suppose $u \in R$ is an idempotent modulo J; that is, $\pi(u) = \bar{u}$ is an idempotent of $\bar{R} = R/J$. Then the right ideal $A_1 = \bar{u} \bar{R}$ is a summand of $\bar{R}$, so that $\bar{R} = A_1 \oplus A_2$, where $A_2 = (1 - \bar{u}) \bar{R}$.

Let $P_i \xrightarrow{f_i} A_i \to 0$ be projective covers with $K_i = \ker f_i$, $i = 1, 2$. *Assertion:* $P_1 \oplus P_2 \xrightarrow{f_1 \oplus f_2} A_1 \oplus A_2$ *is a projective cover of* $\bar{R}$. Since $P_1 \oplus P_2$ is projective and $f = f_1 \oplus f_2$ is an epimorphism, it remains only to show that $K_1 \oplus K_2$, the kernel of f, is superfluous in $P_1 \oplus P_2$. To wit, let S be a submodule such that $P_1 \oplus P_2 = S + (K_1 \oplus K_2)$. For notational simplicity, identify P_1, P_2, K_1, K_2 with their images in $P_1 \oplus P_2$. Now $P_1 \supseteq K_1$ implies

$$P_1 = (K_1 + K_2 + S) \cap P_1 = K_1 + (K_2 + S) \cap P_1$$

and $P_1 \bigcirc K_1 \Rightarrow P_1 = (K_2 + S) \cap P_1$, hence $P_1 \subseteq K_2 + S$. By symmetry, $P_2 \subseteq K_1 + S$. Hence,

$$P_1 \oplus P_2 = (S + K_2 + P_2) = S + P_2 = S + (K_1 + S) = S + K_1$$

and

$$P_1 = (P_1 \oplus P_2) \cap P_1 = (S + K_1) \cap P_1 = (S \cap P_1) + K_1.$$

But $P_1 \supseteq K_1$, so $P_1 = S \cap P_1$, and $P_1 \subseteq S$. Similarly, $P_2 \subseteq S$, and so $P_1 + P_2 = P_1 \oplus P_2 = S$. This proves the contention that $K_1 \oplus K_2$ is superfluous in $P_1 \oplus P_2$.

Now R is a projective cover of $\bar{R} = R/J$ by 22.12, so by the uniqueness of the projective cover (22.11) there is a commutative diagram

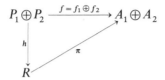

where h is an isomorphism. This shows that direct decomposition of $\bar{R}$ can be "lifted" to direct decomposition of R (since $R = h(P_1) \oplus h(P_2)$, and $\pi h(P_i) = A_i$, $i = 1, 2$), but (as E. Willard pointed out) does not state that the idempotent $\bar{u}$ can be lifted. We now do this. Since $R = h(P_1) \oplus h(P_2)$, $h(P_i)$ is generated by an idempotent e_i, $i = 1, 2$, and $1 = e_1 + e_2$. Therefore,

$$\bar{u} + (1 - \bar{u}) = 1 = \pi(1) = \pi(e_1 + e_2) = f h^{-1}(e_1 + e_2)$$
$$= f_1(h^{-1} e_1) + f_2 h^{-1}(e_2).$$

Since $\bar{R} = A_1 \oplus A_2$, the representation of $\bar{R}$ as a sum of elements of A_i, $i = 1, 2$, is unique; consequently

$$\bar{u} = f_1(h^{-1} e_1) = (f_1 h^{-1}) e_1 = \pi(e_1)$$

and e_1 is the desired idempotent.

22.20   Corollary. (a) *If R is a ring, and if $P_i \to A_i$, $i = 1, \ldots, n$, are finitely many projective covers in mod-R, then the coproduct mapping is a projective cover:*

$$\coprod_{i=1}^{n} f_i \colon \coprod_{i=1}^{n} P_i \to \coprod_{i=1}^{n} A_i$$

where $f_i \colon P_i \to \coprod_{i=1}^{n} A_i$ is the composition of $P_i \to A_i$, and the canonical injection $A_i \to \coprod_{i=1}^{n} A_i$.

(b) *A ring R is semiperfect if and only if every finitely generated right R-module has a projective cover.*

(c) *The (semi)perfect property is a Morita invariant.*

Proof. (a) The assertion in the proof of 22.19 is the $n=2$ case, and the general case follows by induction. (b) is an immediate consequence. (c) was discussed in the introduction to this chapter. $\square$

22.21 Lemma. *Suppose idempotents of R can be lifted modulo an ideal $I \subseteq \mathrm{rad}\, R$, and let π denote $R \xrightarrow{\ \text{canon}\ } R/I$. If e is an indecomposable idempotent of R, then $\pi(e)=e+I$ is an indecomposable idempotent of R/I.*

Proof. Suppose $\pi(e)=u+v$, where u, v, are orthogonal idempotents of R/I, and $u \neq 0$. Then since e is orthogonal to $1-e$, u is orthogonal to $\pi(1-e)$, and so (22.16) u lifts to an idempotent f orthogonal to $1-e$. Then $fe=f=ef$, and $e=f+(e-f)$ is a sum of orthogonal idempotents f, $e-f$. Since e is indecomposable, and $f \neq 0$, necessarily $e=f$. Thus, $\pi(e)=\pi(f)=u$, and so $v=0$; that is, $\pi(e)$ is indecomposable. $\square$

Definition. A right ideal I of a ring R is a **radical right ideal** provided that $I \subseteq \mathrm{rad}\, R$. A **minimal nonradical** right ideal I is one minimal in the set of non-radical right ideals. In this case $I \cap \mathrm{rad}\, R$ is the unique maximal submodule of I.

22.22 Proposition. *Let R be a lift/rad semilocal ring with radical J. Then:*

(a) *R has an Azumaya diagram and 18.23.1–5 hold.*

(b) *A right ideal I is nonradical iff I contains a nonzero idempotent. Thus, a minimal nonradical right ideal is generated by an idempotent.*

Proof. For (a), consult 18.26 and 18.23. Furthermore, for (b) it follows from (b) of 22.12 A that I contains a nonzero idempotent e. Then, eR is a nonradical right ideal contained in I. $\square$

(b) is a theorem of Köthe [30, p. 168, Satz 7].

22.23 Structure Theorem for Projective Modules. *A ring R is a lift/rad semilocal ring iff R is semiperfect. Moreover, a ring R is semiperfect iff R is left semiperfect. When this is so, then the conditions (a)–(c) hold:*

(a) *Let V be any simple right R-module. Then $V \approx eR/eJ$ for some indecomposable idempotent $e \in R$, and the natural map $eR \to V$ is a projective cover of V.*

(b) *Write $R=\sum_{i=1}^{n} \oplus\, e_i\, R$, where, $\forall i$, e_i is an indecomposable idempotent (22.22), and let $J=\mathrm{rad}\, R$. If A is a right R-module, then A/AJ is semisimple and a direct sum $\sum_{k \in K} \oplus\, e_{i_k} R/e_{i_k} J$, where, for each k, $e_{i_k} \in \{e_1, \ldots, e_n\}$. The module $P=\sum_{k \in K} \oplus\, e_{i_k} R$ is projective. Let $f: P \to A/AJ$ denote the coproduct of the canonical projections $p_{i_k}: e_{i_k} R \to e_{i_k} R/e_{i_k} J$, $\forall i_k$ and $\forall k \in K$. Then, there is a commutative diagram*

Furthermore, $P \xrightarrow{\ g\ } A$ is a projective cover under either of the following two conditions:

(1) *If B is a module requiring no more generators than A, then B=BJ if and only if B=0.*

(2) *A is finitely generated.*

(c) *If A is any projective module such that A · rad R is superfluous (for example, when A is a B-object 18.4) then A is isomorphic to a direct sum of right prindecs.*

Proof. We shall prove (a), (b), and (c) under the assumption that R is lift/rad semilocal. Then, (2) of (b) implies that R is semiperfect. Moreover, any semiperfect ring is lift/rad semilocal by 22.19. This proves the first equivalence. Since "lift/rad semilocal" is a right-left symmetric property, the second equivalence follows. Thus, it remains only to deduce (a)–(c) assuming that R is lift/rad semilocal, so we have by 18.23.1–5 and (a) follows from 18.23.4. Then, if A is any projective module, A/AJ is semisimple, hence isomorphic to a direct sum $\sum\oplus_{k\in K} e_{i_k} R/e_{i_k} J$ of right prindecs of $\bar{R}=R/J$. (This follows immediately from the Wedderburn-Artin theorems 8.8ff.) Then, $P=\sum_{k\in K}^k \oplus e_{i_k} R$, a direct sum of projective modules, is projective, and the commutative diagram under (b) results.

To complete the proof of (b), let C be a module denoting variously A or P, and let S be a submodule such that $S+CJ=C$. Since C requires no more generators than A, and since $B\cdot J=B$, where $B=C/S$, the hypothesis (1) implies that $B=C/S=0$, so $C=S$ and CJ is a superfluous submodule of C (for either, $C=P$ or A). Since AJ is superfluous in A, and since im g maps onto A/AJ under the canonical map $A\to A/AJ$, then $A=\text{im } g+AJ$, and $A=\text{im } g$, so $g: P\to A$ is an epimorphism. Also, since $\ker g\subseteq\ker f=PJ$, and since PJ is superfluous then $\ker g$ is superfluous, so $P\to A$ is a projective cover. By 18.4, any finitely generated module A satisfies (1), and, hence, has a projective cover. This proves (b).

If A is projective and AJ is superfluous, then the canonical map $A\to A/AJ$ is a projective cover. By the uniqueness of the projective cover, we see that $A\approx P$ has the desired structure (c).

By taking A to be the simple module eR/eJ in (a) (as we can by (a) of 22.22), we see that (a) is a consequence of (b). □

(b) of 22.23 again proves that every (finitely generated) projective module over a local ring is free.

22.24   Corollary[3]. *If R is semiperfect and if R/rad R is a prime ring, then R ≈ A_n, where A is a local ring.*

Proof. There exists an indecomposable idempotent e of R, and $A=eRe$ is a local ring 22.22. Since $R/\text{rad } R$ is simple by the Wedderburn-Artin theorem 8.8,

$$(ReR)+\text{rad } R=R.$$

Since rad R is superfluous, $ReR=R$, so $\text{trace}_R eR=R$, and therefore eR is a generator, hence a progenerator. Then eR is a balanced R-module; that is,

$$R\approx \text{End}_A eR$$

and eR is finitely generated projective over the local ring $A=eRe\approx \text{End } eR_R$ (7.3). By the preceding corollary, eR is a free A-module, say $eR\approx{}_A A^n$, and then $R\approx \text{End}_A A^n\approx A_n$. □

---

[3] Wedderburn [5], Artin [27], Noether [29], Köthe [30]. Cf. Lambek [65] and Chapter 18 Notes.

Perfect Rings Characterized

22.26   Theorem (Bass [60]). *A semiperfect ring R is right perfect if and only if* rad R *is left vanishing.*

Proof. Any perfect ring R has left vanishing radical 22.8. Conversely, let R be semiperfect, and assume that $J = \text{rad } R$ is left vanishing. By (2) of 22.23(b) in order to show that every module has a projective cover, it suffices to show for any non-zero right R-module M, that $MJ \neq M$.

Suppose $M = MJ$ and $x \in M$, $x \neq 0$, and write

$$x = \sum_{i_1} x_{i_1} a_{i_1}, \qquad\qquad\qquad x_{i_1} \in M, \quad a_{i_1} \in J.$$

For each i_1,

$$x_{i_1} = \sum x_{i_1 i_2} a_{i_1 i_2}, \qquad\qquad\qquad x_{i_1 i_2} \in M, \quad a_{i_1 i_2} \in J$$

and continuing,

$$x_{i_1 \dots i_{n-1}} = \sum x_{i_1 \dots i_n} a_{i_1 \dots i_n}.$$

Then, for each n,

$$x = \sum_{i_1, i_2, \dots, i_n} x_{i_1 \dots i_n} a_{i_1 \dots i_n} \cdots a_{i_1 i_2} \cdot a_{i_1}.$$

Since $x \neq 0$, there is a sequence $i_{10}, i_{20}, \dots, i_{n0}$ such that

$$b_n = a_{i_{10} i_{20} \dots i_{n0}} \cdots a_{i_{10} i_{20}} \cdot a_{i_{10}} \neq 0.$$

Letting such sequences be vertices of a tree in which an edge corresponds to adjoining one new index, we see that there exist paths of arbitrary length. By König's graph theorem (see Foreword Exercise 40), there exists a path of infinite length, that is, a sequence $i_{10}, i_{20}, \dots, i_{n0}, \dots$ such that for all n the element b_n, just displayed, is nonzero. This contradicts the left vanishing of rad R.   □

22.27   *Exercises.* (a) Any right Artinian ring, more generally any semi-primary ring, is right and left perfect.

(b) Any commutative semiperfect ring is a direct product of a finite number of local rings.

(c) The ring of lower triangular row finite matrices in the ring of (finite or) infinite matrices over a field is a perfect ring.

The next lemma is used in the proof of the following theorem characterizing perfect rings.

22.28   Lemma. *Let R be any ring. The following propositions are equivalent:*

(a) *R contains no infinite set of orthogonal (nonzero) idempotents;*

(b) *R_R satisfies the d.c.c. on summands;*

(c) *R satisfies the d.c.c. on principal right ideals of the form eR, where $e = e^2 \in R$;*

(d) *The right-left symmetry of (b) or (c).*

Proof. A summand of R_R has the form eR, where $e \in R$ is idempotent, so (b) ⇔ (c) is evident.

(c) $\Rightarrow$ (a). Let $\{e_n \,|\, n = 1, 2, \ldots\}$ be an infinite set of orthogonal nonzero idempotents. Then $f_n = e_1 + \cdots + e_n$, $n > 1$, and $1 - f_n$, are idempotent. Since

$$R e_1 \subset R f_2 \subset \cdots \subset R f_n \subset \cdots$$

then

$$(1 - e_1) R \supset (1 - f_1) R \supset \cdots \supset (1 - f_n) R \supset \cdots.$$

Thus not (a) implies not (c).

(a) $\Rightarrow$ (c). Let $f_1 R \supset f_2 R \supset \cdots \supset f_n R \supset \cdots$ be a strictly decreasing sequence, where $f_n R$ is idempotent, $n = 1, 2, \ldots$. If e is an idempotent, and if a right ideal A is a summand of eR, $eR = A \oplus B$; then, since $eRe \approx \operatorname{End} eR_R$, there exist orthogonal idempotents $e_1, e_2 \in eRe$ such that $e = e_1 + e_2$, $e_1 eR = e_1 R = A$ and $e_2 R = B$. Applying this fact to the above sequence, for each n, there exists a pair of nonzero orthogonal idempotents

$$g_{n+1}, h_{n+1} \in f_n R f_n$$

such that

$$f_n = g_{n+1} + h_{n+1}, \quad \text{and} \quad f_{n+1} R = h_{n+1} R.$$

If $m > n$, then $g_m \in h_m R h_m \subseteq h_n R h_n$, and $g_n h_n = h_n g_n = 0$ implies $g_m g_n = g_n g_m = 0$. Therefore $\{g_n \,|\, n = 1, 2, \ldots\}$ is an infinite set of orthogonal nonzero idempotents. $\square$

The Roman numerals I through IV used below are consistent with earlier use in the introduction to this chapter.

22.29 Theorem. *The following conditions on a ring R are equivalent:*

I. *R is right perfect.*

II. *$R/\operatorname{rad} R$ is semisimple, and $J = \operatorname{rad} R$ is left vanishing.*

III. *R is left socular, and R has a bounded number of orthogonal idempotents.*

IV. *R satisfies the d.c.c. on principal left ideals.*

V. *R satisfies the d.c.c. on finitely generated left ideals.*

VI. *Every left R-module satisfies the d.c.c. on cyclic submodules.*

VII. *Every left R-module satisfies the d.c.c. on finitely generated submodules.*

Proof. I $\Leftrightarrow$ II. Assuming I, then R is semiperfect, so $R/\operatorname{rad} R$ is semisimple, and $\operatorname{rad} R$ is left vanishing by 22.26. The converse also follows from this. For II implies that $\operatorname{rad} R$ is nil, so idempotents lift, and therefore R is semiperfect. By 22.26, R is then perfect.

II $\Rightarrow$ III. Any semisimple ring R has a bounded number of orthogonal idempotents (bounded by the number of minimal right ideals in a direct sum decomposition of R; cf. 21.6), and since nonzero orthogonal idempotents of R map onto nonzero orthogonal idempotents of $R/\operatorname{rad} R$, then R has a bounded number of orthogonal idempotents.

We wish to show that every nonzero M has a nonzero semisimple submodule. Since R is semisimple·modulo $J = \operatorname{rad} R$, it is equivalent to show that M has a nonzero R/J submodule. The sublemma below does this.

Sublemma 1. *Let R be any ring with left vanishing radical J. Then every nonzero left R-module M contains a nonzero R/J submodule, or equivalently, for every left module M and proper submodule L, there exists $y \in M$ not in L such that $J y \subseteq L$.*

Proof. The first is the $L=0$ case of the second, while the second is obtained from the first by letting M/L play the role of M. The proof of the first statement follows from the observation that the denial of the conclusion means that $Jy \neq 0$ for all $y \neq 0$ in M, hence the existences of an infinite sequence $\{n_i\}_{i=1}$ of elements of J such that for all i,

$$n_i\, n_{i-1} \dots n_2\, n_1\, y \neq 0$$

which denies the left vanishing of J. □

III $\Rightarrow$ II. First $J = \text{rad } R$ is left vanishing by 22.10. We wish to show that $R/\text{rad } R$ is semisimple. Since J is a nil ideal, idempotents can be lifted modulo J (22.15). Since $\bar{R} = R/J$ is semiprime, every simple left ideal is generated by an idempotent (8.5). Every submodule M of socle $\bar{R}$ is semisimple, and if M is finitely generated, then M is generated by an idempotent e (19.25). Since $M = e\bar{R}$ is semisimple, there exist orthogonal idempotents $\{e_i \mid i=1, \dots, n\}$ in $e\bar{R}e$ such that $e = \sum_{i=1}^{n} e_i$, and such that $e_i \bar{R}$ is simple, $i=1, \dots, n$. Furthermore, by 22.17, orthogonal idempotents of $\bar{R}$ can be lifted to orthogonal idempotents of R. Since R has only a bounded number of orthogonal idempotents, this shows that $S = \text{socle}_{\bar{R}}\, \bar{R}$ must be a direct sum of only finitely many simple left ideals, so S is generated by an idempotent f. The complementary summand $\bar{R}(1-f)$ has zero socle, and (since R is left socular) is therefore zero. Hence $\bar{R} = S$ is semisimple, proving that III $\Rightarrow$ II.

IV $\Rightarrow$ II. Let R satisfy the d.c.c. on principal left ideals. Then rad R is left vanishing by 22.9. In order to deduce II, we need to show that $\bar{R} = R/\text{rad } R$ is semisimple. Since $\bar{R}$ inherits the hypothesis on R, by 22.28, $\bar{R}$ satisfies the d.c.c., hence the a.c.c., on summands. Since each finite sum of minimal left ideals of $\bar{R}$ is generated by an idempotent (see the proof of III $\Rightarrow$ II), hence is a summand of $\bar{R}$, this implies that $\bar{R}$ has left socle S of finite length, and that $S = \bar{R}e$, where e is an idempotent. The d.c.c. on principal left ideals implies that every nonzero left ideal contains a minimal left ideal, that is, has nonzero intersection with S. Thus $\bar{R}(1-e)=0$, and $\bar{R} = S$ is semisimple.

The following implications are trivial:

$$\text{VII} \Rightarrow \text{VI} \Rightarrow \text{IV}$$
$$\text{VII} \Rightarrow \text{V} \Rightarrow \text{IV}.$$

Thus, we have proved the implications indicated by the solid arrows:

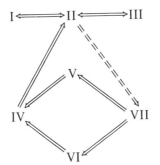

We propose to prove II $\Rightarrow$ VII which will complete the cycle. Before doing so, however, it is instructive to look at the proof of a theorem of Björk which has the implication II $\Rightarrow$ VII as a corollary.

22.30 Theorem (Björk). *If, over a ring R, a module M satisfies the* d.c.c. *on cyclic submodules, then it satisfies the* d.c.c. *on finitely generated submodules.*

Proof. The set $F = F(M)$, consisting of all submodules L which satisfy the d.c.c. on finitely generated submodules, is inductively ordered by inclusion. Let L be a maximal element of F. We wish to show $L = M$. Suppose not. Then there exists an element $y \in M$ such that Ry is minimal in the set of those cyclic left R-modules not contained in L. Clearly, L must be a maximal submodule of $Ry + L$.

Sublemma 2. *If L_1 is a finitely generated[4] submodule of $Ry + L$ not contained in L, then there exists an element $x \in L$ such that $y + x \in L_1$, and there exists a finitely generated submodule L'_1 of $L \cap L_1$ such that*

$$L_1 = R(y + x) + L'_1.$$

Proof. Let $z \in L_1$, $z \notin L$. Since $L_1 \subseteq Ry + L$, then $z = ry + x'$, with $r \in R$, and $x' \in L$. Then $z \notin L$ implies that $ry \notin L$, so that $Rry + L = Ry + L$, since L is a maximal submodule of the right side. Let $a \in R$ be such that $ary = y + m$, with $m \in L$. Then, $az = y + (m + ax') \in L_1$, and $x = m + ax' \in L$ as required.

Let $y_1, \ldots, y_n$ be a finite set of generators of L_1. Since $y' = y + x \notin L$, then $Ry + L = Ry' + L$, so we can write

$$y_i = a_i y' + m_i, \qquad\qquad a_i \in R, \quad m_i \in L.$$

$i = 1, \ldots, n$. Then $m_i = y_i - a_i y' \in L \cup L_1$, for every i. Let L'_1 be the submodule of L_1 generated by the m_i. Then

$$L_1 = \sum_{i=1}^n R y_i = \sum_{i=1}^n R(a_i y' + m_i) \subseteq Ry' + L'_1 \subseteq L_1,$$

that is, $L_1 = Ry' + L'_1$. $\square$

Now let

(1) $L_1 \supseteq L_2 \supseteq \cdots \supseteq L_n \supseteq \cdots$

be a descending sequence of finitely generated submodules of $Ry + L$. We will contradict the maximality of L in $F(M)$ by showing that $Ry + L \in F(M)$, that is, that (1) terminates. If $L_n \subseteq L$ for some n, this is trivial, since $L \in F(M)$. Thus, we will assume that no L_n is contained in L. Then by the last sublemma, for each n, we may choose $y_n = y + x_n \in L_n$, $x_n \in L$, and a finitely generated submodule L'_n of L such that

(2) $L_n = R y_n + L'_n.$

Furthermore, the hypothesis VI allows us to choose y_n such that $R y_n$ is minimal in the set of cyclic submodules for which (2) holds for any L'_n.

---

[4] The proof shows that if L_1 is not finitely generated, then L'_1 requires no more generators than the number of generators required for L_1, assuming only that L is a maximal submodule of $Ry + L$.

Sublemma 3. *Let* X *be any submodule of* $L \cup L_n$ *such that*

$$L_n = R\, y_n + X\,.$$

Then

$$L_n = R\, y_{n+1} + X\,.$$

Proof. Since $y_{n+1} \in L_{n+1} \subseteq L_n$, there exists $b \in R$ such that

$$y_{n+1} \equiv b\, y_n \,(\text{mod } X)\,.$$

Since $X \subseteq L$, and $R\, y_{n+1} + L = R\, y_n + L = R\, y + L$, we have

$$R b\, y_n \equiv R\, y_n\,(\text{mod } L)\,.$$

Write

$$y_n = a b\, y_n + m\,, \qquad\qquad a \in R,\quad m \in L\,.$$

Then $m \in L_n \cup L$, and

$$R\, y_n + X = R\,(a b\, y_n + m) + X \subseteq R b\, y_n + R m + X \subseteq R\, y_n + X\,.$$

Thus, by the minimality of $R\, y_n$, then $R b\, y_n = R\, y_n$. Since $y_{n+1} \equiv b\, y_n\,(\text{mod } X)$, we have that

$$R\, y_{n+1} + X = R\, y_n + X = L_n$$

as desired.

Completion of the Proof of Bjørk's Theorem. Write $L_1 = R\, y_1 + L_1'$ as in Sublemma 2.

Since $L_1 = R\, y_2 + L_1'$ by Sublemma 3, as in the proof of Sublemma 2, we can write

$$L_2 = R\, y_2 + \bar{L}_2\,,$$

where $\bar{L}_2$ is a finitely generated submodule of L_1. By induction, we get $L_n = R\, y_n + \bar{L}_n$, where $\bar{L}_n$ is a finitely generated submodule of $\bar{L}_{n-1} \subseteq L$. Since $L \in F(M)$, $\bar{L}_n = \bar{L}_{n+1}$ for some n, and then

$$L_n = R\, y_{n+1} + \bar{L}_n = R\, y_{n+1} + \bar{L}_{n+1} = L_{n+1}\,,$$

completing the proof of Bjørk's theorem.

II $\Rightarrow$ VII (Bjørk). As in the proof of Bjørk's theorem, we have a maximal element L in the set $F = F(R)$ of left ideals of R which satisfies the d.c.c. on finitely generated left subideals, and we wish to prove that $L = R$. Otherwise, by Sublemma 1, there exists $y \in R$, $y \notin L$ such that $J y \subseteq L$, where $J = \text{rad } R$. Since R/J is semisimple, and $JR y = J y \subseteq L$, then $R y + L$ is semisimple modulo L. Thus, we may choose y such that $R y + L$ is simple modulo L. Furthermore, by (22.22), the identity element of R is a sum of finitely many orthogonal indecomposable idempotents $\{e_i\}_{i=1}^n$, so we may assume that there exists an i, say $i=1$, such that $y = e_1 y$ is the element with these properties. For simplicity write $e = e_1$.

Since L is maximal in $R y + L$, by Sublemma 2, for every finitely generated left subideal L_i of $R y + L$, there corresponds an element $x_i \in L$ and a finitely generated left subideal $\hat{L}_i \subseteq L \cap L_i$ such that $y_i' = y + x_i \in L$, and

$$L_i = R\, y_i' + \hat{L}_i\,.$$

Since $e y_i' = y + e x_i$, and $e x_i \in L$, we may assume that $y_i' = e y_i'$, and that $x_i = e x_i$.

Sublemma 4. *If* $L_1 \supseteq L_2$, *then there exists an element* $e x_2 \in \hat{L}_1$, *and a finitely generated submodule* $\hat{L}_2$ *of* $\hat{L}_1$ *such that*

$$L_2 = R(y + e x_1 + e x_2) + \hat{L}_2.$$

Proof. By the results above, there exists an element z' in L such that $y + z' \in L_2$. Since $L_1 = R(y + e x_1) + \hat{L}_1$, we can write

$$y + z' = r(y + e x_1) + z, \qquad\qquad r \in R, \quad z \in \hat{L}_1.$$

Since $y \notin L$, then $r y \notin L$, so $R y \equiv R r y \pmod{L}$, and hence $R e \equiv R r e \pmod{J}$, since $r e \notin J$. Since $J e = J \cap R e$ is the unique maximal left subideal of $R e$, it follows that $R e = R r e$. Write $e = a r e$, with $a \in R$. Then

$$a y + a z' = a r y + a r e x_1 + a z$$
$$= y + e x_1 + e x_2 \in L_2,$$

with $x_2 = a z \in \hat{L}_1$. Then, the rest of the lemma follows from the proof of Sublemma 2.

Applying Sublemma 4, if $L_1 \supseteq L_2 \supseteq \cdots \supseteq L_n \supseteq \cdots$ is any chain of finitely generated left subideals of $R y + L$, we can write

$$L_2 = R(y + e x_1 + e x_2) + \bar{L}_2$$

where $\bar{L}_2$ is a finitely generated submodule of L_1, and $e x_2 \in L_1$. Inductively, we can write

$$L_n = R y + (e x_1 + \cdots + e x_n) + \bar{L}_n,$$

where $e x_n \in \bar{L}_{n-1}$ and $\bar{L}_n$ is a finitely generated submodule of $\bar{L}_{n-1}$. Since each $\bar{L}_n \subseteq L$, and since $L \in F(R)$, we get $\bar{L}_n = \bar{L}_{n+1}$ for some n. Then, necessarily $e x_{n+1} \in \bar{L}_n = \bar{L}_{n+2}$, so that $L_{n+1} = L_{n+2}$. This completes the proof of 22.29. $\square$

22.31 A   Theorem (Bass [60]). *A ring R is right perfect iff every flat right R-module is projective.*

Proof. Assume that R is right perfect, and let M be a flat right R-module, and let $F \to M \to 0$ be the projective cover, with superfluous kernel K. Now, by the proof of 11.32 (I, p. 438) K is the direct limit of finitely generated projectives $\{P_j\}_{j \in J}$ contained in K, and direct summands of F. Since K is superfluous, so is $P_j \,\forall j \in J$, which is impossible for direct summands, unless the summands are $= 0$. Thus, $K = 0$, and $M \approx F$ is projective.

The converse is much more difficult, and Bass's proof requires the functor Tor which is not even discussed in the text. For this reason, we leave the proof for an exercise, or refer to Bass [60, p. 475]. $\square$

When Products of Projectives are Projective

The preparation for the next theorem has been laid in Chapters 11 and 20.

22.31 B   Theorem (Chase [60]). *For any ring R, the following statements are equivalent:*

(a) *The direct product of any family of projective left R-modules is projective.*

(b) *The direct product of any family of copies of R is projective as a left R-module.*

(c) *R is left perfect and right coherent.*

Proof. (a) $\Rightarrow$ (b) is trivial.

(b) $\Rightarrow$ (c). Since a projective module is a summand, and hence a pure submodule, of a free module, it follows from the assumptions that the hypotheses of 20.21 are satisfied, and therefore R is left perfect. Furthermore, since the direct product of any family of copies of R is a projective—and hence flat—left R-module, by 11.34 (I, p. 439), R is right coherent.

(c) $\Rightarrow$ (a). Since any projective module is flat (I, 11.22, p. 433), then by 11.34, the direct product P of any family of projective left R-modules is flat. But since R is left perfect, every flat left R-module is projective (22.31 A), so the product is projective. $\square$

Loewy Series and Modules

Let M be a module, and inductively define a well ordered sequence $\{M_i\}_{i \in I}$, where I is a set of ordinals preceded by 0, and where

$$M_0 = 0$$
$$M_1 = \text{socle } M$$
$$M_{b+1}/M_b = \text{socle } M/M_b$$
$$M_a = \bigcup_{b < a} M_b \qquad (a = \text{limit ordinal}).$$

The set $\{M_i\}_{i \in I}$ is called the **ascending Loewy chain** (or series) of M. The module M is called a **Loewy module** when there is an ordinal a such that $M = M_a$, and then the least such ordinal is called the **Loewy length** of M. Note that the Loewy length of a module with a finite composition series is $\leq$ the length. For example, a semisimple module has Loewy length ≤ 1.

Dual to ascending Loewy chains, there is the concept of a descending Loewy chain.

If M is a Loewy module of finite length n, and if $J = \text{rad } R$, then $M_1 J = 0$, and $M_2 J \subseteq M_1$, so $M_2 J^2 = 0$, and inductively, $MJ^n = 0$. Conversely, if R is semilocal, and $MJ^n = 0$, then M has finite Loewy length $\leq n$. Similarly, in this case M has a descending Loewy chain of length $\leq n$. These are special cases of the next proposition.

22.32 Proposition. (a) *A ring R is left socular if and only if every left R-module is a Loewy module and iff R is a left Loewy module.*

(b) *A ring R is a right B-ring if and only if every right R-module has a descending Loewy chain.*

Proof (Ascending). Let $M_0 = \text{socle } M$. If $\beta \in I$ and $\beta = \gamma + 1$, define M_β such that $M_\beta/M_\gamma = \text{socle } M/M_\gamma$, and otherwise let $M_\beta = \sum_{\delta < \beta} M_\delta$. Thus, M_β is defined by transfinite induction $\forall \beta \in I$. Since R is left socular, there exists an ordinal α such that $M = M_\alpha$. Then $\{M_\beta\}$ is an ascending Loewy series. The converse is trivial, and also so is the last equivalence. The proof for (b) is dual. $\square$

Simple Factors

If $L = \{M_i \,|\, i \in I\}$ is an ascending Loewy series, then a **simple factor** of L is defined to be a simple module V that is isomorphic to a summand of $M_{\gamma+1}/M_\gamma$ for some γ. A simple left R-module V is isomorphic to Re/Je for some indecomposable idempotent e of R, where $J = \mathrm{rad}\, R$. In fact, if f is an indecomposable idempotent of R, then $V \approx Rf/Jf \Leftrightarrow fV \neq 0$. The following corollary shows that a simple factor of L is a simple factor of L' for any Loewy series L' of M. For this reason, we shall henceforth speak of **the simple factors of M**.

22.33   Corollary. *Let R be a right perfect ring, M a left (right) R-module, and e be an indecomposable idempotent of R. Let $\{M_\beta \,|\, \beta \in I\}$ be an ascending (descending) Loewy series for M and let $J = \mathrm{rad}\, R$. Then the following statements are equivalent:*

(a) $eM \neq 0 \, (Me \neq 0)$.

(b) $eM_\beta \neq 0 \, (M_\beta e \neq 0)$ *for some* $\beta \in I$.

(c) *There exists* $\gamma \in I$ *such that the semisimple module* $M_{\gamma+1}/M_\gamma \,(M_\gamma/M_{\gamma+1})$ *has a summand isomorphic to the simple R-module* $Re/Je\,(eR/eJ)$.

(d) $Re/Je\,(eR/eJ)$ *is a simple factor of the Loewy series* $\{M_\beta \,|\, \beta \in I\}$.

Proof. (c) $\Leftrightarrow$ (d) follows from the definition of simple factor. (a) $\Leftrightarrow$ (b). If $x \in M$ is such that $ex \neq 0$, then $ex \in M_\beta$ for some β, and then $e(ex) = ex \in e M_\beta$, so $eM_\beta \neq 0$.

(b) $\Rightarrow$ (c). If $eM_\beta \neq 0$, say $ex \neq 0$ for $x \in M_\beta$, then there exists an ordinal number γ such that $ex \in M_{\gamma+1}$ and $ex \notin M_\gamma$. Then $e(M_{\gamma+1}/M_\gamma) \neq 0$, since $e(ex + M_\gamma) \nsubseteq M_\gamma$. Now $W = M_{\gamma+1}/M_\gamma$ is a semisimple module, so there is a simple summand V of W such that $eV \neq 0$. Then by (a) of 22.22, $V \approx Re/Je$.

(c) $\Rightarrow$ (a). (c) implies by 22.22 that $eM_{\gamma+1} \neq 0$, whence $eM \neq 0$. (The proof for the parenthetical statements is dual.) $\square$

Linked Ideals

In this paragraph, we assume the notation, hypotheses, and results of the structure theorem for semiperfect rings (22.22), except that, in addition, we assume A is a perfect ring. Thus

$$(1) \qquad A = \sum_{i=1}^n e_i A = \sum_{i=1}^n A e_i$$

where $\{e_i \,|\, i = 1, \cdots, n\}$ are sum-1 orthogonal indecomposable idempotents.

Any left ideal of A isomorphic to some left ideal $A e_i$ appearing in (1) is called a **principal indecomposable left ideal**, and similarly for right ideals. From the structure theorem (22.22) for semiperfect rings, one deduces that a left ideal L is principal indecomposable if and only if $L = Af$ for some indecomposable idempotent $f \neq 0$.

Two principal indecomposable left ideals Ae and Af are said to *link*, notationally $Ae \sim Af$ or $e \sim f$, provided that there is a finite sequence

$$e_1 = e, \quad e_2, \ldots, e_n = f$$

of indecomposable idempotents such that each two successive left A-modules $A\,e_i$ and $A\,e_{i+1}$ have a simple factor in common, $i=1,\ldots,n-1$. We also say that $A\,e$ **is linked to** $A\,f$ **by** $e_1,\ldots,e_n$. Linkage is an equivalence relation on the set of principal indecomposable left ideals of A. We let $\{A\,e\}$ denote the equivalence class of left ideals linked to $A\,e$. Then $A\,f\in\{A\,e\}$ if and only if $A\,e\sim A\,f$.

Blocks

A (left) **block** of A is defined to be the sum of the left ideals in $\{A\,e\}$, where e is an indecomposable idempotent. This is called the **block determined by** e, or by $A\,e$, and is denoted $[A\,e]$. Thus

$$[A\,e]=\sum\nolimits_{A\,f\in\{A\,e\}} A\,f.$$

We will prove that $[A\,e]$ is an ideal of A; it is obviously a left ideal. Since $A=\sum_{i=1}^{n}\oplus A\,e_i$, obviously

$$A=\sum\nolimits_{i=1}^{n}[A\,e_i]$$

that is, A is a sum of finitely many blocks. Actually, A is a direct sum of its blocks, something else that will be proved.

An ideal I of a ring A is **indecomposable** if I is not a direct sum of two nonzero ideals of A. A nonzero ideal I of A is **simple** provided that I and O are the only ideals of A contained in I. Clearly any simple ideal of A is indecomposable. An ideal I of A is simple if and only if it is a minimal ideal of A.

22.34 Proposition (Brauer). *Let A be a right perfect ring, or equivalently, suppose A satisfies the d.c.c. on principal left ideals.*

(a) *If two blocks of A, considered as left ideals, have a common simple factor, then the blocks are equal. Distinct blocks annihilate each other.*

(b) *The blocks of A are ideals, and A is the direct sum of its blocks. Write $A=\sum_{i=1}^{n}\oplus A\,e_i$, where e_i is an indecomposable idempotent, $i=1,\ldots,n$. Renumber, if necessary, so that $[A\,e_1],\ldots,[A\,e_s]$, $s\leq n$, constitute all the distinct blocks in the set $\{[A\,e_i]\,|\,i=1,\ldots,n\}$. Then $[A\,e_s],\ldots,[A\,e_s]$ constitute all the blocks of A, and A is their direct sum,*

$$A=\sum\nolimits_{i=1}^{s}\oplus[A\,e_i].$$

(c) *Any indecomposable (ring) direct factor of A is a block, and every block of A is an indecomposable direct factor.*

(d) *Two principal indecomposable right ideals belong to a block if and only if they are linked.*

Proof. (a) Let $[A\,e]$ and $[A\,f]$ be two blocks of A. If $[A\,e]\neq[A\,f]$, then $A\,f$ has no simple factor isomorphic to $A\,e/J\,e$, where $J=\operatorname{rad} A$, so $e\,A\,f=0$. Thus $e'A\,f'=0$ and $A\,e'A\,f'=0$ for indecomposable idempotents e' and f' such that $A\,e'\in\{A\,e\}$ and $A\,f'\in\{A\,f\}$. This proves

(i) $$[A\,e][A\,f]=[A\,f][A\,e]=0.$$

Since A is a sum of blocks, e.g., $A = \sum_{i=1}^{n} [A\,e_i]$, (i) implies

(ii) $[A\,e_j]\,A = \sum_{i=1}^{n} [A\,e_j][A\,e_i] = [A\,e_j]^2 \subseteq [A\,e_j]$.

By the unique decomposition theorem (21.2), every block has the form $[A\,e_j]$ for some j. Thus (ii) shows that each block of A is an ideal, which by (i) annihilates every other block.

Suppose $A\,e'/J\,e'$ is a common simple factor of blocks $[A\,e]$ and $[A\,f]$. Then $e'\,[A\,e] \neq 0$ and $e'\,[A\,f] \neq 0$, so there exist $A\,e'' \in \{A\,e\}$, $A\,e''' \in \{A\,f\}$ such that $e'A\,e'' \neq 0$ and $e'A\,e''' \neq 0$. This implies that $A\,e$ and $A\,f$ are linked (to $A\,e'$), so $[A\,e] = [A\,f]$. This proves (a).

(b) We now prove that

$$A = \sum_{i=1}^{s} \oplus\, [A\,e_i].$$

To do this, using $A = \sum_{i=1}^{n} \oplus A\,e_i$, we only need to prove that $A\,e_j \subseteq [A\,e_i]$ if and only if $A\,e_j \sim A\,e_i$. If $A\,e_j \sim A\,e_i$, then $A\,e_j \subseteq [A\,e_j] = [A\,e_i]$. Conversely, suppose that $A\,e_j \subseteq [A\,e_i]$. Since $A\,e_j$ is a summand of A, it is a summand of $[A\,e_i]$, so $[A\,e_j]$ and $[A\,e_i]$ have a simple factor in common, e.g., $A\,e_j/J\,e_j$. Then $[A\,e_i] = [A\,e_j]$ by (1), so $A\,e_j \sim A\,e_i$ as needed. This proves (2).

(c) Since A is semiperfect A is a direct sum of finitely many indecomposable ideals, and hence

$$A = B_1 \oplus \cdots \oplus B_h$$

for finitely many indecomposable ideals $B_1, \ldots, B_h$. If e is an idempotent of A, then $A\,e = B_1\,e \oplus \cdots \oplus B_h\,e$. If e is indecomposable, then for some i, $A\,e = B_i\,e \subseteq B_i$, and $B_j\,e = 0$, $j \neq i$. Similarly, $e\,B_j = 0$. Since B_i is an ideal, $B_i \supseteq A\,e\,A$. If f is also an indecomposable idempotent, then $A\,f\,A \subseteq B_j$, and if $i \neq j$, then $B_j\,e = e\,B_j = 0$, so $f\,A\,e = e\,A\,f = 0$. This proves that if $B_i \supseteq A\,e$, then B_i contains any $A\,f$ such that $f\,A\,e \neq 0$ or $e\,A\,f \neq 0$. Next suppose $A\,e$ is linked to $A\,f$ by idempotents $e_1 = e, \ldots, e_t = f$. Then $A\,e_q$ and $A\,e_{q+1}$ have a common simple factor idomorphic to $A\,e'/J\,e'$ for some indecomposable idempotent e'. Thus, $e'A\,e_q \neq 0$ and $e'A\,e_{q+1} \neq 0$, so e_q, e_{q+1}, and e' are contained in the same B_k. By induction, $e_1 = e$ and $e_t = f$ belong to the same B_i. This proves that B_i contains the block $[A\,e]$. Since A is a direct sum of its blocks, $[A\,e]$ is a summand of B_i. Since B_i is indecomposable, this proves that $[A\,e] = B_i$ is indecomposable, and hence that $B_1, \ldots, B_h$ are the blocks of A and the only indecomposable direct factors of A.

(d) This was proved in the course of proving (b), since the proof of (b) shows that a right prindec $f\,A \subseteq [e_i\,A]$ iff $f\,A \sim e_i\,A$. (This follows since $f\,A \approx e_j\,A$ for some j.) Thus, two right prindecs $f\,A$ and $g\,A$ are contained in $[e_i\,A]$ iff they are linked.

22.35 Corollary. *Let A be right perfect, let B be a (left) block, and let $\bar{B}$ be the image of B under the canonical map $A \rightarrow \bar{A} = A/\mathrm{rad}\,A$. Then $\bar{B}$ is a direct sum of the simple ideals of $\bar{A}$ that are contained in $\bar{B}$, and the number of these simple ideals equals the number of the isomorphism classes of principal indecomposable left ideals of A contained in B.*

Proof. Since $\bar{A} = A/J$, where $J = \operatorname{rad} A$ is semisimple, $\bar{A}$ is a direct sum of simple (or minimal) ideals, and every ideal of $\bar{A}$ is a direct sum of some of these. Write $\bar{B} = \bar{S}_1 \oplus \cdots \oplus \bar{S}_t$, where $\bar{S}_i$ is a simple ideal of $\bar{A}$. All the minimal (or simple) ideals contained in $\bar{S}_i$ are isomorphic, since $\bar{S}_i$ is a ring isomorphic to a full matrix ring over a field. Now B is a direct sum of indecomposable left ideals, $A e_i$, say $i = 1, \ldots, h$, so

$$\bar{B} = \sum_{i=1}^{h} \oplus \overline{A e_i} = S_1 \oplus \cdots \oplus S_t.$$

But $\overline{A e_i} = A e_i / J e_i$, and by 22.5, $A e_i \approx A e_j \Leftrightarrow A e_i / J e_i \approx A e_j / J e_j$. Since there are precisely t nonisomorphic $\overline{A e_i}$, we conclude that there are precisely t non-isomorphic $A e_i$. Since every principal indecomposable left ideal contained in B is isomorphic to some $A e_i$, the proof is complete. $\quad\square$

22.36 Corollary (same hypotheses as in Proposition 22.34). (a) *If B is a block of A, any simple factor of B is isomorphic to $A e_i / J e_i$ for some $A e_i \subseteq B$.*

(b) *Let $B_i = [A e_i]$ denote a block of A, $i = 1, \ldots, s$. Then $A = B_1 \oplus \cdots \oplus B_s$. Let M be any left A-module. Then*

$$M = B_1 M \oplus \cdots \oplus B_s M$$

and each simple factor of $B_i M$ is a simple factor of B_i, $i = 1, \ldots, s$.

Proof. (a) Any simple factor V of B is one of A, so $V \approx A e_i / J e_i$ for some i, whence $e_i B \neq 0$, which shows that $A e_i \in B$.

(b) $AM = M$ implies $M = \sum_{i=1}^{s} B_i M$. Let e_i denote the unit element of B_i. If $x \in B_i M \cap \sum_{j \neq i} B_j M$, then $x = e_i x = e_i (\sum_{j \neq i}^{s} e_j x) = 0$, so $M = \sum_{i=1}^{s} \oplus B_i M$. Since $B_i M$ is annihilated by all $e_j \notin B_i$, it follows that if V is a simple factor of $B_i M$, then $V \approx A e_k / J e_k$ for some $e_k \in B_i$. Then $A e_k \subseteq B_i$, so $V \approx A e_k / J e_k$ is a simple factor of B_i too. $\quad\square$

Exercises of Chapter 22

1. There exist left socular right B-rings which are not semiperfect.

2. The class of (semi)perfect (resp. basic) rings is closed under finite products.

3. The ring A consisting of all 2×2 lower triangular matrices over a field k is a basic ring not a product of local rings.

*4. (D. W. Jonah [70] A ring is left perfect if and only if every left R-module has the a.c.c. on cyclic submodules.

5. If P is a projective Artinian R-module, then $\operatorname{rad}(\operatorname{End} P_R)$ is right vanishing.

6. (a) By dualism to quasiinjective modules, define quasiprojective modules. Show that a module P having a projective cover $E \to P$ is quasiprojective if and only if $\ker(E \to P)$ is a fully invariant submodule of E. — (b) If R is perfect, the indecomposable quasiprojective modules are the modules P/PA, where P is an indecomposable summand of R, and A is an ideal of R (Jans-Wu [67]).

7. (Sandomierski [64]) The ring R is a test module for injectivity (in the sense discussed in Chapter 3). If R is perfect, show that R is a test module for projectivity, that is, a module P is projective if and only if every row exact diagram

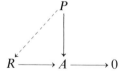

can be completed.

8. (Sandomierski [64]) If P is quasiprojective, and $S = \operatorname{End} P_R$, then

$$\operatorname{rad} S = \{f \mid P \bigcirc \operatorname{im} f\}$$
$$= \{f \mid \operatorname{im} f \text{ is superfluous}\}.$$

Furthermore, if for each $f \in S$ there exists a submodule M such that $\operatorname{im} f + M = P$ and $\operatorname{im} f \cap M$ is superfluous, then $S/\operatorname{rad} S$ is a regular ring. (If P is Artinian, choose M minimal such that $\operatorname{im} f + M = P$. If P is semisimple, M can be chosen so that $\operatorname{im} f \cap M = 0$.)

9. Let R and S be semilocal lift/rad ($=$ semiperfect) rings. Then $R \approx S$ if and only if R and S have isomorphic basic rings.

10. (Bass [60]) The following conditions on a ring R are equivalent: (a) R is left perfect. (b) The flat or weak dimension of a left R-module equals the projective dimension. (c) A direct limit of modules of proj. dim. $\leq n$ has proj. dim. $\leq n$. (d) Every flat left R-module is projective.

11. (Courter [69]) The following conditions on a ring R are equivalent: (a) Every module in mod-R, and in R-mod, is rationally complete. (b) R is a product of finitely many full matrix rings over right and left perfect local rings. (c) R is right perfect and every right R-module is rationally complete. (d) R is right and left perfect, and every right or left module is corationally complete (in a sense dual to rational completeness).

12. If R is a local ring with nilpotent radical, then every right or left module is rationally complete. In particular, $R = \bar{R} = \operatorname{Bien} \hat{R}_R$ (cf. 19.34).

13. (Chase [60]) A commutative ring R such that R^a is projective for any cardinal a is Artinian, and conversely.

14. A commutative Artinian ring (over which every product of projectives is projective by 15) need not have the property that products of free modules are free.

Notes for Chapter 22

Since perfect rings generalize Artinian (also semiprimary) rings, it is clear that no attempt to make complete references will ever succeed. (As did Alice, one runs faster and faster in the same place. Thus, we leave future-perfect rings for posterity, and concentrate on the past-perfect ones!)

See Bass [60] for the background to the idea of projective cover, and also for many theorems on finitistic homological dimension not mentioned here. Eilenberg [56] is an ideological forerunner.

Cozzens [70] gave examples of Noetherian domains (*V*-domains), not fields, possessing a simple injective cogenerator. Thus, these rings are right and left *B*-rings, but not perfect, answering a question of Bass [60] (see Exercise 22.7 A (a)).

Rentschler [67] gave a non-homological proof of the implication II ⇒ IV of 22.29, answering a problem posed by Bass, and also solved by Björk's theorem (22.30). However, this advance was made at the expense of part of Bass's theorem (the part stated in 21.31 A) which still does not have a non-homological proof.

Fuchs [70 b] and Osofsky [71 b] affirmed another problem of Bass [60] when they proved that for any pair of infinite cardinal numbers a and b there is a right and left perfect ring of left and right Loewy lengths respectively $a+1$ and $b+1$.

Camillo and Fuller [74] proved that any ring of finite left Loewy length n has finite right Loewy length $\leq 2^n$.

Woods [71] and Renault [70] proved that a group ring RG is perfect iff R is perfect and G is finite, thereby generalizing Connell's theorem for Artinian rings (Connell [63]) (cf. also the corresponding result for QF-rings 24.28).

The structure of G when RG is semiperfect is an open problem for the general case, however consecutive contributions have been made by Connell [63], Burgess [69], Valette and others. Consult Goursaud [73] and Lawrence [75] for certain improvements and references. Gupta [68] characterized orders in perfect rings (compare 18.48).

Perfect and semiperfect modules have been characterized by Mares [63] and Kasch-Mares [66].

Wu-Jans [67] studied quasiprojective covers, and showed that if a right perfect ring has infinite lattice of ideals, then there are infinitely many non-isomorphic indecomposable quasiprojective principal cyclic left modules. Ditto for quasi-injective modules. (This shows that a ring of finite module type [= FFM ring in Chapter 20] has finite lattice of ideals.)

Koehler [70], Fuller [69 b], and Golan [71] proved that the existence of quasiprojective covers implies the ring is perfect!

Mueller [70] (an alias for Müller) and Warfield [72] extended the structure theorem for projective modules over perfect rings to semiperfect rings: they are direct sums of prindecs. (See 22.23. Also see Notes for Chapters 20 and 21 where other "direct summand" theorems are discussed.) Warfield's theorem similarly determines projectives over any lift/rad ring which is von Neumann regular modulo radical, simultaneously extending the theorem of Kaplansky [58] and Albrecht [61] for regular rings. See Chapter 21, Exercises 6–8.

Sandomierski [64] showed that over a perfect ring R, that R is a "test module" for projectivity in a sense dual to the requirement for injectivity of a module M that maps of submodules of R into M can be lifted to maps of $R \to M$ (Baer's Criterion for Injectivity 3.41 (I, p. 157)). The characterization of all such rings is still an open problem.

Courter and Jonah (Jonah [70]) characterize left perfect rings by the property that for each integer n, each left module M, satisfies the acc-n, that is, the a.c.c. on submodules generated by n elements.

Brauer's theory of blocks (for Artinian rings) is contained in Brauer-Weiss [64]. (For an exposition and simplifications in group algebras, consult Rosenberg-Zelinsky [61].)

Michler [69c] showed that an idempotent ideal in a perfect ring is generated by an idempotent which is central modulo radical.

We mentioned Courter's theorem (Exercise 11) on rings similar to finite products of perfect local rings (cf. Dlab [70]). Teply [70] has classified those rings R for which every hereditary torsion theory (=hereditary radical as defined in Volume I, p. 528) is such that the torsion functor (=radical) on mod-R is exact: this happens iff R is a finite direct product of left [sic!] perfect rings each having a unique maximal ideal. In this case, R is a right B-ring (in our terminology) iff rad R is left vanishing (Teply [70, p. 200, Cor. 3.4]; cf. 22.7B. Also see Dlab [70] in this connection.)

Certain other aspects of (semi)perfect rings are discussed in Notes for Chapter 18.

Note on Loewy Series

Loewy introduced what are now called Loewy series in his paper in 1905 on representations of matrix groups, and later in 1917 in his study of complexes of matrices representing linear systems of differential equations. The group or set (complex) of matrices generates an algebra, and the matrices corresponding to linear transformations of a vectors space. Loewy obtained lower triangular matrix representations with irreducible blocks down the diagonal, and the essential uniqueness of these amount to statements about the factors of the Loewy series of the algebra, and invariant subspaces of the vector space.

Krull [26] defined the terms "Loewy series" and "invariants or elementary divisors" in use today, and applied them to "generalized rings of integers" in his paper written in 1928. In the latter paper, Krull was even aware of the possibility of transfinite Loewy series (as we have defined them in this chapter) as indicated by a footnote on page 64, but he made no use of this idea.

Two papers in Crelle's journal (much better named after it's founder!), Fuchs [70] and Shores [74], go into the historical background in some detail.

References

Bass [60], Björk [69, 70], Brauer and Weiss [64], Camillo and Fuller [74], Chase [60], Connell [63], Courter [69], Cozzens [70], Dickson and Kelly [70], Dlab [70], Eilenberg [56], Fuchs [70], Fuller [69b], Fuller and Hill [70], Golan [70, 72], Gupta [68], Hinohara [62], Jonah [70], Kasch and Mares [66], Koehler [70], Köthe [30], Krull [28], Loewy [05, 17, 26], Mares [63], Michler [69], Müller [70], Osofsky [71b], Rant [72], Renault [70], Rentschler [67], Rosenberg and Zelinsky [61], Sandomierski [64], Shores [71, 72, 74], Swan [60, 68], Teply [70], Warfield [72], Wu and Jans [67], Woods [70], Zöschinger [73], Albrecht [61], Azumaya [75], Kaplansky [58a], Krull [26].

Chapter 23. Morita Duality

In this chapter we introduce Morita duality. Roughly speaking, these theorems are dual to the Morita theorems on category equivalence (Chapter 12).

If C and D are categories, then a **duality** $C \leadsto D$ is an ordered pair (T, S) of contravariant functors $T: C \leadsto D$ and $S: D \leadsto C$ such that

$$ST \approx 1_C \quad \text{and} \quad TS \approx 1_D.$$

[Then (S, T) is a duality $D \leadsto C$.]

23.1 *Exercise.* If (T, S) and (T, S') are dualities $C \leadsto D$, then S and S' are naturally equivalent functors.

If (T, S) is a duality $C \leadsto D$, we abbreviate this by saying $T: C \leadsto D$ is a duality. The exercise then shows that we may speak of S as the **inverse** of T. (Then T is the inverse of S.) We say that T and T' are equivalent dualities provided that $T \approx T'$ as functors. (Then their respective inverses are equivalent functors.)

Let A and B be rings, let C be a full subcategory of mod-A, and let D be a full subcategory of B-mod. A duality (T, S) of $C \leadsto D$ is called a U-duality provided that there exists a (B, A)-bimodule U such that $\mathrm{Hom}_A(\ , U)$ induces a functor naturally equivalent to T, and $\mathrm{Hom}_B(\ , U)$ induces a functor naturally equivalent to S. We then say that $\mathrm{Hom}_A(\ , U)$ induces T, and that T is a U-**duality**. For brevity, we also say that U induces T. Thus, "U induces a U-duality" is meaningful. A theorem of Morita states that any duality between the categories C and D as just defined is equivalent to a U-duality provided that $A \in \mathrm{obj}\, C$ and $B \in \mathrm{obj}\, D$ (23.29). A necessary and sufficient condition on U in order that U defines a U-duality is that U is an injective cogenerator both in mod-A and in B-mod such that $B \approx \mathrm{End}\, U_A$ and $A \approx \mathrm{End}\, U_B$ (23.16). When A is right Artinian, this condition can be cut in half: U is a finitely generated injective cogenerator in mod-A and $B \approx \mathrm{End}\, U_A$ (23.25). A necessary condition on A for the existence of a U-duality is that A be semiperfect (23.18). Then, if A is perfect, A must be right Artinian, and by symmetry, B left Artinian (23.20). Quasifrobenius rings have an A-duality

$$\text{fin. gen. mod-}A \leadsto \text{fin. gen. } A\text{-mod}$$

and many other characterizations given in Chapter 23. (Here, of course, fin. gen. mod-A is the full subcategory of mod-A, consisting of the finitely generated modules.)

For some semiperfect rings, dualities are common. For example, 23.32 states that if A is a finite dimensional algebra over a field k, then $\mathrm{Hom}_k(\ , k)$ induces a duality

$$\text{fin. gen. mod-}A \rightsquigarrow \text{fin. gen. } A\text{-mod}.$$

Let $X' = \text{Hom}_k(X, k)$ for any $X \in \text{obj mod-}A$. By adjoint associativity, we have

$$\text{Hom}_A(X, A') = \text{Hom}_A(X, \text{Hom}_k(A, k))$$
$$\approx \text{Hom}_k(X \otimes_A A, k)$$
$$\approx \text{Hom}_k(X, k)$$
$$= X'.$$

It follows that the duality defined by $\text{Hom}_k(\ , k)$ is an A'-duality in the sense just discussed; namely, A' is an (A, A)-bimodule that induces an A'-duality

$$\text{fin. gen. mod-}A \rightsquigarrow \text{fin. gen. } A\text{-mod}.$$

This raises the question: when does $\text{Hom}_k(\ , k)$ or $\text{Hom}_A(\ , A')$ induce an A-duality? Obviously, this will be the case if and only if $A \approx A'$ as right A-modules. (The subject of A-duality is discussed in the next chapter on quasi-Frobenius rings.)

U-Duals

For any object U in an additive category C, we have designated the symbol h^U to represent the functor $C \rightsquigarrow \text{ABELGRPS}$ defined by $\text{Mor}_C(U,)$, and h_U the contravariant functor defined by $\text{Mor}_C(\ , U)$. If U is a (B, A)-bimodule, then for each $X \in \text{mod-}A$, the object $h_U(X)$ is a left B-module, called the U-**dual** of X, under the operation

$$bf = h_U(b_s) f$$

$\forall b \in B$, $f \in h_U(X)$, and where b_s is the homothet, or equivalently

$$(bf)(x) = b \cdot f(x) \qquad\qquad\qquad \forall x \in X.$$

Thus, h_U defines a functor $\text{mod-}A \rightsquigarrow B\text{-mod}$, called the U-**dual functor**. We also denote $\text{Hom}_B(\ , U)$ by h_U. Then h_U^2 denotes the composite functor $h_U \circ h_U$, in either direction. For each X, $h_U^2(X)$ is called the U-**bidual** of X, and h_U^2 is the U-**bidual functor**.

Abbreviated Notation

When U is fixed, and understood, we let $(\ )^*$ denote the functor h_U. Thus, $X^* = h_U(X)$, and for a morphism $f: X \to Y$, we let $f^* = h_U(f)$. Other abbreviations are:

$$h_U^2(X) = X^{**} = (X^*)^* \qquad\qquad h_U^3(X) = X^{***}$$
$$h_U^2(f) = f^{**} = (f^*)^* \qquad\qquad h_U^3(f) = f^{***}.$$

The U-dual functor $h_U: \text{mod-}A \rightsquigarrow B\text{-mod}$ is denoted by $(\ )_A^*$, or simply $(\ )^*$. The U-bidual functor $h_U^2: \text{mod-}A \rightsquigarrow \text{mod-}B$ is designated by $(\ )_A^{**}$, or $(\ )^{**}$.

Symmetric Notation

For $x \in X$, $f_1 \in X^*$, $f_2 \in X^{**}$, $f_3 \in X^{***}$ let

$$\langle f_1, x \rangle = f(x)$$
$$\langle f_1, f_2 \rangle = (f_1) f_2 \qquad \text{(value of } f_2 \text{ at } f_1)$$
$$\langle f_3, f_2 \rangle = f_3(f_2).$$

Then it is clear that there exists an A-map

$$\pi_U(X) : X \to X^{**}$$

such that

$$\langle f, \pi_U(X)(x) \rangle = \langle f, x \rangle$$

for all $f \in X^*$, $x \in X$. We abbreviate $\pi(X)(x)$ by x^{**}, when both U and X are understood. [The notation x^{**} for $\pi(X)(x)$ can be justified.] Similarly, we have a B-map

$$\pi_U(Y) : Y \to Y^{**}$$

when $Y \in B$-mod.

The Natural Transformation

As in the case $U = A = B$, π defines a natural transformation of functors

$$\begin{cases} \pi_U : 1_{\text{mod-}A} \to h_U^2 = (\ )^{**} \\ \pi_U(X) : X \to h_U^2(X) = X^{**}. \end{cases}$$

U-Torsionless and Reflexive Modules

An A-module X is said to be U-**torsionless** provided that $\pi_U(X) : X \to X^{**}$ is injective (monic), where $X^{**} = h_U^2(X)$. If $\pi_U(X)$ is an isomorphism, then X is called U-**reflexive**. Since the maps $\pi_U(X)$ are natural, then X is U-reflexive (resp. U-torsionless) if and only if every A-module isomorphic to X is.

When $U = A$, we say X is torsionless, or reflexive, instead of A-torsionless, or A-reflexive. Also $(\ )^* = h_A$ is called the **canonical contravariant functor** mod-$A \rightsquigarrow A$-mod.

23.2 Proposition. *Let U be a (B, A)-bimodule.*

(a) *B is U-torsionless if and only if U is a faithful B-module.*

(b) *B is U-reflexive if and only if U is a balanced B-module, that is, if and only if the canonical map $B \to \text{End } U_A$ is an isomorphism.*

(c) *When B is U-reflexive, then U is U-reflexive as an A-module.*

Proof. Now

$$\pi_U(B) : B \to \text{Hom}_A(\text{Hom}_B(B, U), U)$$

where

$$\langle \pi_U(b), f \rangle = \langle b, f \rangle = (b) f = b \cdot (1) f$$

$\forall b \in B, f \in \mathrm{Hom}_B(B, U)$. Since

$$\begin{cases} \mathrm{Hom}_B(B, U) \to U \\ \qquad f \mapsto (1) f \end{cases}$$

is a (B, A)-bimodule isomorphism, it follows that B is U-torsionless if and only if U is a faithful B-module, and that B is U-reflexive if and only if the canonical map

$$\begin{cases} B \to \mathrm{Hom}_A(U, U) \approx \mathrm{Hom}_A(\mathrm{Hom}_B(B, U), U) \\ b \mapsto b_s \end{cases}$$

is an isomorphism, that is, if and only if U is a balanced B-module. This proves (a) and (b).

(c) The natural B-isomorphism

where
$$h: U \to \mathrm{Hom}_B(B, U)$$
$$(b)\, h(u) = b\, u \qquad\qquad\qquad\qquad\qquad \forall b \in B, \quad u \in U$$

is an A-map, which the natural isomorphism $B \approx \mathrm{Hom}_A(U, U) = U^*$ shows corresponds to

$$\pi_U : U \to \mathrm{Hom}_B(U^*, U) = U^{**}.$$

Thus, $B \approx U^*$ yields $U^{**} \approx \mathrm{Hom}_B(B, U) \approx U$; that is, U is U-reflexive. $\square$

The main fact about U-torsion, to be proved, is that X is U-torsionless if and only if X can be embedded in a product of copies of U. We first establish a lemma.

23.3 Lemma. *The class of U-torsionless modules is closed under the taking of products and submodules.*

Proof. Let $X = \prod_{i \in I} X_i$ be a product of U-torsionless modules, and let $p_i: X \to X_i$ be the projections. If $x \in X$, and $x \neq 0$, then $x' = p_i(x) \neq 0$ for some i. Now $x' \in X_i$, so there exists $f \in X_i^*$ such that

$$\langle f, x' \rangle = \langle f, p_i x \rangle = \langle f p_i, x \rangle \neq 0.$$

Since $f p_i \in X^*$, X is therefore U-torsionless.

Next let X be torsionless, and let $h: M \to X = \prod_{i \in I} X_i$ be monic. If $m \neq 0 \in M$, then there exists $f \in X^*$ such that

$$\langle f, h(m) \rangle = \langle f h, m \rangle \neq 0.$$

Then $f h \in M^*$. $\square$

23.4 Proposition. *Let U be a (B, A)-bimodule, and let B be U-reflexive. For any A-module X, the following statements are equivalent:*

(a) *X is U-torsionless.*

(b) *There exist a monic $X \to U^I$ in a product of copies of U.*

(c) *The canonical map*

$$\begin{cases} F: \ X \to U^{\mathrm{Hom}_A(X,\,U)} = \prod_{f \in \mathrm{Hom}_A(X,\,U)} U_f \\ F: \ x \mapsto (\dots, f(x), \dots) \end{cases}$$

is injective.

Proof. (a) $\Rightarrow$ (c). If $x \neq 0$ in (c), then by (a), there exists $f \in X^*$ such that

$$\langle f, \pi_U(x) \rangle = \langle f, x \rangle = f(x) \neq 0$$

and then $F(x) \neq 0$. (c) $\Rightarrow$ (b) is trivial. (b) $\Rightarrow$ (a). By (c), of 23.2, U is U-reflexive, hence, by the lemma, any submodule of U^I is U-torsionless. Since X is isomorphic to a torsionless module, X is torsionless. $\square$

23.5 Proposition. *For each $X \in \mathrm{mod}$-A, the sequence*

(1) $0 \to X^* \xrightarrow{\ \pi_U(X^*)\ } X^{***}$

is split exact.

Proof. The functor ( )* converts $\pi_U: X \to X^{**}$ into $\pi_U^*: X^{***} \to X^*$. We will show that (1) is split exact by showing $\pi_U(X^*) \circ \pi_U^* = 1_{X^*}$. Since X^* and X^{***} are left B-modules, we write π_U^* and $\pi_U(X^*)$ on the right. Let $x \in X$, $g \in X^*$, $f \in X^{***}$. Then

$$\langle (f) \pi_U^*, x \rangle = \langle f, \pi(x) \rangle$$

so

$$\langle g (\pi_U(X^*) \circ \pi_U^*), x \rangle = \langle (g) \pi_U(X^*), \pi(x) \rangle$$
$$= \langle g, \pi(x) \rangle$$
$$= \langle g, x \rangle.$$

Thus, $\pi_U(X^*) \circ \pi_U^* = 1_{X^*}$. $\square$

23.6 Corollary. *Any U-dual module X^* is torsionless.*

Proof. $\pi_U(X^*): X^* \to X^{***}$ is monic. $\square$

The Category of Short Exact Sequences

Let S-EX SEQ denote the category of all short exact sequences

$$\{A_i\}: \ 0 \to A_1 \to A_2 \to A_3 \to 0$$

of morphisms and objects in an Abelian category C.

A morphism $\gamma: \{A_i\} \to \{B_i\}$ in the category is an ordered triple $\gamma = (\gamma_1, \gamma_2, \gamma_3)$, where γ_1, γ_2, and γ_3 are morphisms of C such that

$$
\begin{array}{ccccccccc}
\{A_i\}: & 0 \longrightarrow & A_1 & \longrightarrow & A_2 & \xrightarrow{\ \alpha\ } & A_3 & \longrightarrow 0 \\
& & \downarrow{\scriptstyle \gamma_1} & & \downarrow{\scriptstyle \gamma_2} & & \downarrow{\scriptstyle \gamma_3} & \\
\{B_i\}: & 0 \longrightarrow & B_1 & \xrightarrow[\ \beta\ ]{} & B_2 & \longrightarrow & B_3 & \longrightarrow 0
\end{array}
$$

commutes. Then S-EX SEQ is an additive category. The 5-lemma yields $\operatorname{Ker}\gamma$, namely,

(1) $0 \to \operatorname{Ker}\gamma_1 \to \operatorname{Ker}\gamma_2 \to \alpha \operatorname{Ker}\gamma_3$

and $\operatorname{Coker}\gamma$,

(2) $0 \to \beta \operatorname{Coker}\gamma \to \operatorname{Coker}\gamma_2 \to \operatorname{Coker}\gamma_3$.

23.7 Proposition. *Let C be an abelian category, and suppose*

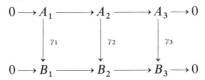

is row exact and commutative. Then

(a) *If γ_2 is monic, then γ_1 is monic.*

(b) *If γ_2 is epic, then γ_3 is epic.*

(c) *If γ_2 is an equivalence, then γ_1 is epic if and only if γ_3 is monic.*

(d) *If γ_2 is an equivalence, then γ_1 is an equivalence if and only if γ_3 is.*

(e) *If γ_1 and γ_3 are equivalences, so is γ_2.*

Proof. If $\gamma = (\gamma_1, \gamma_2, \gamma_3)$ is any morphism in S-EX SEQ, then $\operatorname{Ker}\gamma$ and $\operatorname{Coker}\gamma$ have been constructed, and this yields (a), (b), and (c). Then (d) is an immediate consequence. Finally, (e) is a direct consequence of (c) of the 5-lemma (13.16). □

23.8 Corollary and Definition. *Let C be a category, $T: C \rightsquigarrow C$ a functor, and $h: 1_C \to T$ a natural transformation. An object A in C is said to be h-**reflexive** provided that $h(A): A \to TA$ is an equivalence. Let C be Abelian, and*

$$0 \to A_1 \to A_2 \to A_3 \to 0$$

exact. Assume T is an exact functor. Then

(a) *If A_2 is h-reflexive, then A_1 is h-reflexive if and only if A_3 is h-reflexive.*

(b) *If A_1 and A_3 are h-reflexive, then A_2 is h-reflexive.*

Proof. Take $B_i = T(A_i)$, and $\gamma_i = h(A_i)$, $i = 1, 2, 3$ in the proposition. □

If U denotes a (B, A)-bimodule, then, of course, the h_U-reflexive modules are what we have called U-reflexive modules.

23.9 Proposition. *Let U be a (B, A)-bimodule, which is injective both in mod-A and in B-mod. Let $X_0 \subseteq X$ in mod-A or mod-B. Then*

(a) *If X is U-reflexive, then X_0 is U-reflexive if and only if X/X_0 is U-reflexive.*

(b) *If both X_0 and X/X_0 are U-reflexive, then X is U-reflexive.*

Proof. Since U is injective on both sides, h_U is exact, hence h_U^2 is exact, and so the corollary applies. □

23.10   Corollary. *Any summand of a U-reflexive module is U-reflexive.*

Proof. For notational simplicity, let T denote the U-bidual functor, and let h denote π_U. Let Y be U-reflexive, and let

$$0 \to X \xrightarrow{\ f\ } Y \xrightarrow{\ g\ } Y/X \to 0$$

be split exact. Since $gf = 1_X$, we have $T(g)\,T(f) = 1_{TX}$, so

$$0 \to TX \xrightarrow{\ Tf\ } TY \xrightarrow{\ Tg\ } T(Y/X) \to 0$$

is split exact, and the diagram of 23.7, with $A_1 = X$, $B_1 = TX$, $h(X) = \gamma_1$, and so on, is commutative. Since Y/X is equivalent to a subobject of Y, Y/X is torsionless; that is, $h(Y/X) = \gamma_3$ is monic, and γ_1 is monic for the same reason. Hence, by (c) of 23.7, the fact that $h(Y) = \gamma_2$ is an equivalence implies that γ_1 is also epic, thus an equivalence. Then γ_3 is also an equivalence by (d) of 23.7. $\square$

23.11   Exercise. (a) Let A be a ring. Any finitely generated projective A-module is A-reflexive.

(b) (Freyd) If C is an abelian category, then modulo split exact sequences, the category of short exact sequences in C is abelian (see quotient categories, Chapter 15).

Closed Modules

Let U be a (B, A)-bimodule. If M is a right A-module, and if M^* denotes the U-dual of M, then the mapping

$$\begin{cases} M^* \times M \to U \\ (y, x) \mapsto \langle y, x \rangle = y(x) \end{cases}$$

for all $y \in M^*$, $x \in M$, defines the following annihilator relations between subsets of M and M^*: If $X \subseteq M$, and $Y \subseteq M^*$, define

$$X' = \{y \in M^* \mid \langle y, x \rangle = 0 \ \ \forall x \in X\} = \operatorname{ann}_{M^*} X$$
$$Y' = \{x \in M \mid \langle y, x \rangle \ = 0 \ \ \forall y \in Y\} = \operatorname{ann}_M Y.$$

For a subset X of either M or M^*, X'' is defined to be $(X')'$. A subset X is U-**closed** provided that $X \circ X''$. If $X \subseteq M$, then X' is a left B-submodule of M^*, and if $X \subseteq M^*$, then X' is a right A-submodule of M. If $M = A$, then $A^* \approx U$, and then for a subset X of A, the corresponding subset of U is just the usual annihilation by module elements, namely, $\operatorname{ann}_{M^*} X$. Similarly, if $A = \operatorname{End}_B U = U^*$ is the U-dual of U, and if Y is a subset of U, then the corresponding set is $\operatorname{ann}_Y A$.

Since $\operatorname{Hom}_A(\ , U)$ is a left exact functor, for each exact sequence of right A-modules

$$0 \to X \xrightarrow{\ f\ } Y \xrightarrow{\ g\ } Y/X \to 0$$

there is a dual exact sequence

(1) $$0 \to (Y/X)^* \xrightarrow{\ g^*\ } Y^* \xrightarrow{\ f^*\ } X^*.$$

If f is the inclusion map, g is the canonical map $Y \to Y/X$, and $k \in Y^*$.

commutes, and

$$\ker f = \{k \in Y^* \mid f^*(k) = 0\}$$

or, equivalently,

$$\ker f = \{k \in Y^* \mid kf(x) = k(x) = 0 \ \forall x \in X\}.$$

Thus, $\ker f = X'$. The exactness of the sequence (1) proves that $X' \approx (Y/X)^*$.

23.12 Proposition. *Let U be a (B, A)-bimodule, and let $Y^* = \operatorname{Hom}_A(Y, U)$ for any $Y \in \operatorname{mod-}A$. If X is an A-submodule of Y, then the left B-module $X' = \operatorname{ann}_{Y^*} X$ is isomorphic to the left B-module $(Y/X)^*$.* $\square$

23.13 Proposition. *If U is a cogenerator in $\operatorname{mod-}A$ then every submodule X_0 of a right A-module X satisfies the double annihilator condition:*

$$X_0 = \operatorname{ann}_X \operatorname{ann}_{X^*} X_0$$

where X^* denotes the U-dual of X. If I is any right ideal of A, then

$$I = \operatorname{ann}_A \operatorname{ann}_U I.$$

Proof. We must show, that if $x \in X$ and if $x \notin X_0$, then $x \notin \operatorname{ann}_X \operatorname{ann}_{X^*} X_0$; that is, there exists $f \in X^* = \operatorname{Hom}_A(X, U)$ with $\ker f \supseteq X_0$, and $f(x) \neq 0$. Since $X_0 \cap xA \neq xA$, $C = xA/(X_0 \cap xA)$ is a nonzero cyclic module, so there exists a simple module V, and an epic $C \to V$. Since $C \approx (xA + X_0)/X_0$, we therefore have an epic $g : (xA + X_0) \to V$ such that $\ker g$ contains X_0. Since U is a cogenerator, by 12.2, there is a monic $h : \hat{V} \to U$, where $\hat{V} = \operatorname{inj hull} V_A$. Let $M = h(\hat{V})$. Then $k = hg$ is a nonzero map $(xA + X_0) \to M$, which, since M is injective, can be extended to a map $f : X \to M$. Since $M \subseteq U$, then we can consider $f \in X^*$. Since $\ker f \supseteq X_0$, this completes the proof of the first part. For the second part, take $X = A$, and $X^* = \operatorname{Hom}_A(A, U) \approx U$. $\square$

23.14 Corollary. *Let U be a (B, A)-bimodule, and U an injective cogenerator for $\operatorname{mod-}A$. Then every submodule of a U-reflexive A-module is U-reflexive.*

Proof. Let $f : X_0 \to X$ be an injection, where X is U-reflexive. The diagram

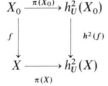

commutes, where $\pi = \pi_U$. Let

$$X' = \pi(X)^{-1} h^2(f) \pi(X_0)(X_0)$$

and let $g \in \mathrm{ann}_{h_U(X)} f(X_0)$; that is, $g: X \to U$ and $gf = 0$. Since $h^2(gf) = 0$, and

$$h^2(g) \pi(X) = \pi(X) g$$

then

$$\begin{aligned} g(X') &= g \pi(X)^{-1} h^2(f) \pi(X_0) \\ &= \pi(X)^{-1} h^2(g) h^2(f) \pi(X_0) \\ &= \pi(X)^{-1} h^2(gf) \pi(X_0) = 0. \end{aligned}$$

This implies that $X' \subseteq \mathrm{ann}_X \mathrm{ann}_{h_U} f(X_0)$, so $X' = \mathrm{Im} f$ by 23.13. If $y \in h_U^2(X_0)$, then

$$\pi(X_0) f^{-1} \pi(X)^{-1} h^2(f)(y) = y \in \mathrm{Im} \pi(X_0).$$

Thus, $\pi(X_0)$ is epic, hence an equivalence. □

Serre Class

If C is a category, a subclass $\mathscr{S}$ of objects of C is a **Serre class** provided that every exact sequence

$$0 \to X' \to X \to X'' \to 0$$

of objects in C it is true that X belongs to C if and only if both X' and X'' belong to C.

If C is an abelian category, and if $\mathscr{S}$ is a Serre class in C, then the full subcategory of C defined by $\mathscr{S}$ is abelian (15.2).

23.15 Theorem. *Let U be a (B, A)-bimodule such that U is an injective cogenerator in mod-A, and injective in mod-B. Then the class of U-reflexive right A-modules is a Serre class.*

Proof. If X is U-reflexive, and if $X_0 \subseteq X$, then X_0 is U-reflexive by the corollary, and hence X/X_0 is U-reflexive by (a) of 23.9. Conversely, if both X_0 and X/X_0 are U-reflexive, then so is X by (b) of 23.9. □

U-Duality

We began this chapter with the definition of duality. We next give necessary and sufficient conditions for a U-duality.

23.16 Theorem and Definition. *Let U be a (B, A)-bimodule, let U_A-Ref denote the class of U-reflexive right A-modules, and $_B U$-Ref the class of U-reflexive left B-modules. The following two conditions are equivalent:*

(a) U_A-Ref is a Serre class containing both A and U, and $_B U$-Ref is a Serre class containing both B and U.

(b) U *is an injective cogenerator both in* mod-A *and* B-mod, *and* A *and* B *are each* U-*reflexive*.

(c) U_A-Ref (*resp.* $_BU$-Ref) *is an abelian full subcategory of* mod-A (*resp.* B-mod) *containing* A *and* U (*resp.* B *and* U), *and* U *induces a* U-*duality* U_A-Ref$\rightsquigarrow$ $_BU$-Ref.

(d) U_A-Ref *is a Serre class containing* A, *and* $_BU$-Ref *is a Serre class containing* B.

(e) U_A-Ref *is an Abelian full subcategory of* mod-A *containing* A, $_BU$-Ref *is an abelian full subcategory of* B-mod *containing* B, *and* U *induces a* U-*duality* U_A-Ref$\rightsquigarrow$ $_BU$-Ref.

When any of the preceding conditions hold, then $_BU_A$ *is called a* **duality context.**

Proof. By 23.2, (a) $\Leftrightarrow$ (d), and (c) $\Leftrightarrow$ (e). Furthermore, (b) $\Rightarrow$ (a) is 23.15.

(a) $\Rightarrow$ (c). The full subcategory of mod-A defined by U_A-Ref is Abelian since it is a Serre class. The functor h_U induces a functor U_A-Ref$\rightsquigarrow$ $_BU$-Ref denoted by h_{U_A}. Dually, $h_{_BU}$: $_BU$-Ref$\rightsquigarrow U_A$-Ref, so

$$h_U^2 \colon U_A\text{-Ref}\rightsquigarrow U_A\text{-Ref}.$$

Moreover, h_U^2 is naturally equivalent to the identity functor on the U-reflexive modules on either side. Hence h_{U_A} and h_{BU} are dualities when (a) holds.

(c) $\Rightarrow$ (b). By (b) of 23.2, the fact that $B\in{}_BU$-Ref implies that the canonical map $B\to \operatorname{End}U_A$ is an isomorphism. Furthermore, U_A is an injective cogenerator in U_A-Ref since B is a projective generator in $_BU$-Ref, and the duality h_U maps B onto $h_U(B)=\operatorname{Hom}_B(B, U)\approx U_A$. By symmetry, U is an injective cogenerator in $_BU$-Ref.

It remains to show that U is an injective cogenerator in mod-A. (The result for $_BU$ will follow by symmetry.) If I is a right ideal of A, then A and I are in U_A-Ref, and exactness of h_U implies that

$$\operatorname{Hom}_A(A, U)\to \operatorname{Hom}_A(I, A)$$

is surjective. Hence, by Baer's criterion U_A is injective in mod-A. Next let V be a simple A-module. Then $V\approx A/I$ for some right ideal I of A. Since $W=A/I\in U_A$-Ref, the cogenerator property of U_A in U_A-Ref gives a monic $W\to U$. Hence, U is an injective object containing as a subobject every simple object in mod-A, so U is a cogenerator of mod-A by 12.7. $\square$

23.17 *Exercise.* Assume that $_BU_A$ is a duality context.

(a) Show that an infinite direct sum of nonzero U-reflexive injective modules is never U-reflexive.

(b) Let X be U-reflexive in mod-A, and let P be one of the following properties of objects: projective, injective, projective cover, injective hull, generator, cogenerator, and so on. For which P is it true, for all X, that X has property P in the category of U-reflexive modules if and only if X has property P in the category of all modules? Show that this is not true for all P and all X (cf. Osofsky [66]).

(c) For any ideal I of A, there is a duality context $_{B/I'}E_{A/I}$ where $E=\operatorname{ann}_U I\approx \operatorname{Hom}_A(A/I, U)$, and $I'=\{b\in B\,|\,bE=0\}$. Show that this duality is induced by the duality context $_BU_A$, that is, $\mathrm{Hom}_{A/I}(X, E) \approx \mathrm{Hom}_A(X, U)$, for any $X \in \mathrm{mod}\text{-}A/I$.

(d) *Let* $T: C \rightsquigarrow D$ *and* $S: D \rightsquigarrow E$ *be contravariant functors of Abelian categories.*

(1) *If* T *is left exact, and if* S *is right exact, then* $ST: C \rightsquigarrow E$ *is right exact.*

(2) *If* $T: C \rightsquigarrow D$, $S: D \rightsquigarrow C$ *are both contravariant, and* $ST \approx 1_C$, *then* T *is faithful and* S *is full. Furthermore, if* S *is faithful, then* T *is exact.*

(3) *If* $T: C \rightsquigarrow D$, $S: D \rightsquigarrow C$ *is a duality, then* T *and* S *are each fully faithful exact functors.*

23.18   Corollary (Osofsky [66]). *Let* $_BU_A$ *be a duality context. Then* A *and* B *are semiperfect rings.*

Proof. Every module in mod-A has an injective hull, but since U_A is injective, the submodules of U^n have an injective hull contained in U^n, hence, have U-reflexive injective hulls. The duality maps a finitely generated B-module Y onto a submodule Y^* of U^n for some n. Let $Y^* \to E$ be the injective hull of Y^* contained in U_A-Ref such that $0 \to E \to U^n$ is exact. Then there exists an exact sequence $B^n \to E^* \to 0$. Since E^* is projective in $_BU$-Ref, this sequence splits, so E^* is projective in B-mod. Then $E^* \to Y$ is a projective cover of Y in B-mod. Since every finitely generated B-module therefore has a projective cover, then B is semiperfect, and by symmetry, so is A. $\square$

Thus, the existence of a duality context places a restriction on a ring. We shall see that not every semiperfect ring possesses a duality context; however, every finite dimensional algebra does (see 23.32).

23.19   Lemma (Ornstein [66] and Osofsky [66]). (a) *If* I *is an ideal of a ring* R *such that* I/I^2 *is Noetherian, then* I^k/I^{k+1} *is Noetherian, for every* $k \geq 1$.

(b) *If* R *is a ring with d.c.c. on finitely generated right ideals* (=*left perfect ring* 22.29), *and if* $J = \mathrm{rad}\,R$ *is finitely generated modulo* J^2, *then* R *is right Artinian.*

Proof. (a) Let $x_1, \ldots, x_t \in I$ generated I modulo I^2. Then $\{x_i x_j\}_{i,j=1}^n$ generated I^2 modulo I^3. For, each element $z \in I^2$ is a finite sum of products of $y_1 y_2$, with $y_i \in I$, $i = 1, 2$, and

$$y_i = \sum_{j=1}^t x_j r_{ij}$$

with $r_{ij} \in R$, $i = 1, 2$, $j = 1, \ldots, t$. Thus, z is a finite sum of the products $x_j r_{ij} x_m r_{km}$. But, since $r_{ij} x_m r_{km} \in I$, it is a linear combination of the x_i's with coefficients in R plus an element of I^2. Thus, $x_j r_{ij} x_m r_{km}$ is a linear combination of the $x_i x_j$'s with coefficients in R plus an element of I^3. Hence, I^2/I^3 is finitely generated, and (a) follows by induction.

(b) Let F be a finitely generated right ideal of R such that $J = F + J^2$. Then $J^2 = FJ + J^3$, and since $FJ \subseteq F \cap J^2$, then $J^2 = F \cap J^2 + J^3$. Since J^2/J^3 is finitely generated, then there is a finitely generated right ideal F_2 of $F \cap J^2$ such that $J^2 = F_2 + J^3$. By induction, there is a sequence of finitely generated right ideals

$$F_1 \supseteq F_2 \supseteq \cdots \supseteq F_n \supseteq \cdots$$

where F_n generates J^n modulo J^{n+1}. Since R satisfies the d.c.c. on finitely generated right ideals, then $F_n = F_{n+1}$ for some n, and then $J^n = J^{n+1}$. However, since R is left perfect, then a left R-module M satisfies $M = JM$ if and only if $M = 0$, since R is a left B-ring 22.8. Thus, $J^n = 0$, and then (a) implies that R has a right composition chain (cf. 18.12–18.13), and hence, is right Artinian. $\square$

23.20 Proposition (Osofsky [66]). *Let $_BU_A$ be a duality context, and let $J = \operatorname{rad} A$.*

(a) *For any U-reflexive module M of* mod-A, MJ^n/MJ^{n+1} *is a semisimple module of finite length, for all $n \geq 0$.*

(b) *If A is either right or left perfect, then A is right Artinian and B is left Artinian.*

(c) *In this case, U has finite length both in* mod-A, *and B-mod.*

Proof. Let $J = \operatorname{rad} A$. By 23.16 the U-reflexive modules contain fin. gen. mod-A. Since U-Ref is a Serre class, exactitude of

$$0 \to MJ/MJ^2 \to M/MJ^2 \to M/MJ \to 0$$

implies that M/MJ and MJ/MJ^2 are U-reflexive modules. Since A is semiperfect by 23.18, then M/MJ and MJ/MJ^2 are semisimple modules. Since no infinite direct sum of nonzero modules is U-reflexive, it follows that M/MJ, MJ/MJ^2, and, by induction, MJ^n/MJ^{n+1} has finite length, for every n.

(b) The last lemma implies (b) in case A is left perfect, since (a) implies that J/J^2 is finitely generated. If A is right perfect, then A is a right B-ring, and hence by 18.3(b) and (c), $\operatorname{rad} M = MJ$ is superfluous for any module M. Now if M is U-reflexive, then M/MJ is finitely generated by (a), $M = F + MJ$, for some finitely generated submodule F of A, then superfluity of MJ implies that $M = F$ is finitely generated. Since every submodule of a U-reflexive module is U-reflexive, this implies that U-Ref consists of Noetherian modules. In particular, A is right Noetherian. Since A is perfect, J is a nil ideal, hence by Levitzki's theorem 9.17, J is nilpotent. Then the theorem 18.13 of Hopkins and Levitzki implies that A is right Artinian.

(c) Since U is U-reflexive, then (a), and the fact (b) that $J^n = 0$ for some n, imply that U has finite length in mod-A. Then, the U-duality shows that $B = \operatorname{Hom}_A(U, U)$ has finite length in B-mod, that is, that B is left Artinian. By symmetry, then U has finite length in B-mod. $\square$

Remark. This shows that an arbitrary (semi) perfect ring does not have a duality context.

Let I be a right ideal of a ring A, and let U be a right A-module. Then U is **I-complete** provided that each A-homomorphism $f \colon I \to U$ is induced by an element $m \in U$, that is,

$$f(x) = mx \qquad\qquad\qquad\qquad\qquad\qquad \forall\, x \in I.$$

Expressed otherwise,

$$\operatorname{Hom}_A(A, U) \to \operatorname{Hom}_R(I, U) \to 0$$

is exact. By 3.41, U is injective if and only if U is I-complete for all right ideals I of A.

Consider conditions (a), (b), (c), and (d):

(a) U is I-complete for every right ideal (Baer's criterion 3.41).

(b) $\operatorname{ann}_U (I \cap J) = \operatorname{ann}_U I + \operatorname{ann}_U J$ for all right ideals I and J.

(c) $\operatorname{ann}_U \operatorname{ann}_A S = US$ for every left ideal S of A, where $\operatorname{ann}_A S = S^\perp$.

(d) If $B = \operatorname{End} U_A$, then $Y = \operatorname{ann}_U \operatorname{ann}_A Y$ for every B-submodule Y of U.

If $x = a, b, c$ or d, then (x^*) denotes the condition (x) with the onesided ideal, or submodule, in the statement restricted to those conditions that are finitely generated, and (x^{**}) denotes the restriction to those conditions that are principal (cyclic). Thus, (d^*) asserts that $Y = \operatorname{ann}_U \operatorname{ann}_A Y$ for every finitely generated B-submodule of U.

23.21 **Proposition** (Ikeda and Nakayama [56]). *The following equivalences hold:*

23.21.1 $(a^{**}) \Leftrightarrow (c^{**})$.

23.21.2 $(a^*) \Leftrightarrow (b^*)$ and (c^{**}).

23.21.3 $(a) \Rightarrow (b)$, (c^*) and (d^*).

Proof. Let $^\perp I = \operatorname{ann}_U I$, and $I^\perp = \operatorname{ann}_A I$ for any subset $I \subseteq A$.

(a) $\Rightarrow$ (b). Trivially, $(I_1 \cap I_2) \supseteq {}^\perp I_1 + {}^\perp I_2$. Let $z \in {}^\perp(I_1 \cap I_2)$. If $a \in U$, then

$$\psi_1: x_1 \mapsto a x_1 \qquad\qquad\qquad\qquad x_1 \in I_1$$

$$\psi_2: x_2 \mapsto (a+z) x_2 \qquad\qquad\qquad x_2 \in I_2$$

are maps $\psi_i: I_i \to U$, $i = 1, 2$. Since $\psi_1 = \psi_2$ on $I_1 \cap I_2$, then

$$\psi: x_1 + x_2 \mapsto \psi_1(x_1) + \psi_2(x_2) \qquad\qquad x_1 \in I_1, \ x_2 \in I_2$$

is a map of $I_1 + I_2$ into U. By (a), there exists $m \in U$ such that $\psi(x) = m x \ \forall x \in I_1 + I_2$. Then, since

$$a x_1 = m x_1, \quad (a+z) x_2 = m x_2 \qquad\qquad \forall x_1 \in I_1, \ x_2 \in I_2$$

it follows that $u = (a-m) \in {}^\perp I_1$, $v = (a+z-m) \in {}^\perp I_2$. Then necessarily $z = v - u \in {}^\perp I_1 + {}^\perp I_2$, establishing (b). The proof also establishes $(a^*) \Rightarrow (b^*)$. (a) $\Rightarrow$ (d^*) is a special case of 19.2, whereas $(a^{**}) \Leftrightarrow (c^{**})$ is 6.2.

(b*) and (c**) $\Rightarrow$ (a*). Let $I = x_1 A + \cdots + x_n A$ be any finitely generated right ideal of A, and let f be a map of I into M. Since $(c^{**}) \Leftrightarrow (a^{**})$ shows that U is I-complete in the case $n = 1$, by induction we can assume that U is I_1-complete, where $I_1 = x_1 A + \cdots + x_{n-1} A$. Let f_i be restriction of f on I_i, $i = 1, 2$, where $I_2 = x_n A$. Since U is I_i-complete, $i = 1, 2$, (by (a**) and the induction assumption), there exist $m_i \in U$ such that $f_i(x) = m_i(x) \ \forall x \in I_i$, $i = 1, 2$. Since $f_1 = f_2 = f$ on $I_1 \cap I_2$, clearly $m_1 - m_2 \in {}^\perp(I_1 \cap I_2)$. Since (b*) implies that $^\perp(I_1 \cap I_2) = {}^\perp I_1 + {}^\perp I_2$, there exist $b_i \in {}^\perp I_i$ such that $m_1 - m_2 = b_1 - b_2$. Then, $m_1 - b_1 = m_2 - b_2$; call this element m. If $w = w_1 + w_2 \in I_1 + I_2 = I$, $w_1 \in I_1$, $w_2 \in I_2$, then $m w_i = m_i w_i$, $i = 1, 2$, so that for all $w \in I$

$$m w = m_1 w_1 + m_2 w_2 = f_1(w_1) + f_2(w_2) = f(w).$$

Thus, U is I-complete, establishing 23.21.2.

Finally, (a) $\Rightarrow$ (c*). Using (a) $\Rightarrow$ (b), and (a) $\Rightarrow$ (c**), if $x_1, \ldots, x_n$ in S generate S, then

$$\operatorname{ann}_U \operatorname{ann}_A S = \operatorname{ann}_U (\bigcap_{i=1}^n \operatorname{ann}_A x_i) = \sum_{i=1}^n \operatorname{ann}_U \operatorname{ann}_A x_i$$
$$= \sum_{i=1}^n \operatorname{ann}_U \operatorname{ann}_A (A x_i) = \sum_{i=1}^n U(A x_i)$$
$$= U(\sum_{i=1}^n A x_i) = US. \quad \square$$

23.22 Corollary. *If every right ideal of A is finitely generated, then any two of the following three conditions on an R-module M imply the other.*

(a) *M is injective.*

(b) *$\operatorname{ann}_M I + \operatorname{ann}_M J = \operatorname{ann}_M (I \cap J)$ for all right ideals I and J.*

(c) *$\operatorname{ann}_M \operatorname{ann}_R a = M a$ for every $a \in A$.* $\quad \square$

23.23 Corollary. *Let U be a (B, A) bimodule, let $B \approx \operatorname{End} U_A$ canonically, and let θ_A and η_A be the mappings*

$$\theta_A = \begin{cases} \theta: \text{ right ideals of } A \to B\text{-submodules of } U \\ \theta: I \mapsto {}^\perp I = \operatorname{ann}_U I \end{cases}$$

$$\eta_A = \begin{cases} \eta: B\text{-submodules} \to \text{right ideals} \\ \eta: Y \mapsto Y^\perp = \operatorname{ann}_A Y. \end{cases}$$

(a) *If U is a cogenerator in mod-A, then θ is injective, η is surjective, and $\eta \theta = 1_A$.*

(b) *If U is quasiinjective in mod-A, then, η induces an injective mapping on the finitely generated B-submodules of U.*

(c) *If U_A is an injective cogenerator, and either A is right Artinian or U is a Noetherian B-module, then θ and η are bijections, and $\theta = \eta^{-1}$.*

Proof. (a) is 23.13, and (b) is 23.21.3.

(c) Since η is bijective on finitely generated submodules, if A is Artinian, then ${}_B U$ satisfies the a.c.c. on finitely generated submodules, hence is Noetherian, so every submodule is finitely generated, η is bijective, and $\theta = \eta^{-1}$. $\quad \square$

When θ_A and η_A are bijections, we say that ${}_B U_A$ satisfies the **double annihilator conditions** for A. When ${}_B U_A$ satisfies these conditions for both B and A, we say that U satisfies the (B, A)-**annihilator conditions.**

23.24 Lemma. *Let A be right Noetherian, and let U be a (B, A)-bimodule. If ${}_B U$ is injective in B-mod, and if A is U-reflexive, then each finitely generated right A-module is U-reflexive.*

Proof. The functor $T = h_U$: mod-$A \rightsquigarrow B$-mod is left exact, and $S = h_U$: B-mod $\rightsquigarrow$ mod-A is exact, so h_U^2: mod-$A \rightsquigarrow$ mod-A is (right) exact. If $A^n \to X \to 0$ is exact, then using reflexivity of A, we get $A^n \to X^{**} \to 0$ exact. Thus, h_U^2 induces a functor

$$h_U^2: \text{fin. gen. mod-}A \rightsquigarrow \text{fin. gen. mod-}A = \mathscr{S}_A.$$

Since the natural transformation π_U of the identity functor to h_U^2 is a natural equivalence on A, by 13.7, $h_U^2 \approx 1_{\mathscr{S}_A}$. $\quad \square$

23.25   Theorem (Morita [58], Azumaya [59]). *Let A be a right Artinian ring, and let U be a (B, A)-bimodule. The following conditions are equivalent:*

(1) $_BU_A$ *is a duality context.*

(2) U *is a finitely generated injective cogenerator both in* mod-A *and B-mod, and A and B are both U-reflexive.*

(3) B *is left Artinian, and every finitely generated module in* mod-A *and in B-mod is U-reflexive.*

(4) (a) U_A *and $_BU$ are faithful modules.*
(b) *Every simple module V in* mod-A, *or in B-mod, is isomorphic to its U-bidual V^{**}.*
(c) B *is left Artinian.*

(5) (a) U_A *and $_BU$ are faithful modules.*
(b) *The U-dual of every simple object, in* mod-A, *or in B-mod, is simple.*
(c) U_A *is finitely generated.*

(6) (a) *Every right ideal I of A, and every B-submodule W of U, satisfies the double annihilator conditions:*

$$I = \mathrm{ann}_A \, \mathrm{ann}_U I$$

$$W = \mathrm{ann}_U \, \mathrm{ann}_A W.$$

(b) *Every left ideal J of B, and every A-submodule V of U satisfies the double annihilator conditions:*

$$J = \mathrm{ann}_B \, \mathrm{ann}_U J$$

$$V = \mathrm{ann}_U \, \mathrm{ann}_B V.$$

(7) (a) B *is U-reflexive; that is, the canonical ring homomorphism $B \to \mathrm{End}\, U_A$ is an isomorphism.*
(b) U_A *is a finitely generated injective cogenerator.*

(8) *There exists a U-duality*

$$\text{fin. gen. mod-}A \rightsquigarrow \text{fin. gen. } B\text{-mod}.$$

Proof. $(1) \Leftrightarrow (2)$. By 23.16, (1) is equivalent to (2) minus the statement that U is finitely generated. When A is right Artinian, however, then by 23.20, (1) implies that U is finitely generated on both sides.

$(1) \Rightarrow (3)$. B is left Artinian by 23.19, and the rest is clear.

$(3) \Rightarrow (4)$ is trivial.

$(4) \Rightarrow (5)$. Let X be a simple A-module. Then $\mathrm{ann}_X X^* = X$ would imply $X^* \subseteq \mathrm{ann}_{X^*} X = 0$, and then $X^{**} = 0$ is not isomorphic to X. Therefore $\mathrm{ann}_X X^* = 0$. This implies that $\pi_U \colon X \to X^{**}$ is monic, and since X^{**} is simple, π_U is an isomorphism. Therefore every simple module (on either side) is U-reflexive. Since B is left Artinian, any proper B-submodule Y of X^* is contained in a maximal submodule Y_1. (B is a B-ring in the terminology of Chapter 22.) Let $X_1 = \mathrm{ann}_X Y_1$. Then $(X^*/Y_1)^* \approx X_1$. Hence $X_1 \neq 0$, since X^*/Y_1 is reflexive by assumption. Hence

$X_1 = X$, so $Y_1 = 0$. This proves that X^* is simple. By symmetry, the dual of a simple B-module is simple.

We now show that U_A is finitely generated. If $X_1 \subset X_2$ are submodules of U such that X_1/X_2 is simple, then $V^* = (X_1/X_2)^* \neq 0$, so $X_1^* \neq X_2^*$. But $U^* \approx B$, so

$$0 \to X_i \to U$$

exact implies

$$X_i^* \approx B/\mathrm{ann}_B X_i$$

so $\mathrm{ann}_B X_2 \subset \mathrm{ann}_B X_1$. Thus, since B has a left composition series, it follows that U_A has a composition series.

$(5) \Rightarrow (3)$. If $V \in \mathrm{mod}\text{-}A$ is simple, then V^* and V^{**} are simple. The proof of $(4) \Rightarrow (5)$ shows that $\pi_U: V \to V^{**}$ is an isomorphism, so every simple module is U-reflexive. Now let X be a finitely generated A-module. Since A is right Artinian, X has a composition series of finite length $n \geq 1$. Our induction assumption is as follows:

If X_1 has length $< n$, then X_1 is U-reflexive, and if $f: X_1 \to X$ is a nonzero monic, where X has length $\leq n$, then $h_U^2(f) \neq 0$.

We first note that $\pi_U(X)$ cannot be the zero morphism. Let $t: X \to V$ be epic, where V is simple. Then

$$
\begin{array}{ccc}
& \pi_U(X) & \\
X & \longrightarrow & X^{**} \\
\downarrow{\scriptstyle t} & & \downarrow{\scriptstyle h^2(t)} \\
V & \longrightarrow & V^{**} \\
& \pi_U(V) &
\end{array}
$$

commutes, and $h^2(t)\,\pi_U(X) = \pi_U(V)\,t = 0$ would imply, by the fact that $\pi_U(V)$ is an isomorphism, that $t = 0$, a contradiction.

Now let $X_1 = \ker \pi_U(X)$, and let $f: X_1 \to X$ be the injection. Then

$$h_U^2(f)\,\pi_U(X_1) = \pi_U(X)\,f = 0.$$

Since $X_1 = \ker \pi_U(X) \neq X$, then length $X_1 < n$, so $\pi_U(X_1)$ is an equivalence, and $h_U^2(f) = 0$. By the induction assumption, then $f = 0$, which proves that $\pi_U(X)$ is monic. Now let X_0 be a simple submodule. Then

$$0 \to X_0 \to X \to X/X_0 \to 0$$

exact implies by left exactness of h_U^2 that

$$0 \to X_0^{**} \to X^{**} \to (X/X_0)^{**}$$

is exact, so

$$\text{length } X^{**}/X_0^{**} \leq \text{length}\,(X/X_0)^{**}.$$

Since X_0 and X/X_0 are U-reflexive, this proves that length $X^{**} \leq n$. Since $\pi_U(X): X \to X^{**}$ is monic, we must have $\mathrm{im}\,\pi_U(X) = X^{**}$, so $\pi_U(X)$ is epic, and X is U-reflexive.

(3) $\Rightarrow$ (2). By (3) $\Rightarrow$ (5), U_A is finitely generated. Since B and A are U reflexive, $B \approx \operatorname{End} U_A$ and $A \approx \operatorname{End}_B U$ canonically, so U_A is faithful. Hence, by 12.6, there is a monic $0 \to A \to U^n$ for some n. Using $A^* \approx U$ and $U^* \approx B$, we get $B^n \to U \to 0$ epic, so $_B U$ is finitely generated. Hence $\mathscr{S}_A = \operatorname{fin. gen. mod-}A$ is a Serre class containing A and U, and symmetrically for $_B \mathscr{S}$. It follows from 23.16 that U is an injective cogenerator in mod-A, and in B-mod.

(2) $\Rightarrow$ (6). Since B is left Artinian, and $_B U$ finitely generated, then $_B U$ is Noetherian. Also U_A is Noetherian. When $_B U$ and U_A are Noetherian, then (a) $\Rightarrow$ (c*) of (3) in 23.21 is the implication (2) $\Rightarrow$ (6).

(3) $\Rightarrow$ (8). When (3) holds, then U_A and $_B U$ are finitely generated, and then h is a duality as claimed.

(8) $\Rightarrow$ (3). For this part we need not assume that A is right Artinian. This will follow from the assumption (8). The proof (c) $\Rightarrow$ (b) of 23.16 shows that U is a cogenerator on both sides. Hence, by 23.13, for B-modules $Y_0 \subseteq Y$, we have

$$Y_0 = \operatorname{ann}_Y \operatorname{ann}_{Y^*} Y_0.$$

If Y is finitely generated, then Y^* is a finitely generated right A-module. Furthermore, $Y^*/\operatorname{ann}_{Y^*} Y_0$ is then finitely generated, and

$$(Y^*/\operatorname{ann}_{Y^*} Y_0)^* \approx \operatorname{ann}_{Y^{**}} \operatorname{ann}_{Y^*} Y_0 \approx \operatorname{ann}_Y \operatorname{ann}_{Y^*} Y_0 = Y_0.$$

Since every U-dual of a finitely generated module is finitely generated, this proves that Y_0 is finitely generated, and that Y is Noetherian. By the duality, Y^* is Artinian, and by symmetry, Y is Artinian. Thus, B is left Artinian, and A is right Artinian. The rest is clear.

(7) $\Rightarrow$ (5). Since U_A is a cogenerator, U_A is faithful. Since $B \approx \operatorname{End} U_A$, then $_B U$ is faithful too.

Since A is Artinian, by 23.21.3, the mapping

$$\eta : B\text{-submodules of } U \to \text{right ideals of } A$$

is bijective, so $_B U$ has finite length. Since $A^n \to U \to 0$ is exact for some n, and since $A^* \approx U$, $U^* \approx B$, applying $(\ )^*$ we get $0 \to B \to U^n$ exact, so B is left Artinian too. We can now apply 23.24 to conclude that every finitely generated left B-module is reflexive. Since every simple B-module is therefore reflexive, the proof of (4) $\Rightarrow$ (5) shows that the dual of every simple B-module is simple.

It remains to show that the dual of every simple A-module V is simple. Since U_A is an injective cogenerator, V is embedded in U, and contains a copy of each simple A-module. Each simple A-module is isomorphic to eA/eJ, where e is a principal indecomposable idempotent, and $J = \operatorname{rad} A$. Let $e_1, \ldots, e_t$ correspond to a complete set $V = V_1 \approx e_1 A/e_1 J, \ldots, V_t = e_t A/e_t J$ of nonisomorphic simple A-modules. Since $B \approx \operatorname{End} U_A$, the principal indecomposable idempotents of B are in 1-1 correspondence with indecomposable summands of U_A. Hence, if $\hat{V}_i$ denotes an injective hull of V_i contained in U_A, let e_i' denote the corresponding projection idempotent in B such that $e_1', \ldots, e_t'$ are mutually orthogonal. Since U_A is injective, by 19.27

$$N = \operatorname{rad} B = \{b \,|\, \ker b \text{ is an essential submodule}\}.$$

Since the socle $S(U)$ of U is essential, it follows that

$$N = \operatorname{rad} B = \operatorname{ann}_B S(U).$$

Since

$$\operatorname{ann}_B(e_i' S(U)) = \{b \in B \mid b\, e_i' \in N\} = B(1 - e_i') + N e_i'$$

then injectivity of U, and $U^* \approx B$, yields

$$(e_i' S(U))^* \approx B/\operatorname{ann}_B(e_i' S(U)) \approx B e_i'/N e_i'.$$

Since

$$e_i' S(U) = \operatorname{socle}(e_i' U) = \operatorname{socle} \hat{V}_i = V_i \approx e_i A/e_i J$$

we have $(V)^* \approx B e_1'/N e_1'$. Since e_1' is a principal indecomposable idempotent of B, then $B e_1'/N e_1'$ is simple, showing that a dual of a simple A-module is simple.

(2) $\Rightarrow$ (7). (7) is one-half of (2).

(6) $\Rightarrow$ (5). The lattice anti-isomorphism

$$\begin{cases} \text{right ideals of } A \to B\text{-submodules of } U \\ \qquad\qquad I \mapsto \operatorname{ann}_U I \end{cases}$$

given by (a) implies that U_A is faithful; also $\operatorname{ann}_U I$ is a minimal left B-submodule if and only if I is a maximal right ideal. Thus, if I is a maximal right ideal, then

$$(A/I)^* = \operatorname{Hom}_A(A/I, U) \approx \operatorname{ann}_U I$$

implies that the U-dual of every simple right A-module is simple. This proves 5(b). The rest of (5) follows by symmetry. $\quad\square$

23.26 *Exercise* (Morita [58]). (a) Show that any commutative Artinian ring has a finitely generated injective cogenerator E, hence an E-duality.

(b) Let R be a local commutative Artinian ring, and let E be the injective hull of $R/\operatorname{rad} R$. Then $R \approx \operatorname{End} E_R$ canonically, that is, there is an E-duality fin. gen. mod-R into itself, hence an involution (cf. end-of-chapter Exercise 13).

(c) (K. Fuller [69]). If R is right Artinian, and if $e = e^2 \in R$ is such that eR is injective, and if f is an idempotent R such that the top of fR is isomorphic to the direct sum of an equivalence class of simple modules in the socle of eR, then eRf defines a (eRe, fRf) duality context.

(It suffices to assume that eR is indecomposable injective, and in this case f is an idempotent such that the top of fR is the bottom of eR. In any case, eRf is an injective cogenerator in eRe-mod, and in mod-fRf. Compare end-of-chapter Exercise 8.)

23.27 **Proposition.** *Let T and T' be right exact additive functors $T: C \rightsquigarrow D$ of Abelian categories. Assume that C has a generator U, and that every object in C is finitely generated with respect to U. If $h: T \to T'$ is a natural transformation, and if $h(U)$ is an equivalence, then h is a natural equivalence of functors.*

Proof. Additive functors preserve finite coproducts. Since the functors T and T' preserve cokernels, then, as in the proof of 5.37, the 5-lemma shows that h is a functor equivalence. $\quad\square$

23.28 *Exercise. Let A and B be rings, let $\mathscr{S}_A$ and $_B\mathscr{S}$ denote respective full Abelian subcategories of* mod-*A and B*-mod *such that $A \in \mathscr{S}_A$ and $B \in {}_B\mathscr{S}$. If $T: \mathscr{S}_A \rightsquigarrow {}_B\mathscr{S}$ is a duality, then T is representable; that is, $T \approx h_U$, where $U = TA$. Furthermore, two contravariant functors $h_{U_i}: \mathscr{S}_A \rightsquigarrow {}_B\mathscr{S}$ are naturally equivalent, $i = 1, 2$, if and only if $U_1 \approx U_2$ as (B, A)-bimodules.*

23.29 **Theorem** (Morita [58]). *Let A and B be rings, let $\mathscr{S}_A$ and $_B\mathscr{S}$ be Abelian full subcategories of* mod-*A and B*-mod, *respectively, such that $\mathscr{S}_A$ contains A, and $_B\mathscr{S}$ contains B. Then any duality*

$$\mathscr{S}_A \underset{S}{\overset{T}{\approx}} {}_B\mathscr{S}$$

is equivalent to a U-duality; that is, $U = TA$ defines a U-duality, and there exist natural equivalences of functors

$$T \approx \mathrm{Hom}_A(\ , U), \qquad S \approx \mathrm{Hom}_B(\ , U).$$

Proof. (Following Cohn [66, p. 56].) By the exercise, there exist equivalences $T \approx \mathrm{Hom}_A(\ , U)$, where $U = TA$, and $S \approx \mathrm{Hom}_B(\ , V)$, where $V = SB$. In the situation $X_A, {}_BY_A, Z_B$, there is a homomorphism natural in all variables

$$\psi: X \otimes_A \mathrm{Hom}_B(Y, Z) \to \mathrm{Hom}_B(\mathrm{Hom}_A(X, Y), Z)$$

where

$$(f_1)\, \psi(x \otimes f_2) = (f_1(x))\, f_2$$

for all $x \in X$, $f_1 \in \mathrm{Hom}_A(X, Y)$, and $f_2 \in \mathrm{Hom}_B(Y, Z)$. For $U = Y$ and $V = Z$, the natural homomorphism ψ defines a natural transformation t of functors

$$t': \otimes_A \mathrm{Hom}_B(U, V) \to \mathrm{Hom}_B(\mathrm{Hom}_A(\ , U), V).$$

Now $ST \approx 1_{\mathscr{S}_A}$ implies a right A-module isomorphism $\mathrm{Hom}_B(\mathrm{Hom}_A(X, U), V) \approx X$, and, for $X = A$, we get an A-module, isomorphism

$$g: A \approx \mathrm{Hom}_B(U, V)$$

and a commutative diagram,

$$
\begin{array}{ccc}
A \otimes_A \mathrm{Hom}_B(U, V) & \xrightarrow{\; a_s \otimes 1 \;} & A \otimes_A \mathrm{Hom}_B(U, V) \\
{\scriptstyle t(A)} \downarrow & & \downarrow {\scriptstyle t(A)} \\
\mathrm{Hom}_B(U, V) \approx A & \xrightarrow{\quad a_s \quad} & A
\end{array}
$$

where, $t(A) = g^{-1} t'(A)$. Thus,

$$t(A)[(a_s \otimes 1)(1 \otimes f)] = a_s [t(A)(1 \otimes f)]$$

for all $a \in A$, $f \in \mathrm{Hom}_B(U, V)$. Using the natural isomorphism

$$A \otimes_A \mathrm{Hom}_B(U, V) \approx \mathrm{Hom}_B(U, V),$$

we can write this

$$t(A)(af) = a\, t(A)(f)$$

that is, $t(A)$ is an (A, A)-bimodule isomorphism. Let g denote the (A, A)-isomorphism

$$g\colon A \to \operatorname{Hom}_B(U, V)$$

defined by $t(A)^{-1}$. By symmetry, we have a (B, B)-bimodule isomorphism

$$h\colon B \to \operatorname{Hom}_A(V, U).$$

Let $e = g(1)$, and let $f = h(1)$. Let Φ denote the natural homomorphism

$$\Phi\colon \operatorname{Hom}_A(V, U) \otimes_B U \to \operatorname{Hom}_A(\operatorname{Hom}_B(U, V), U)$$

where

$$\Phi(f_1 \otimes u)(f_2) = f_1((u) f_2)$$

for $f_1 \in \operatorname{Hom}_A(U, V)$, $u \in U$, and $f_2 \in \operatorname{Hom}_B(U, V)$. Then Φ induces a map $U \to U$

$$\begin{cases} U \approx B \otimes_B U \xrightarrow{h \otimes 1} \operatorname{Hom}_A(V, U) \otimes_B U \xrightarrow{\Phi} \operatorname{Hom}_A(\operatorname{Hom}_B(U, V), U) \approx U \\ u \mapsto 1 \otimes u \to f \otimes u \xrightarrow{\Phi} \Phi(f \otimes u) \to f e u. \end{cases}$$

The homomorphism

$$\begin{cases} U \to U \\ f e(u) \mapsto u \end{cases}$$

is an isomorphism, and this implies that f is epic and e monic. Similarly, f is monic and e is epic, so f and e are isomorphisms. Now f is an A-map by definition, and we check that f is a B-map. The (B, B)-isomorphism $h\colon B \approx \operatorname{Hom}_A(V, U)$ under which $1 \mapsto f$ yields

(*) $h(b) = h(b \cdot 1) = b\, h(1) = b f$

whence

$$f(b v) = f b(v) = b f(v) = b(f(v))$$

for all $b \in B$, $v \in V$, so f is a (B, A)-isomorphism $V \to U$. Therefore, we have functor equivalences

$$T \approx \operatorname{Hom}_A(\ , U) \approx \operatorname{Hom}_A(\ , V)$$

and

$$S \approx \operatorname{Hom}_B(\ , V) \approx \operatorname{Hom}_B(\ , U).$$

Since any duality functor is exact and faithful, we conclude that U is an injective cogenerator both in mod-A and in B-mod. By 23.16, in order to conclude that U defines a duality, it remains to show that B and A are each U-reflexive, and by 23.2, this will be the case if and only if the canonical homomorphisms $B \to \operatorname{End} U_A$ and $A \to \operatorname{End}_B U$ are isomorphisms. We do this for $B \to \operatorname{End} U_A$. By 23.28, U and V are isomorphic (B, A)-bimodules. If we identify V with U, then we have the (B, B)-bimodule isomorphism

$$h \begin{cases} B \to \operatorname{Hom}_A(U, U) = \operatorname{End} U_A \\ 1 \mapsto f. \end{cases}$$

Since $f: U_A \to U_A$ is an isomorphism, f is unit in $\operatorname{End} U_A$. Since $f^2 = f$, then f is the identity element of $\operatorname{End} U_A$, and the calculation $(*)$ shows that for $b \in B$, $u \in U$,

$$h(b)\,u = (bf)\,u = b(fu) = b\,u.$$

This proves that h is the canonical map $B \to \operatorname{End} U_A$, which is thereby an isomorphism. $\square$

The Dual of a Free Module

For any right A-module M, the A-dual $M^* = \operatorname{Hom}_A(M, A)$ is a left A-module. Any contravariant functor h_U converts coproducts into products 13.8*. In case M is the free A-module $A^{(I)}$, then $M^* \approx A^I$, which is not always a free module (e.g., $\mathbb{Z}^\omega$ is not free). However, if M is the free module A^n, where n is an integer > 0, then the fact that $(A_A)^* \approx {}_A A$ yields $M^* \approx A^n$, a free left A-module.

23.31   Dual Basis Lemma for Free Modules. *Let A be a ring, and n a natural number.*

(a) If M is a free A-module $\approx A^n$, then M is reflexive, and M^* is a free left A-module $\approx A^n$.

(b) If $x_1, \ldots, x_n$ is a basis of M (resp. of M^*), then there exists a basis $y_1, \ldots, y_n$ of M^* (resp. of M) such that $\langle x_i, y_j \rangle = \delta_{ij}$ (Kronecker-δ) (resp. $\langle y_i, x_j \rangle = \delta_{ij}$), $i, j = 1, \ldots, n$.

Proof. M^* is a free left A-module by the preceding remarks, but this is reproved in the proof of (b). To prove (b), note that there exist elements $y_1, \ldots, y_n \in M^*$ such that $\langle x_i, y_j \rangle = \delta_{ij}$, $i, j = 1, \ldots, n$. Let $y = \sum_{j=1}^{n} r_j y_j$ be an element of the left A-submodule S of M^* generated by $y_1, \ldots, y_n$, where $r_1, \ldots, r_n \in A$. If $y = 0$, then $\langle x_i, y \rangle = \sum_{j=1}^{n} r_j \delta_{ij} = r_i = 0$, $i = 1, \ldots, n$. This shows that S is a free left A-module with a basis $y_1, \ldots, y_n$. Let u be any element of M^*, and let $r_i = \langle x_i, u \rangle$, $i = 1, \ldots, n$. Since $\langle x_i, u \rangle = \langle x_i, w \rangle$, $i = 1, \ldots, n$, where $w = \sum_{j=1}^{n} r_j y_j \in S$, then $u = w$, so $M^* = S$. This proves that M^* is a free left A-module $\approx A^n$. For any $x \in M$, let x^{**} denote the image of x under $\pi_A(M): M \to M^{**}$. Since M is torsionless, $x_1^{**}, \ldots, x_n^{**}$ are n distinct elements of M^{**}, and

$$\langle x_i^{**}, y_j \rangle = \langle x_i, y_j \rangle = \delta_{ij}, \qquad\qquad i, j = 1, \ldots, n.$$

Applying the left-right symmetry of the result just proved to the left A-module $N = M^*$, we obtain that $N^* = M^{**}$ is a free right A-module with a basis $x_1^{**}, \ldots, x_n^{**}$, so $\pi_A(M): M \to M^{**}$ is epic, hence an isomorphism. This proves (a). The proof of the parenthetical statement in (b) remains. If $u_1, \ldots, u_n$ is a free basis of M^*, then by the first part of (b), there exists a basis $\{v_1, \ldots, v_n\}$ of $(M^*)^* = M^{**}$ such that $\langle v_i, u_j \rangle = \delta_{ij}$, $i, j = 1, \ldots, n$. Since M is reflexive, there exists $x_i \in M$ such that $x_i^{**} = v_i$, $i = 1, \ldots, n$. Then $\langle x_i, u_j \rangle = \langle x_i^{**} u_j \rangle = \delta_{ij}$, $i, j = 1$. This completes the proof. $\square$

If A and B are algebras over a commutative ring k, then a duality between subcategories of mod-A and B-mod is understood to be k-linear functors as defined in Chapter 12.

23.32   Corollary. *If A is an algebra over a field k, and finite dimensional, then* $\mathrm{Hom}_k(\ ,k)$ *induces a duality*

$$\text{fin. gen. mod-}A \rightsquigarrow \text{fin. gen. } A\text{-mod}.$$

Proof. Let $\mathscr{S}_A$ and $_A\mathscr{S}$ denote the respective categories of finitely generated modules. If $X \in \mathscr{S}_A$, then $h_k(X) = \mathrm{Hom}_k(X, k) \in {}_A\mathscr{S}$, and $h_k \circ h_k(X) = h_k^2(X) \in \mathscr{S}_A$. The natural transformation $\pi_k \colon 1_{\mathscr{S}_A} \rightsquigarrow h_k^2$ is an equivalence because: (a) finite dimensional vector spaces are reflexive by 23.1; (b) A-homomorphisms are k-linear; (c) if $f \colon X \to X'$ is a right A-map, then $h_k(f) \colon h_k(X') \to h_k(X)$ is a left A-map; and (4) $\pi_k(X)$ is a right A-map. $\square$

23.33   *Exercises.* (a) For finite dimensional algebras A and B over a field k, the following conditions are equivalent:
(1) A and B are Morita equivalent algebras.
(2) There exists a duality from fin. gen. mod-$A \rightsquigarrow$ fin. gen. B-mod.
(3) A and B have isomorphic basic algebras (rings).

(b) In 23.32, show that $\mathrm{Hom}_k(\ ,k) \approx \mathrm{Hom}_A(\ ,A')$, where A' is the k-dual of A.

(c) (Nagao-Nakayama [53].) Let A be a finite dimensional algebra over a field k. A left A-module X is injective if and only if X is isomorphic to a direct sum of k-duals of principal indecomposable right A-modules.

The Nakayama Automorphism

For any right A-module X, and $a \in A$, let a_X be the homothetic $X \to X$ defined by a. If U and V are A-modules, then a **semilinear** transformation $f \colon U \to V$ is a pair of mappings

(a) A ring automorphism $\theta \colon A \to A$.

(b) A group homomorphism $f \colon U \to V$ such that

$$(u\,a)\,f = (uf)\,\theta(a)$$
or
$$\theta(a)_V = f^{-1}\,a_U\,f$$

for all $a \in A$, $u \in U$. If θ is an automorphism, then f is called a semilinear (A, θ)-isomorphism $U \to V$.

23.34   Proposition. *Let $_B U_A$ and $_B V_A$ be duality contexts.*

(a) *Any B-isomorphism $f \colon U \to V$ is a semilinear (A, θ)-isomorphism for a unique automorphism θ of A, called the **Nakayama automorphism** defined by f.*

(b) *Assuming (a), then U and V define equivalent dualities if and only if the Nakayama automorphism θ is an inner automorphism.*

(c) *If φ is any ring automorphism of A, then there exist a duality context $_B W_A$, and a B-isomorphism $g \colon U \to W$ with Nakayama automorphism φ. The module W will be denoted by (U, φ).*

Proof (Cohn [66]). (a) Since $A \approx \mathrm{End}_B V$ canonically (23.16), for each $a \in A$, there exists an element a' in A such that $a'_V = f^{-1}\,a_U\,f$. The mapping $a \mapsto a'$ is the desired automorphism.

(b) If $\theta(a) = x^{-1} a x$, for all $a \in A$, and some $x \in A$, then

$$\theta(a)_V = x_V^{-1} a_V x_V = f^{-1} a_U f$$

and $f x_V^{-1} \colon U \to V$ is a (B, A)-isomorphism of V. Hence U and V define equivalent dualities (23.1).

Conversely, if $h \colon U \to V$ is a (B, A)-isomorphism, then $g = f h^{-1}$ is a B-auto-morphism of U; that is, g is a unit in $\operatorname{End}_B U$. Let $x \in A$ be such that $x_U = g$. Then $f = x_U h$, and

(iii) $\theta(a)_V = f^{-1} a_U f = h^{-1} x_U^{-1} a_U x_U h.$

Since h is an A-isomorphism, then $h^{-1} b_U h = b_U \ \forall b \in A$, so from (iii) we obtain $\theta(a)_V = x_U^{-1} a_U x_U \ \forall a \in A$, and θ is the inner automorphism determined by x.

(c) If U is the A-module defined by $t \colon A \to \operatorname{End}_{\mathbb{Z}} U$, let W be the A-module

$$t \varphi \colon A \to \operatorname{End}_{\mathbb{Z}} U$$

that is, $u \cdot a = u \theta(a) \ \forall a \in A$, $u \in U$. Clearly, $B = \operatorname{End} W_A$. Furthermore, $_B W_A$ is a duality context because W_A (resp. $_B W$) is an injective cogenerator in mod-A, and A and B are both W-reflexive (23.16). $\square$

When there exists an (A, θ)-isomorphism $f \colon U \to V$, and if f is the identity mapping, then V is denoted by (U, θ). By (c) of the proposition, the U and (U, θ) dualities are equivalent if and only if θ is an inner automorphism.

23.35 *Exercises* (Morita). (a) If $_B U_A$ is a duality context, then there is a ring isomorphism center $A \approx$ center B. Conclude that commutative rings are linked by dualities only if they are isomorphic.

(b) Let A be a commutative ring, and let $\mathcal{S}_A$ be a Serre class of mod-A containing A, and let T be a duality $\mathcal{S}_A \rightsquigarrow \mathcal{S}_A$.
(1) T induces a ring automorphism θ of A of order ≤ 2.
(2) If $U = TA$, then T induces a (A, θ)-semilinear isomorphism $U \to U$.
(3) T is equivalent to the (U, θ)-duality.

(c) Let A be a commutative Artinian ring. Show that there exists a bijection between the classes of equivalent dualities $\mathcal{S}_A \rightsquigarrow \mathcal{S}_A$ and the set of ring auto-morphisms θ of A of order ≤ 2. [Hint: It suffices to assume A is local. Show that the injective hull U of $A/\operatorname{rad} A$ is finitely generated. Then U defines an A-duality, and every duality is an (U, θ)-duality for an appropriate θ.]

Exercises for Chapter 23

1. If a ring A has a duality context $_B U_A$, then any factor ring $A' = A/I$ has a duality context $_{B'} U'_{A'}$, where $U' = \operatorname{ann}_U I$, and $B' = B/K$, where $K = \{b \in B \mid b U' = 0\}$.

[A module M is (algebraically) **linearly compact (l.c)** (i.e. linearly compact in the discrete topology—see Chapter Notes) provided that any finitely solvable system of congruences $x \equiv x_a \pmod{M_a}$ has a simultaneous solution, where the M_a are submodules. Thus, a valuation ring R is linearly compact iff R is a maximal valuation ring. Cf. *sup.* 24.46.]

2. Müller [70]) A ring A has a Morita duality context $_BU_A$ iff the minimal injective cogenerator E of mod-A is linearly compact. In this case, then E defines a duality context. Moreover, for any duality context, the U-reflexive right modules coincide with the linearly compact modules. In particular, A is linearly compact in mod-A.

3. (Onodera [72]) Any linearly compact right A-module M has finite Goldie dimension. Any submodule or quotient module of a l.c. module is l.c.

4. (Onodera [72]) Any linearly compact module M is complemented, that is, for any submodule N there is a submodule N' minimal with respect to $M = N + N'$.

5. (Osofsky-Onodera) Any linearly compact ring R is semiperfect. Moreover, if R is right l.c., and right socular, then R is right Artinian (compare 23.20).

6. (Onodera [72]) A ring R is a twosided injective cogenerator iff R is a onesided linearly compact cogenerator.

7. There exists a left Artinian ring R, with finitely generated injective cogenerator U, but R is not right Artinian. (See Osofsky [68e]. Compare Examples 24.34.)

8. (Müller [70]) If a commutative ring A has a Morita duality, context $_BU_A$, then there is a duality context $_AW_A$.

9. (Müller, loc.cit.) A commutative local ring R has a Morita duality iff R is U-complete (with respect to the finite topology defined by the minimal injective cogenerator U), and R/I is an injective cogenerator ring for every completely co-irreducible ideal I (cf. PF rings, 24.32).

10. Let A be a finite dimensional algebra over a field k. Show that A is right (injective) σ-cyclic if and only if every indecomposable (projective) left A-module embeds in $A' = \mathrm{Hom}_k(A, k)$.

11. (Fuller [69a]) If e is an idempotent element in a left Artinian ring R, then the following are equivalent:

(a) Re is injective.

(b) For each e_i in a basic set of idempotents for e there is a primitive idempotent f_i in R such that socle $Re \approx Rf_i/Jf_i$ and socle $f_i R \approx e_i R/e_i J$.

(c) There exists an idempotent f in R such that (i) $\mathrm{ann}_{fR} Re = 0$ and $\mathrm{ann}_{Re} fR = 0$. (ii) The functors $\mathrm{Hom}_{fRf}(\_, fRe)$ and $\mathrm{Hom}_{eRe}(\_, fRe)$ define a duality between the category of finitely generated left fRf-modules and the category of finitely generated right eRe-modules.

Moreover, if Re is injective, then the $f_i R$ of (b) and fR of (c) are also injective.

12. (Morita [58] and Dlab-Ringel [72a]) Over any ring R, any finendo injective cogenerator is balanced.

13. (Cohn [66]) Let U be a (B, A)-bimodule, where $B = \mathrm{End}\, U_A$ is left, and $A = \mathrm{End}_B U$ is right, Noetherian. (a) If U is a cogenerator (resp. injective) in mod-A, then U is injective (resp. cogenerator) in B-mod. (b) Conclude that U is an injective cogenerator in A-mod if and only if it is B-mod, and the $_BA_A$

is a duality context. Conclude that U is finitely generated, and A (B) is right (left) Artinian.

14. An infinite direct sum of U-reflexive injective modules cannot be U-reflexive.

15. (Erdös-Kaplansky) If V is an infinite dimensional vector space over a field A of dimension b, then the dimension of the A-dual of V is 2^b. Formulate a generalization of the exercise.

16. (Matlis [58]) Let R be a commutative Noetherian local ring with radical P, let $E = \widehat{R/P}$, and $B = \operatorname{End} E_R$. Then B is the P-adic completion of R (cf. 18.9.7), E is Artinian, and there is a duality between Artinian R-modules and finitely generated B-modules induced by $(-, E) = \operatorname{Hom}_R(-, E)$.

Notes for Chapter 23

The Pontryagin [39] duality Char on the category LCAb of locally compact Abelian groups to itself defined by sending a group $X \in \mathrm{LCAb}$ into its character group:
$$\operatorname{Char} X = \operatorname{Hom}(X, \mathbb{R}/\mathbb{Z})$$
where $\mathbb{R}/\mathbb{Z}$ is the circle group (the "reals mod-1"), and the morphisms are continuous (cf. I, 2.6.6, p. 94).

Pontryagin's theorem states that the category Ab of (discrete) Abelian groups is dual to the category of compact Abelian groups under the functor $X \mapsto \operatorname{Char} X$, giving the pointwise topology when X is discrete (when $X \in \mathrm{Ab}$), and the uniform topology in case X is compact. (Thus, X has the compact open topology in either case.) Morita [58] proved that if D is any duality on LCAb, then D is equivalent to a U-duality induced by $U = D(\mathbb{Z}) \approx \mathbb{R}/\mathbb{Z}$. Thus,
$$D(X) = \operatorname{Hom}(X, U) \approx \operatorname{Hom}(X, \mathbb{R}/\mathbb{Z}) = \operatorname{Char} X$$
that is, every duality is equivalent to the Pontryagin duality. (Here, of course, $\mathbb{Z}$ is endowed with the discrete topology.) [For duality for noncommutative compact groups, consult Hochshild [65].]

In his book, *Algebraic Topology*, Lefschetz extended the duality for finite dimensional vector spaces over a field R to a duality between all abstract and all linearly compact topologized vector spaces. Kaplansky [53] (discussed in Kaplansky [69, pp. 79–80]) did the same for modules over a discrete valuation ring. Here one uses $X^* = \operatorname{Hom}(X, K/R)$ the set of all continuous homomorphisms into the quotient field K modulo R. Then, if X is discrete, X^* is linearly compact (in the topological sense), and conversely, X is the full dual of X^*.

These theorems have been generalized by Leptin, Schöneman, MacDonald, and Müller [71]. Moreover, Müller shows that among the topologies defined by the U-dualities, there is a coarsest and a finest.

Oberst [70] characterized the dual of a Grothendieck category (roughly) as the category of strict complete topological coherent (STC) left R-modules over some STC linearly compact ring R. (Warning: read (2) of Oberst [70, p. 540]). Müller [70] characterizes when a ring R has a Morita duality via the condition that the minimal injective cogenerator E is linearly compact. Then, for any

U-duality of R, the U-reflexive right R-modules coincide with the linear compact modules. (This aspect is gone into in depth in Müller [71]. In another direction, paralleling the duality between compact spaces and commutative C^*-algebras given by the Gelfand and Naimark theorem, Hoffman [70] establishes a complete duality between the category of compact semigroups and the category of C^*-algebras with comultiplication.)

These results still leave unknown the structure that a ring must have in order to possess a duality. In reference to Müller's characterization, how does one know when E is linearly compact just by looking at R? Of course, for any duality context $_BU_R$; R itself must be (right) linearly compact (since it is U-reflexive). (Also see Chapter Exercise 9.) Any commutative Artinian ring has a duality (23.32 and 23.26), but an arbitrary Artinian ring may not. This was shown by a theorem of Rosenberg and Zelinsky [57] which reduced the problem to the existence of a field K having a subfield F such that K had right vector space dimension over F a finite integer not equal to the left vector space dimension. Such an example was adduced by Cohn [61] to answer Artin's problem.

Lambek and Rattray [75] look at duality in cocomplete additive categories, generalizing many classical theorems, including theorems of Matlis [58] (see Exercise 16) and Kaplansky [53]. For a noncommutative Noetherian ring version of Matlis's theorem, see Jategaonkar [74a, b].

In the next chapter Artinian rings with an $_RR_R$ duality context are characterized ideal-theoretically (24.5). More generally, rings possessing an $_RR_R$ duality context are characterized subject to knowing when R is right (left) selfinjective (24.32).

For further references to these rings and other notes on duality, see Notes for Chapter 24.

References

Azumaya [59], Baer [43b], Bass [60], Cohn [61, 66a], Curtis [59], Fuller [69a], Hochschild [65], Hoffman [70], Ikeda [51, 52], Ikeda and Nakayama [54], Jans [61], Jategaonkar [74a, b], Kaplansky [53], Lambek and Rattray [75], Matlis [58], Morita [58, 67, 69], Morita, Kawada, and Tachikawa [57], Müller [70b, 71], Nagoa and Nakayama [53], Nakayama [39, 40, 41], Oberst [70], Onodera [72, 73], Ornstein [66], Osofsky [66, 68e], Pontryagin [39], Rosenberg and Zelinsky [57, 61], Sandomierski [72], Tachikawa [58, 59].

Chapter 24. Quasi-Frobenius Rings

A ring A is quasi-Frobenius (QF) in case A is right and left Artinian, and there exists an A-duality

$$\text{fin. gen. mod-}A \rightsquigarrow \text{fin. gen. } A\text{-mod.}$$

This is characterized 24.4 by the demand that every onesided ideal of A be an annulet (annihilator). Such rings are right and left selfinjective. Conversely, if A is such that A is an injective A-module, and if A is either right or left Noetherian or Artinian [1], then A is QF 24.5. Over a QF-ring A, every faithful module is a generator, and contains the basic module as a summand 24.6. An arbitrary ring A is QF if and only if every injective right A-module is projective 24.12. The dual theorem also holds 24.23 (but the proof is not dual). Over basic QF rings, finitely generated faithful modules are determined by their endomorphism rings 24.26. Remarks about inner automorphisms of basic QF rings in the chapter should be compared with the corresponding theorems of Skolem-Noether for Azumaya algebras 32.42. The exercises indicate numerous related theorems.

Pseudo-Frobenius (PF) rings are defined, generalizing the property which holds in QF rings: any faithful right R-module generates the category mod-R of all right R-modules. Then R is said to be right PF. By 24.32, R is right PF iff R is a semilocal right selfinjective ring with essential socle. (In other words, a right selfinjective essentially right Artinian ring!) If every factor ring is PF, then R is Artinian primary-decomposable serial, and conversely. (This is proved in Chapter 25. See 25.4.6 A.)

We begin with some simple remarks about annulets.

24.1 Proposition. *The following two conditions on a ring R are equivalent:*

(a) *Each cyclic module is contained in a projective module in* mod-R.

(b) *Each right ideal of R is the annihilator of a finite subset of R.*

Proof. This is 20.26. $\square$

24.2 Corollary. *The following two conditions for a ring R are equivalent:*

(a) *Every cyclic R-module can be embedded in a product of copies of R in* mod-R.

(b) *Every right ideal of R is a right annulet.* $\square$

24.3 *Exercise.* (a) If every right ideal (and every left ideal) of R is the annihilator of a finite subset of R, then R is left and right Artinian.

---

[1] It is enough to assume the a.c.c. on right or left annulets 24.25.

(b) Find an example of a ring which every cyclic right R-module is contained in a free R-module in mod-R, but such that R is not left Artinian.

(c) If Q is an ideal of A, then every left ideal of A containing Q is a left annulet iff Q is a left annulet and every left ideal of the factor ring A/Q is a left annulet. Conclude in a semilocal ring A that every (maximal) left ideal of A containing the radical J is a left annulet iff J is a left annulet.

(d) A semiprime ring R is semisimple iff every left ideal is a left annulet.

(e) Show that there exists a nil ideal N of a ring R with the properties that N is the intersection of all ideals H such that R/H is semiprime. [Then N is called the *(Baer) lower nil radical of* R. Cf. 26.11.] Show that the following are equivalent: (1) Every left ideal of R containing N is a left annulet; (2) R/N is semisimple, and N is a left annulet. In this case conclude that $N = \operatorname{rad} R$.

(f) A right selfinjective ring R is a cogenerator in mod-R iff every maximal right ideal of R is a right annulet and iff R is a semilocal ring such that the radical J is a right annulet.

(g) (Faith-Walker [67]) A semilocal ring R which is a cogenerator of mod-R is right selfinjective.

24.4   Theorem and Definition. *A ring A is called* **quasi-Frobenius,** *abbreviated* QF, *provided that A is a left and right Artinian and satisfies any of the following equivalent conditions:*

(a) *A defines an A-duality*

$$\text{fin. gen. mod-}A \rightsquigarrow \text{fin. gen. } A\text{-mod.}$$

(b) *Every simple module in mod-A, or in A-mod, is reflexive.*

(b') *The radical of A is a right and left annulet, and every simple right or left ideal is an annulet.*

(c) *The A-dual of every simple right, or left, A-module is simple.*

(d) *The mapping*

$$\begin{cases} \varphi: \text{ right ideals } A \rightarrow \text{left ideals } A \\ \varphi: \ I \mapsto {}^\perp I = \operatorname{ann}_A I \end{cases}$$

is a lattice isomorphism between the lattice of right ideals of A and the lattice of left ideals of A. Expressed otherwise, every one-sided ideal of A is an annulet.

(e) *A is injective as a right A-module.*

(f) *A_A is an injective cogenerator.*

Proof. (a)–(d) correspond to the $A = U$ case of various parts (23.25). The equivalence of (a)–(d) therefore follows from 23.25, and likewise the implications (a) $\Rightarrow$ (e), and (f) $\Rightarrow$ (a).

(e) $\Rightarrow$ (f). Let e and f be principal indecomposable idempotents of A. Since eA is a summand of A_A, it is injective, and indecomposable, so socle(eA) is a simple A-module, and socle$(eA) \approx$ socle(fA) if and only if $eA \approx fA$. By the results on semiperfect rings, namely 22.22, the result just stated implies that every simple right A-module is isomorphic to the socle of some fA. Since every simple right A-module is therefore isomorphic to a submodule of A_A, then A is an injective antecogenerator in mod-A, and hence, by 3.31, A is an injective cogenerator.

(b') $\Leftrightarrow$ (c). By Exercise (c) of 24.3, every maximal right ideal M is a right annulet. Now the A-dual of the simple module $V = A/M$ is canonically isomorphic to $^\perp M$ under $f \mapsto f(1 + M)$. Then, if W is a simple left ideal contained in $^\perp M$, necessarily $W^\perp = M$. Since W is a left annulet, then $^\perp M = W \approx \mathrm{Hom}_A(V, A)$ is simple. This proves that the A-dual of every simple right and, by symmetry, that of every simple left, module is simple. This proves the non-parenthetical part of (b') $\Rightarrow$ (c), and the parenthetical part is an exercise. Furthermore, (c) $\Rightarrow$ (d), and (d) $\Rightarrow$ (b') are trivial. $\square$

24.5 Theorem. *The following conditions on a ring A are equivalent:*

(a) A *is quasi-Frobenius (QF).*

(b) A_A *is injective and Noetherian.*

(c) A_A *is injective and Artinian.*

(d) A_A *is injective, and* $_A A$ *is Noetherian.*

Proof. (b) $\Rightarrow$ (c). Since A is right Noetherian, A satisfies the d.c.c. on left annulets. Since A_A is injective, every finitely generated left ideal of A is, by 19.3, a left annulet (cf. 23.21). Thus, A satisfies the d.c.c. on finitely generated left ideals, so A is right perfect, and hence right Artinian by lemma 23.19.

(c) $\Rightarrow$ (d). Since A_A is Artinian, A satisfies the a.c.c. on left annulets, hence, by 19.3, the a.c.c. on finitely generated left ideals, so A is left Noetherian.

(d) $\Rightarrow$ (a). By Utumi's theorem 19.30, $A/\mathrm{rad}\,A$ is regular, so the Noetherian hypothesis implies semisimplicity of $A/\mathrm{rad}\,A$. Now A_A injective and $_A A$ Noetherian implies that every left ideal of A is a left annulet. In particular, if $J = \mathrm{rad}\,R$, then J^n is a left annulet for each n. Now $(J^n)^\perp$ is an ideal of A, so $_A A$ Noetherian implies that $(J^n)^\perp = (J^{n+1})^\perp$ for large n, and then $J^n = J^{n+1}$. Since $M = J^n$ is a finitely generated left ideal satisfying $JM = J$, then $M = J^n = 0$ by 18.4, so A is therefore semiprimary, and 18.12 implies that A is left Artinian. Furthermore, by the proof of (e) $\Rightarrow$ (f) of 24.4, it follows that A is an injective cogenerator in mod-A. In this case, by (a) of 23.23, or by 24.2, every right ideal of A is an annulet. Consequently, the Noetherian hypothesis on $_A A$ implies that A is right Artinian. Then A is QF by 24.4. $\square$

24.6 Corollary. *If A is QF, the following conditions on an A-module M are equivalent:*

(a) M *is a faithful right A-module.*

(b) M *contains the basic module of A 18.24.*

(c) *The basic module of A is isomorphic to a summand of M.*

(d) M *is a generator of* mod-A.

Proof. (a) $\Rightarrow$ (d). By 19.13A, there exists $n > 0$ such that M^n contains a copy of A. Since A_A is injective, A is a summand of M^n, hence there exists an epic $M^n \to A$, so M is a generator. (d) $\Rightarrow$ (c) is 18.24.

(b) $\Rightarrow$ (a). The basic module is a generator, hence is faithful. $\square$

24.7 Corollary. *Any finitely generated faithful projective A-module P over a QF ring A is a progenerator for* mod-*A, and* $S = \operatorname{End} P_A$ *is QF.* □

24.8 *Exercise.* (a) A ring A is QF if and only if every injective module is projective, and every projective module is injective.

(b) If A is QF, then the basic module B of A is the injective hull, and projective cover, of socle B.

(c) If A is QF, then A is the injective hull of $A/\operatorname{rad} A$.

(d) The basic module of a QF ring A is a minimal injective cogenerator in mod-A.

(e) (Baer [43] and Ikeda [51]) If A is right Artinian, and if there exists a lattice anti-isomorphism of the lattice of right ideals to the lattice of left ideals, then A is QF.

(f) (Kawada [57], Morita and Tachikawa [56], Utumi [60]) Let $P(M)$ denote the condition on a right R-module M which states for any two submodules A and B that $M/A \approx M/B$ only if $A \approx B$. Furthermore, let (l_n) be the condition $P(R^n)$ in R-mod, and (r_n) the same condition in mod-R. Then:
(1) R is QF $\Leftrightarrow (r_n)$ and (l_n) $\forall n$.
(2) If R is a f.d. algebra over an algebraically closed field, then R is QF $\Leftrightarrow (r_1)$ or (l_1) holds.
(3) If R satisfies (l_2) or (r_2), then R is QF.
(4) If the total matrix ring R_2 of degree 2 satisfies (r_1) or (l_1), then R is QF.

(g) Show that the property that every faithful right A-module generates mod-A characterizes QF rings among right or left Artinian rings.

(h) (Tachikawa [69]) Prove Exercise (g) with "finitely generated" modifying "faithful right A-module".

(i) Every group ring RG of a finite group over a QF ring is QF.

Projective Cogenerators Over Semilocal Rings

A QF ring A is semilocal, and A is a projective cogenerator in mod-A. We now establish a theorem on general projective cogenerators over semilocal rings, which we need in the determination of QF rings as those rings over which every injective module is projective 24.12.

24.9 Proposition. *If P is a finitely generated projective cogenerator in* mod-*R, and if R is semilocal, that is, if R/*rad*R is semisimple, then P and R are injective R-modules.*

Proof. The proof uses the following two results (see 22.5 and 22.6).

I. Let P and Q be finitely generated projective modules in mod-R and $J = \operatorname{rad} R$. Then $P/PJ \approx Q/QJ$ if and only if $P \approx Q$.

II. Let P be a finitely generated projective module in mod-R, let $\Lambda = \operatorname{End}_R P$, $Q = \operatorname{rad} \Lambda$, and $J = \operatorname{rad} R$. Then the ring $\operatorname{End}_R P/PJ$ is isomorphic to the quotient ring Λ/Q.

Let $U_1, \ldots, U_n$ denote a complete set of nonisomorphic simple right R-modules, and let $\hat{U}_i$ denote the injective hull, $i = 1, \ldots, n$. Since P is a cogenerator, we may assume $\hat{U}_1, \ldots, \hat{U}_n$ are contained in P. Since these modules are injective, they are summands of P, hence are finitely generated and projective along with P. Since $U_1, \ldots, U_n$ are nonisomorphic simple submodules of P, they are independent submodules; that is, the sum $U_1 + \cdots + U_n$ is direct. Consequently, the sum $\hat{U}_1 + \cdots + \hat{U}_n$ is direct. Furthermore, $C = \hat{U}_1 \oplus \cdots \oplus \hat{U}_n$, being a direct sum of projective injective modules, is projective and injective. Since U_i is simple, $\hat{U}_i$ is indecomposable (and injective), so $\Lambda_i = \operatorname{End}_R \hat{U}_i$ is a local ring. By II, $\Lambda_i/Q_i \approx \operatorname{End}_R \hat{U}_i/\hat{U}_i J$, where $Q_i = \operatorname{rad} \Lambda_i$, $i = 1, \ldots, n$. Since $\hat{U}_i/\hat{U}_i J$ is a semisimple module whose endomorphism ring is a division ring $\approx \Lambda_i/Q_i$, $\hat{U}_i/\hat{U}_i J$ must be a simple module. Since $\hat{U}_i$ is finitely generated, I implies that $\hat{U}_i/\hat{U}_i J$ is isomorphic to $\hat{U}_j/\hat{U}_j J$ if and only if $\hat{U}_i \approx \hat{U}_j$. Since U_i is the unique simple submodule of $\hat{U}_i$, and since isomorphic modules have isomorphic injective hulls, we see that this occurs if and only if $U_i \approx U_j$, that is, if and only if $i = j$. Thus, $\hat{U}_1/\hat{U}_1 J, \ldots, \hat{U}_n/\hat{U}_n J$ is a complete set of n nonisomorphic simple modules. Expressed otherwise, each simple module is an epimorph of $C = \hat{U}_1 \oplus \cdots \oplus \hat{U}_n$. Since C is projective, this implies that C is a generator in mod-R. Therefore, there exists an epimorphism $\sum C_\alpha \to R$ of a direct sum of copies of C onto R. Since R is projective, this epimorphism splits, so R is (isomorphic to) a summand of $\sum C_\alpha$. Since R is finitely generated, R is a direct summand of finitely many copies of C. Then injectivity of C implies that of R. Since P is finitely generated and projective, the same argument shows that P is also injective in mod-R. $\square$

24.10 **Corollary.** *If R is a cogenerator in* mod-R, *and if $R/\operatorname{rad} R$ is semisimple, then R is right selfinjective.* $\square$

24.11 *Exercise* (Osofsky [66]). If R is an injective cogenerator in mod-R, then R is semilocal. (Converse of 24.10.)

24.12 **Theorem** (Faith-Walker [67]). *A ring R is QF if and only if each injective right R-module is projective.*

Proof. First assume that R is QF, and let M be any injective right R-module. Since R is right Artinian, it is right Noetherian, so M is a direct sum of indecomposable injective modules. Since a direct sum of projective modules is projective, we are reduced to the case where M itself is indecomposable (and injective). Then M is the injective·hull $\hat{C}$ of any nonzero cyclic submodule C. By the first proposition, C is contained in a finitely generated free module R^n. Since R_R is injective, so is R^n; therefore, the imbedding of C into R^n can be extended to an imbedding of $M = \hat{C}$ into R^n. But M, being injective, is a summand of R^n, and therefore is projective.

Conversely, if each injective module is projective, then each module is contained in a free module, in particular, is contained in a direct sum of cyclic modules, so R is right Artinian by theorem 20.18. But, since each M in mod-R is contained in a direct product, in fact, direct sum, of copies of R, R is a cogenerator. Now $R/\operatorname{rad} R$ is semisimple, so 24.10 implies that R is right selfinjective, so R is QF by 24.5. $\square$

24.13 Corollary. *An Artinian ring R is a cogenerator in mod-R if and only if R is QF.*

Proof. Apply 24.10 and 24.4(f). □

24.14 Theorem. *A ring R is QF if and only if each injective right R-module is a direct sum of cyclic modules that are isomorphic to principal indecomposable right ideals of R.*

Proof. If R is QF, and M is injective, then M is a direct sum of indecomposable injective modules $\{M_i | i \in I\}$. But each M_i is projective by 24.12, and hence isomorphic to a principal indecomposable right ideal by 20.15. This proves one part. Since a direct sum of principal indecomposable right ideals is projective, the converse follows from 24.12. □

24.15 Corollary. *A ring R is QF if and only if every right R-module is contained in a free right R-module.* □

24.16 Corollary. *If every (injective) right R-module is contained in a direct sum of right ideals, then R is QF.*

For then every injective module is projective.

Injective Projectives

The rest of this chapter is mainly devoted to the proof of the dual to 24.12; namely, a ring R is QF iff every projective right R-module is injective. This implies that every free module is injective, and hence that the injective right R-module R is $\sum$-injective. Thus, the results from Chapter 20 on $\sum$-injectivity are available:

24.17 Corollary. *If the injective hull of R in mod-R is countably $\sum$-injective, then R satisfies the a.c.c. on right annulets.* □

24.18 Theorem (Faith [66a]). *The following conditions on a ring R are equivalent:*

(a) *Any countably generated projective module in mod-R is injective.*

(b) *R is right selfinjective, and satisfies the a.c.c. on right annulets.*

(c) *Any projective module in mod-R is injective.*

Proof. (a) $\Rightarrow$ (b). R_R is injective, and so is $R^{(\omega)}$. Thus, $\hat{R}$ is countably $\sum$-injective, so R satisfies the a.c.c. on right annulets by 20.3 A.

(b) $\Rightarrow$ (c). Since $\mathscr{A}_r(\hat{R}, R)$ coincides with the set of right annulets of R, 20.3 A implies that any free, hence, any projective, module in mod-R is injective. □

24.19 Lemma. *If R is right perfect, and if R satisfies the a.c.c. on right annulets, then R is semiprimary.*

Proof. Let $J = \text{rad} R$. Then R/J is semisimple. By the chain condition,

$$(J^n)^\perp = (J^{n+1})^\perp$$

for some n. If $R = (J^n)^\perp$, then $J^n = 0$, which is what we want. Otherwise, since R is right perfect, the left R-module $R/(J^n)^\perp$ has nonzero socle $T/(J^n)^\perp$, where T is a left ideal $\supset (J^n)^\perp$, and $T \neq (J^n)^\perp$. Since J annihilates this module, then $JT \subseteq (J^n)^\perp$, so $T \subseteq (J^{n+1})^\perp = (J^n)^\perp$, a contradiction. $\quad\square$

24.20   Theorem. *The following conditions on a ring R are equivalent:*

(a) *R is quasi-Frobenius.*

(b) *Each injective right R-module is projective.*

(c) *Each projective right R-module is injective.*

(d) *R is right selfinjective and satisfies the* a.c.c. *on right annulets.*

Proof. (a) $\Leftrightarrow$ (b) is theorem 24.12. (a) $\Rightarrow$ (c) follows from the fact that over a Noetherian ring every module is $\sum$-injective. Hence, every free, that is, every projective, is injective.

(c) $\Leftrightarrow$ (d) is 24.18.

(d) $\Rightarrow$ (a). By the lemma, R is semiprimary, hence by 18.14, R satisfies the d.c.c. on finitely generated right ideals. Let $A_1 \supseteq \cdots \supseteq A_n \supseteq \cdots$ be any descending sequence of right annulets of R. By the following lemma, there is a corresponding sequence $A_1' \supseteq \cdots \supseteq A_n' \supseteq \cdots$ of finitely generated right ideals such that $^\perp A_i' = {}^\perp A_i$, $i = 1, 2, \ldots$. Consequently, there is an integer n such that $A_n' = A_m'$ $\forall m \geq n$. Then $^\perp A_n = {}^\perp A_m$ $\forall m \geq n$, proving that R satisfies the d.c.c. (resp. a.c.c.) on right (resp. left) annulets. Since each finitely generated left ideal is a left annulet, R satisfies the a.c.c. on finitely generated left ideals. Consequently, each left ideal is finitely generated. Then R is QF by theorem 24.5. $\quad\square$

24.21   Lemma. *If R is right selfinjective, and if R satisfies the* a.c.c. *on right annulets, and A, B are right ideals such that $A \supseteq B$, then there exist finitely generated subideals $A' \supseteq B'$ such that $^\perp A = {}^\perp A'$ and $^\perp B = {}^\perp B'$.*

Proof. Since R_R is injective, by 23.21,

$$^\perp(A \cap B) = {}^\perp A + {}^\perp B.$$

Let $A \supseteq B$, and let A' be the finitely generated subideal of A, given by 20.2 A such that $^\perp A = {}^\perp A'$. Then

$$^\perp(A' \cap B) = {}^\perp A + {}^\perp B = {}^\perp B.$$

The last equality holds since $^\perp A \subseteq {}^\perp B$. Hence, by 20.2 A again, there exists a finitely generated subideal B' of $A' \cap B$ such that $^\perp B' = {}^\perp B$. $\quad\square$

24.22   Corollary. *If R is right or left selfinjective, and satisfies the* a.c.c. *on right annulets, then R is QF.*

Proof. If R is left selfinjective, then by 19.11 every finitely generated right ideal is a right annulet, and hence the hypothesis implies that R is right Noetherian, and that R is QF by 24.5. Hence, we may assume that R is right selfinjective, and then R is QF by 24.20 $\quad\square$

Isomorphism of Faithful Modules Over Quasi-Frobenius Rings

If ω is a semilinear (A, θ)-isomorphism $U \to V$, then the mapping

$$\varphi: \operatorname{End} U_A \to \operatorname{End} V_A$$
$$\varphi: f \mapsto \omega f \omega^{-1} \qquad\qquad (f \in \operatorname{End} U_A)$$

is a ring isomorphism $\operatorname{End} U_A \approx \operatorname{End} V_A$ (cf. 23.34).

24.23 Definition. *A finitely generated faithful A-module U is said to be determined by its endomorphism ring provided that to each finitely generated faithful A-module V, and ring isomorphism $\varphi: \operatorname{End} U_A \to \operatorname{End} V_A$, there corresponds a semilinear (A, θ)-isomorphism $\omega: U \to V$ such that*

$$\varphi(f) = \omega f \omega^{-1} \qquad\qquad f \in \operatorname{End} U_A.$$

Special cases of the next theorem are theorems of Jacobson [64], Shoda (1929) (cited in Morita [62]), Asano [39], and Baer (1943).

24.24 Theorem (Morita [62]). *Any finitely generated, faithful module U over a basic QF-ring A is determined by its endomorphism ring*[2].

Proof. By 24.6, U and V are proper generators in mod-A (see 16.17). Hence, by 16.20, the isomorphism $\varphi: \operatorname{End} U_A \to \operatorname{End} V_A$ implies an equivalence $T: \operatorname{mod-}A \rightsquigarrow \operatorname{mod-}A$ such that $TU = V$, and such that T induces φ. If $\varphi = T(1_U)$, then

$$
\begin{array}{ccc}
U & \xrightarrow{\ \omega\ } & V = TU \\
f \downarrow & & \downarrow T(f) = \varphi(f) \\
U & \xrightarrow[\ \omega\ ]{} & V = TU
\end{array}
$$

commutes, where $\omega: U \to V$ is the natural isomorphism defined by T. Therefore, $\varphi(f) = \omega f \omega^{-1}$. $\square$

If f is an inner automorphism of a ring R, there exists $x \in R$ such that $f(a) = x a x^{-1} \ \forall a \in R$. Thus f induces the identity mapping on center R. A ring R has the *inner automorphism property* if a ring automorphism f is necessarily inner if f induces the identity mapping on center R. Besides commutative rings, examples of such rings are simple algebras of finite dimension (Theorem of Skolem-Noether), and Azumaya algebras over local rings. (See Bass [67, p. 109].)

24.25 Theorem. *Let A be any self-basic QF-ring with the inner automorphism property, and let U and V be finitely generated faithful right A-modules. Then any ring isomorphism*

$$\varphi: B = \operatorname{End} U_A \to C = \operatorname{End} V_A$$

such that

$$\varphi(b) = b \qquad\qquad \forall b \in \operatorname{center} B$$

---

[2] Actually, Morita proves that modules are determined by "strongly dense subrings" of endomorphism rings.

has the form

$$\varphi(f) = \omega f \omega^{-1} \qquad\qquad \forall f \in B$$

for some A-isomorphism $\omega: U \to V$.

Proof. By 24.24, we have a semilinear (A, θ)-isomorphism $\omega': U \to V$ which determines φ. Now

$$\begin{cases} h: \text{ center } A \to \text{ center } B \\ h: a \mapsto a_d \end{cases}$$

is a ring isomorphism. If $a \in$ center A, and $u \in U$, then

$$\omega'(u)\,\theta(a) = \omega'(u\,a) = \varphi(a)\,\omega'(u) = a\,\omega'(u) = \omega'(u)\,a.$$

Thus, $\theta | \text{center } A = 1_A$, so θ is inner by assumption. If θ is the inner automorphism effected by $a' \in A$,

$$\theta: a \mapsto a'\,a(a')^{-1} \qquad\qquad \forall a \in A$$

then $\omega = \omega'\,a_d'$ is the desired isomorphism $U_A \to V_A$. $\square$

24.26 Corollary. *Let A be a self-basic QF-ring with the inner automorphism property, let U be a finitely generated faithful right A-module, and let $B = \operatorname{End} U_A$. If A has the inner automorphism property, so does B.* $\square$

Proof. Put $V = U$ in the theorem. Then $\omega \in B = \operatorname{End} U_A$, and φ is the inner automorphism determined by ω. $\square$

Next is an immediate corollary to theorem 24.25 and its proof.

24.27 Corollary. *Let A be a QF-ring, let $A_0 = e_0 A e_0$ denote its basic ring, and let U and V be finitely generated faithful right A-modules. Then there exists a ring isomorphism $\operatorname{End} U_A \approx \operatorname{End} V_A$ if and only if there exists a semilinear (A_0, θ)-isomorphism of the right A_0-modules $U e_0$ and $V e_0$.* $\square$

Group Rings Over Selfinjective Rings

In this section, we prove that any group ring AG of a finite group G is a right selfinjective ring iff A is right selfinjective. An immediate corollary is that AG is QF iff A is QF, a fact that supplies an inexhaustible source of nonsemisimple QF rings since any finite group G defines a nonsemisimple QF group ring AG over any field A of characteristic dividing $|G|$.

We begin with a lemma for group rings over any ring.

24.28 Trace Lemma for Group Rings. *Let G be any group, A any ring, and $R = AG$ the group ring. The trace function*

$$\operatorname{Tr}_G \begin{cases} AG \to A \\ \sum_{g \in G} a_g g \mapsto \sum_{g \in G} a_g \end{cases} \qquad\qquad (a_g \in A \ \ \forall g \in G)$$

is an element of $\text{Hom}_A(AG, A)$, *and for any right AG-module M there is a mono-morphism of Abelian groups*

$$t: \text{Hom}_{AG}(M, AG) \to \text{Hom}_A(M, A)$$

where

$$t(f)(m) = \text{Tr}_G(f(m))$$

for all $f \in \text{Hom}_{AG}(M, AG)$ *and all* $m \in M$. *Moreover, if G is finite, this is an iso-morphism.*

Proof. We first show that

(1) $f(m) = \sum_{g \in G} \text{Tr}(f(m)(g)) \, g^{-1} = \sum_{g \in G} t(f)(m\,g) \, g^{-1}$

so that f is determined uniquely by $t(f)$. To do this, if $r \in AG$, let $c_g[r] \in A$ be defined by

$$r = \sum_{g \in G} c_g[r] \, g^{-1}$$

then

$$r\,h = \sum_{g \in G} c_g[r] \, g^{-1} h \qquad\qquad (\forall\, h \in G)$$

so, setting $k^{-1} = g^{-1} h$, we have the formula:

(2) $c_{rh}[k] = c_g[r] = c_{hk}[r]$.

Therefore,

$$t(f)(m\,g) = c_{f(mg)}[1]$$
$$= c_{f(m)g}[1], \quad \text{since } f \text{ is an } R\text{-homomorphism,}$$
$$= c_{f(m)}[g], \quad \text{using (2),}$$

that is,

(3) $c_{f(m)}[g] = t(f)(m\,g)$ $\forall\, m \in M,\; g \in G$.

This proves the formula (1), and shows that t is an injective homomorphism of groups.

In order to show that t is an isomorphism when G is finite, let $p \in \text{Hom}_A(M, A)$, and using (2) defined $f: M \to R$ by the formula:

$$c_{f(m)}[g] = p(m\,g).$$

Then, $f(m) \in R \; \forall\, m \in M$, and clearly, $t(f) = p$, that is, t is epic. □

24.29   Theorem (Connell [63]). *Let G be a finite group, and A a ring. Then, the group ring AG is right selfinjective iff A is right selfinjective.*

Proof. Sufficiency: Put $R = AG$. Since $\text{Hom}_R(R, R) \approx R$ by (I, 3.63, p. 121) then 24.28 gives the isomorphism

$$t: R \to \text{Hom}_A(R, A) = H$$

where $t(r)(r') = c_{rr'}(1)$. It is easily checked that t is an R-module isomorphism. By (I, 5.62, p. 281), we conclude that H is an injective right R-module, since A

is an injective right module, and R is a flat left A-module. Thus, R is right self-injective as claimed. The necessity is an exercise.   □

24.30 Theorem. *Let A be any QF ring, and G a finite group. Then the group ring $R = AG$ is QF.*

Proof. Since A is right (and left) Artinian, and Noetherian, and since R is a finitely generated A-module, then R has finite (Jordan-Hölder) length as an A-module. However, any onesided ideal of R is canonically an A-module, so that A has finite length as an A-module (both sides). Hence, by 24.5, it suffices to prove that R is selfinjective (either side). However, this follows from 24.29.   □

The **augmentation ideal** of a group ring AG is the kernel of the trace function $AG \to A$.

24.31 *Exercise.* (a) (Jennings [41]) The group ring AG of a finite group G over a field A of characteristic dividing $|G|$ is not semisimple. If G is a p-group, and A has characteristic p, then $\mathrm{rad}(AG)$ is the augmentation ideal, and is nilpotent.

(b) (Connell [63]) A group ring AG is semisimple iff A is a semisimple ring, G is a finite group, and $|G|$ is a unit of A (cf. Maschke's theorem (I, p. 475)).

(c) (Jennings [41], Formanek [70]) Let G be a finitely generated torsionfree nilpotent group, and A be any ring. Then, the augmentation ideal N of AG is **residually nilpotent** in the sense that $\bigcap_{n < \omega} N^n = 0$.

(d) (Connell, Farkas [73], Renault [70], Jain) If AG is right selfinjective, then G must be finite.

Injective Cogenerator Rings

By a theorem of Morita, in order for a ring A to have an A-duality, it is necessary for A to be an injective cogenerator both in A-mod and mod-A. (See 23.16.) As we have noted in this chapter, QF rings are examples of such rings, including the group ring AG of a finite group G over any QF ring A, in particular, over any field (24.30). Until 1966, QF rings were the only known examples of injective cogenerating rings, but then Osofsky [66] provided other examples (see 24.33). We now consider a theorem envisioning a ring R which is an injective cogenerator as a right R-module even though up to now all known examples are twosided injective cogenerators! (However, see Example 24.34.2.)

For references to the next proposition, consult Azumaya [66], Osofsky [66], and Utumi [67].

24.32 Definition and Proposition. *A ring R is said to be **pseudo-Frobenius** (**right** PF) provided that the equivalent conditions hold:*

(a) *R is an injective cogenerator in mod-R.*

(b) *Every faithful right R-module is a generator of mod-R.*

(c) *R is injective in mod-R, with finitely generated essential right socle.*

(d) *R is a semiperfect right selfinjective ring with essential right socle.*

(e) *R is right selfinjective and essentially right Artinian.*

Proof. (a) $\Rightarrow$ (b). (a) implies by 23.23 that every right ideal of R is a right annulet. Thus, if M is any faithful right R-module with $\text{trace}_R M = I \neq R$, then there is a nonzero element $a \in R$ such that $aI = 0$ and so $af(m) = 0 \; \forall f \in M^* = \text{Hom}_R(M, R)$, and $m \in M$. This implies that $aM^* = 0$. Since M is faithful, then, in mod-R, R is a submodule, hence a summand of a product M^I, say $M^I = R \oplus X$. Then, applying the functor $(\ )^* \approx \text{Hom}_R(\ , R)$, and using the isomorphism $(M^I)^* \approx (M^*)^I$, we obtain that R is a summand of $(M^*)^I$. Thus, M^* is faithful, so $aM^* = 0$ implies $a = 0$. This contradiction shows that $\text{trace}_R M = R$, so by 3.26, M is a generator of mod-R.

(b) $\Rightarrow$ (c) Let G be the direct sum of the injective hulls of representatives of each equivalence class of simple right R-modules, that is, let G be the smallest cogenerator of mod-R (see 3.55). Then, G is faithful, and so G is a generator, and $G^n = R \oplus X$, for some integer $n > 0$. By the unique decomposition theorem 18.18, R is a direct sum of finitely many indecomposable injective modules with simple socles. This proves (c).

(c) $\Rightarrow$ (d). (c) implies that R is the direct sum of the injective hulls of simple right ideals. Then R has an Azumaya diagram, and so R is lift/rad semilocal by 18.26, and semiperfect by 22.23.

(d) $\Rightarrow$ (a). If eR and fR are prindecs, then injectivity implies that eR and fR have simple socles, and that $\text{socle}\, eR \approx \text{socle}\, fR$ if and only if $eR \approx fR$. The corresponding statement 22.5 for the tops of eR and fR holds in general by projectivity. Thus, every simple right R-module is isomorphic to the socle of some right prindec, that is, that R is an antecogenerator. Since R is injective, then by 3.41, R is a cogenerator.

This proves the equivalence of (a)–(d); and (c) $\Leftrightarrow$ (e) follows from 19.13A. $\square$

24.33 *Exercise.* (a) (Faith [71d]) A ring R is right PF if R is right self-injective, and every faithful ([quasi]injective) right R-module is finendo.

(b) (Osofsky [66]) There exist right PF-rings which are neither right nor left Artinian nor Noetherian.

(c) (Onodera [68]) If R is right and left cogenerating, then R is right and left selfinjective.

(d) (Morita [58], Kato [68]) R is right and left PF iff every finitely generated module is reflexive.

(e) (Osofsky [66]) If R is right and left PF, and if R is right or left perfect, then R is QF. [Hint: Apply 23.20.]

By 24.5, and 24.32, QF rings are right and left PF. The next examples show that PF rings need not be QF.

24.34 *Examples* (Osofsky [66])

24.34.1   An injective cogenerator without chain conditions.

Let $\mathbb{Z}_{(p)}$ denote the p-adic integers for some prime p.

Define a ring R by $(R, +) = \mathbb{Z}_{(p)} \oplus \mathbb{Z}_{p^\infty}$, and for $(\lambda, x), (\mu, y) \in R$, $(\lambda, x)(\mu, y) = (\lambda \mu, \lambda y + \mu x)$. This multiplication is associative and distributes over addition (the verification uses the facts that $\mathbb{Z}_{(p)} = \text{Hom}_\mathbb{Z}(\mathbb{Z}_{p^\infty}, \mathbb{Z}_{p^\infty})$, and $\mathbb{Z}_{(p)}$ is a commutative ring).

Let I be a proper ideal of R. I may be any additive subgroup of $\mathbb{Z}_{p^\infty}$. If not, let $(\lambda, x) \in I$, $\lambda \neq 0$. Then $(\lambda, x)\mathbb{Z}_{p^\infty} = \mathbb{Z}_{p^\infty} \subseteq I$, and $I/\mathbb{Z}_{p^\infty}$ is an ideal of $\mathbb{Z}_{(p)}$. Such an ideal is of the form (p^i) for some $i \geq 0$, so $I = ((p^i, 0))$.

Thus R is a local ring with maximal ideal $((p, 0))$, and R contains a copy of its only simple module, namely the subgroup of $\mathbb{Z}_{p^\infty}$ of order p. Thus by (I, 3.31, p. 148), if R is injective, it will be cogenerator in mod-R.

Let f be a map from an ideal I of R into R.

If $I \subseteq \mathbb{Z}_{p^\infty}$, f maps I into the torsion subgroup of $(R, +)$, namely $\mathbb{Z}_{p^\infty}$. Since $\mathbb{Z}_{p^\infty}$ is an injective group, f extends to an element $\lambda \in \operatorname{Hom}_\mathbb{Z}(\mathbb{Z}_{p^\infty}, \mathbb{Z}_{p^\infty}) = \mathbb{Z}_{(p)}$. Then $f(0, x) = (\lambda, 0)(0, x)$ for all $(0, x)$ in I.

If $I = ((p^i, 0))$, and $f(p^i, 0) = (0, x)$, then there is a $y \in \mathbb{Z}_{p^\infty}$ such that $p^i y = x$. Then $f(p^i, 0) = (0, y)(p^i, 0)$, so $f(\lambda, z) = (0, y)(\lambda, z)$ for all $(\lambda, z) \in ((p^i, 0)) = I$.

If $I = ((p^i, 0))$, and $f(p^i, 0) = (\lambda, x)$, then $(0 : (p^i, 0))$, the additive subgroup of order p^i, is annihilated by λ. Hence p^i divides λ. Then $f(p^i, 0) = (\lambda/p^i, y)(p^i, 0)$, where y is defined above. Then $f(\mu, z) = (\lambda/p^i, y)(\mu, z)$ for all $(\mu, z) \in I$.

Hence, in all cases, f is given by left multiplication, and R_R is injective by Baer's criterion (I, 3.41, p. 157).

24.34.2   A non-injective cogenerator in mod-R without a.c.c. on direct summands.

Let R be an algebra over a field F with basis $\{1\} \cup \{e_i | i = 0, 1, 2, \ldots\} \cup \{x_i | i = 0, 1, 2, \ldots\}$ such that:

(i) 1 is a twosided identity of R.

(ii) For all i and for all j, $e_i x_j = \delta_{i,j} x_j$ and $x_j e_i = \delta_{i, j-1} x_j$, $e_i e_j = \delta_{i,j} e_i$ and $x_i x_j = 0$.

Here $\delta_{i,j}$ is the Kronecker δ let $J = \operatorname{rad} R$ and $\pi: R \to R/J$ be canonical.

One easily verifies that this multiplication of basis elements associates, and that $J = (\{x_i | i \geq 0\})$. Moreover, $(R/J, +) = \sum \pi(e_i) F + 1 F$, and the simple R-modules are precisely $\{\pi(e_i) R\}$ and $R/\sum e_i R$. Since these are isomorphic to $\{x_{i+1} R\}$ and $x_0 R$ respectively, R will be a cogenerator if each $e_i R$ is injective.

Let I be a right ideal of R, $f: I \to e_i R$. We observe that $(e_i R, +) = e_i F + x_i F$, so that $f = 0$ on $I \cap R(1 - e_i - e_{i-1})$, where $e_{-1} = 0$. Hence f can be nonzero only on $e_i F + e_{i-1} F + x_i F + x_{i+1} F$. Moreover, $x_{i+1} R$ is simple, but not isomorphic to $x_i R$, so f must be 0 on it. We conclude that f can be extended to $I + (1 - [e_i + e_{i-1}])R$ by defining it to be 0 on the last summand. If $e_{i-1} \notin I$, define $f(e_{i-1}) = 0$. Then f is extended to $I + (1 - e_i)R$. If $e_i \in I$, f is extended to a map from $R \to e_i R$; if $x_i \notin I$ or $f(x_i) = 0$, define $f(e_i) = 0$; if not, $f(x_i) = x_i v$ for some $v \in F$, and we define $f(e_i) = e_i v$. In all cases, we have extended f to R, so $e_i R$ is injective by Baer's criterion. $\square$

24.35   *Example* (Levy [66]). Let R be the ring of all formal power series in a variable x indexed by the family W of all well ordered sets of nonnegative real numbers. Thus, an element r of R has the form $r = \sum_{i \in w} a_i x^i$, with $a_i \in \mathbb{R}$. The only nonzero ideals of R are: the principal ideals (x^b), and those ideals of the form

$$(x^{>b}) = \{x^c u \mid c > b, \text{ and } u \text{ a unit of } R\}.$$

Then, if I is any nonzero ideal, then R/I is completely selfinjective (and non-Noetherian) (see Levy [66] for proof). Moreover, R is a local ring, and it is easily verified that the unique simple module $R/\text{rad } R$ embeds in R/I, for any nonzero ideal I, so R/I is an injective cogenerator. (This example is much more intuitive than the ones of 24.34, because any factor ring of a polynomial ring $k[x]$ (over any field k) modulo any nonzero ideal is QF. (Proof?))

24.36 Proposition. *Let R be a basic ring with radical J.*

(a) *Then R is PF if and only if R is isomorphic to the injective hull of its top, R/J, in mod-R.*

(b) *Let A be an ideal of a right PF ring R. Then the following conditions are equivalent:*

(1) *R/A is right PF.*

(2) *The left annihilator $^{\perp}A$ of A is a cyclic right and cyclic left ideal of R.*

(3) *There exists $z \in R$ such that $^{\perp}A = zR = Rz$.*

(4) *There exists $z \in R$ with $z^{\perp} = A$, and $zR = Rz$.*

(5) *There is an isomorphism $^{\perp}A \approx R/A$ in mod-R.*

Proof. If $R \approx \widehat{R/J}$, then R is right PF by (c) of 24.32. Conversely, when R is basic and right PF, the proof of (d) $\Rightarrow$ (a) of 24.32 implies that $R \approx \widehat{R/J}$.

(b) The top of any module M is isomorphic to a summand of the top of M/MA, for any ideal A, since top $M = MJ$ is semisimple, and there is an exact sequence

$$0 \to (MJ + MA)/MJ \to M/MJ \to \text{top } M/MA \to 0.$$

(1) $\Rightarrow$ (3). Since $R = \widehat{R/J}$ by (a), and since the top of R/A embeds in R/J, then there is an embedding of the top of R/A in to R. Again by (a), and the assumption (1), R/A is an essential extension of its top, and hence R/A also embeds in R (3.57). Then A is the right annihilator of the image z of $[1 + A]$ under this embedding (cf. 24.1). Since R is injective in mod-R, then Rz is a left annulet, and so $Rz = {}^{\perp}A$. Since $zR \approx R/A$ is an injective cogenerator in mod-R/A, then every simple submodule W of $^{\perp}A$ in mod-R/A is isomorphic to a submodule V of zR. However, since R is basic, and right selfinjective, necessarily $V = W$, so that socle $^{\perp}A = $ socle zR. Since R has essential right socle, so does $^{\perp}A$. Since the right socles of zR and $^{\perp}A$ are equal, this proves that $^{\perp}A$ is the injective hull of the socle of zR. But $zR \approx R/A$ is the injective hull of the socle of zR. Therefore, $^{\perp}A = Rz = zR$.

(2) $\Rightarrow$ (3) comes from 19.43. Moreover, (3) $\Rightarrow$ (4) and (4) $\Rightarrow$ (5) are obvious from the proof of (1) $\Rightarrow$ (3). Finally, since R has essential socle, then (5) implies that R/A has essential socle, and, by 19.12, that R/A is injective. Then, by (c) of 24.32, R/A is right PF. Thus, (5) $\Rightarrow$ (1). $\square$

Exercises of Chapter 24

1. If A is a (left and right) principal ideal domain, then A/K is QF for any ideal $K \neq 0$. Conclude that a commutative algebra $A = k[x]$ over a field k is either a polynomial ring, or QF.

2. A commutative integral domain R is a Dedekind ring iff R/K is QF for any ideal $K \neq 0$. Cf. Excercise 8, p. 101.

3. A ring R is an Artinian principal ideal ring iff every factor ring of R is QF (resp. right PF) [see 24.36].

4. Let $A = k[x, y]$ be the polynomial ring over a field k in two indeterminates x and y. Then $\bar{A} = A/(x^2, y^2)$ is a QF local algebra of dimension 4 over k. Furthermore, socle $\bar{A} = \bar{A}\bar{x}\bar{y}$, rad $\bar{A} = \bar{A}\bar{x} + \bar{A}\bar{x}$, and $\bar{A}/$socle $\bar{A}$ is not QF.

5. A commutative local Artinian ring is QF if and only if socle A is simple.

6. The truncated polynomial ring in n-variables is the ring

$$A = k[x_1, \dots, x_n]/(x_1^{r_1+1}, \dots, x_n^{r_n+1})$$

where $k[x_1, \dots, x_n]$ is the polynomial ring in n independent indeterminates. Show that A is QF by showing that socle A is simple, in fact, generated by the product $x_1^{r_1} \dots x_n^{r_n}$.

7. (Bass [62]) If A is right and left Noetherian, and if

$$\text{inj. dim } _A A \le 1,$$

and if x is a nonunit non divisor of zero in center A, then $A/(x)$ is QF (see Exercise 1).

8. Let $_B U_A$ be a Morita duality context. Then A is QF if and only if B is QF. Investigate what other ring properties are similarly preserved.

9. Let A be QF. Then rings A and B are Morita equivalent if and only if there exists a duality

$$\text{fin. gen. mod-}A \rightsquigarrow \text{fin. gen. } B\text{-mod}.$$

10. (Baer [43b] and Ikeda [51]) If A is right Artinian, and if there exists a lattice anti-isomorphism

$$\begin{cases} \text{right ideals } A \rightarrow \text{left ideals } A \\ \qquad\qquad I \mapsto I' \end{cases}$$

then A is QF. Moreover, $I' = \text{ann}_A I$ for every I.

11. (Ikeda-Nakayama) A finite dimensional algebra A over a field k is QF if and only if each right ideal is a right annulet if and only if simple right ideals and rad A are annulets.

12. (Nakayama [39], Brauer-Nesbit [37]) An algebra A over a field k is said to be **Frobenius** if $n = \dim_k A$ is finite, and if the following equivalent conditions hold:

(a) The k-dual module A^* of A is isomorphic to A in mod-A.

(b) Every right ideal I is a right annulet, and there holds the dimension relation: $\dim_k(^\perp I) + \dim_k I = \dim^k A$.

(c) A is QF, and the dimension relation for right and left ideals holds.

(d) There exists a hyperplane of the vector space A over k ($=$ a vector subspace of dimension $n-1$) that does not contain a nonzero ideal.

(e) There exists $f \in A^*$ such that ker f contains no nonzero ideal. [By (c), and (e), these conditions are left-right symmetric; and one may assume that ker f is a hyperplane.]

13. (Nakayama [41]) The **Nakayama automorphism** Nak of a Frobenius algebra A over a field k is defined by the linear function f defined by 12(e) such that $\ker f$ contains no nonzero ideal and the identity

$$f(\mathrm{Nak}(x)\,y) = f(y\,x) \quad \forall\, x,\, y \in A,$$

that is, $\mathrm{Nak}(x)\,y - yx$ lies in the hyperplane $H = \ker f$. Moreover, H is invariant under Nak, that is, $\mathrm{Nak}(x) \in H\ \forall\, x \in H$. The Nakayama automorphism is uniquely determined up to an inner automorphism of the algebra A (cf. 23.34).

14. A finite dimensional algebra A over a field k is quasi-Frobenius if and only if the equivalent conditions hold: (a) The basic module of A in mod-A is a summand of $A^* = \mathrm{Hom}_k(A, k)$. (b) A^* is a generator in mod-A.

15. (Tachikawa [69]) If R is left perfect, and if every finitely generated faithful right module M is a generator of mod-R, then R is right PF. Thus, any right or left Artinian ring with this property is QF. (Also see #25 and #26.)

16. If A is semiprimary and an injective cogenerator in A-mod, is A necessarily QF? (Conjecture: No. Cf. Kato [68b]. Cf. also Exercise 24.)

17. Let A be a selfbasic QF ring. Let U be a generator in A, which is not a progenerator (e.g., $U = A \oplus X$, where X is not finitely generated projective). Show that U is finitely generated faithful and projective over $B = \mathrm{End}\, U_A$, but is not a generator in B-mod. Hence, B is not QF even though U_A is faithful and possibly finitely generated.

18. (Faith-Walker [67]) Let A be right selfinjective, and let B be the endomorphism ring of the free module on an infinite set. Then B is right selfinjective if and only if A is QF (cf. Sandomierski [70]).

19. If U is a generator in mod-R, then the endomorphism ring of U is right selfinjective only if R is right selfinjective.

20. (Kato [71]) Over a twosided injective cogenerator ring R (= right and left PF), any finitely generated faithful module M is injective over $\mathrm{End}_R M$ iff M is reflexive.

21. The following conditions on a ring A are equivalent: (a) Every cyclic right module (and every cyclic left A-module) is contained in a projective A-module. (b) Every right ideal (and every left ideal) is the annihilator of a finite subset of A. (c) A is QF.

22. (Faith-Walker [67]) A ring A is QF if and only if each injective right A-module is isomorphic to a direct sum of principal indecomposable right ideals of A.

23. (Kawada [57]) Call a module **pseudoinjective** provided that every isomorphism of a submodule of M into M is induced by an endomorphism of M (cf. S.K.Jain and S.Singh [67, 70]). A finite dimensional algebra A over an algebraically closed field is pseudoinjective as a right A-module iff A is QF, hence iff pseudoinjective as a left A-module.

24. If A is QF, then gl. dim $A = 0$, or ∞.

25 A. (Faith [76 d]) A ring R is said to be **right FPF** if every finitely generated faithful right R-module is a generator of mod-R (see problem 15). A right FPF ring R is *right bounded* in the sense that if I is any essential right ideal, then I contains an ideal $\neq 0$. Moreover, any primitive right FPF ring is simple Artinian, and any prime right FPF ring is right nonsingular. A right and left Noetherian prime ring is (right and left) FPF iff R is a hereditary bounded ring with no idempotent ideals. A commutative domain is FPF iff semihereditary. If G is a finite group of order n, which is a unit in R, and if R is right (F)PF, then so is the group ring RG. (This fails for FPF for $R = \mathbb{Z}$.)

25 B. A semiperfect right FPF ring R is a direct sum of uniform right ideals, and in fact each right prindec is uniform. Moreover the basic ring R_0 of R is *strongly right bounded* in the sense that every nonzero right ideal contains a nonzero ideal. Any right selfinjective strongly right bounded ring R is right FPF. (Partial converse.) Any right semiperfect right FPF ring with nil Jacobson radical is right selfinjective. (Compare the theorem of Tachikawa [69] stated in 24.8(h).)

26 A. (Faith [76 b, c, d]) A semiprime right FPF ring R without infinite orthogonal sets of idempotents is a finite product of prime right FPF rings. A (twosided) Goldie prime ring R is right FPF iff every right ideal $\neq 0$ generates mod-R and R is right bounded. A bounded Goldie prime ring in which finitely generated ideals $\neq 0$ are generators (both sides) is FPF (both sides).

26 B. A Noetherian semiperfect ring R is FPF iff R is a finite product of semiperfect bounded "Dedekind prime rings" ($=$HNP rings with no nontrivial idempotent ideals) and QF rings. (For commutative R drop the semiperfect assumption.)

.27. A finite dimensional algebra A over a field k is **symmetric** provided there exists a nondegenerate bilinear form $f : A \times A \to k$ such is associative, $f(ab, c) = f(a, bc)$, and symmetric, $f(a, b) = f(b, a)$ $\forall$ $a, b, c \in A$. Every group algebra kG of a finite group over a field k is symmetric, and every symmetric algebra over k is Frobenius.

28. Prove the inclusions are proper:

group algebras (finite groups) $\subset$ symmetric algebras $\subset$
Frobenius algebras $\subset$ QF algebras

29. (Nakayama [42]) A finite dimensional algebra A is QF iff the lattices of onesided ideals are selfdual.

30. (Nakayama [39, 41]) Let R be a selfbasic Artinian ring, with radical J, and right prindecs $e_1 R, \dots, e_n R$. Recall that eR/eJ is called the top, and socle eR the bottom of the module eR, for any $e = e^2$. Show that R is QF iff there is a permutation p of $\{1, \dots, n\}$ such that

$$\text{top } e_i R = \text{bottom } e_{p(i)} R \qquad\qquad (i = 1, \dots, n),$$

and similarly for the left-right symmetry of this condition. (Thus, for any Artinian ring R, the condition states that R is QF iff the basic ring of R has this property.)

31. (Boyle's Pajama Game) The following conditions on a ring R are equivalent: *PJ* 1. R is primary decomposable Artinian serial ring; *PJ* 2. The top and bottom of each finitely generated (right and left) module are equal (i.e. isomorphic); *PJ* 3. The top and bottom of each module are equal; *PJ* 4. The injective hull equals the projective cover of each finitely generated module; *PJ* 5. R is Noetherian and the tops and bottoms (resp. injective hulls and projective covers) of indecomposable modules are equal (Boyle [73]).

32. (Auslander) Each prindec of a group algebra of a finite group over a field has $\approx$ top and bottom. (Cf. 31.)

Notes for Chapter 24

The subject of Frobenius and quasi-Frobenius rings had its origins in the study of representations of a finite group G in a field k. Theorem 24.30 implies that the group algebra kG is quasi-Frobenius, but it is, in fact, not only Frobenius but symmetric. Symmetric algebras were introduced by Nesbitt and Thrall (in a Proc. Nat. Acad. Sci. paper in 1937). Basically, the idea of symmetry stems from the classical definition of a Frobenius algebra A over a field k, namely: A is Frobenius iff the right regular representation **R** and the left regular representation **S** are equivalent to each other. Then A is symmetric if **R** can be transformed into **S** by a nonsingular symmetric matrix (cf. Nakayama [39, p. 611] and Curtis-Reiner [62, p. 440]).

No matter which definition of QF ring was employed (see Chapter Exercises 10, 12, 28, and 29), various dualities were evident. The duality

(D_1) fin. gen. mod-$R \rightsquigarrow$ fin. gen. R-mod

defined by $\mathrm{Hom}_R(\ , R)$, however, was not explicitly given until Morita-Tachikawa [56], Dieudonné [58], and Morita [58] (cf. Tachikawa [58]). Dieudonné distinguishes 3 dualities: (D_1), defined above; (D_2) annihilation of onesided ideals; and (D_3) (in case R is an algebra over a field k of finite dimension) the duality

(D_3) fin. gen. mod-$R \rightsquigarrow$ fin. gen. R-mod

induced by the k-duality (D_1) for $R = k$. By Exercise 23.32(b), (D_3) is also induced by the functor $\mathrm{Hom}_R(\ , A')$, where A' is the k-dual module. Thus, the A-duality (D_1), and the A'-duality (D_3), will coincide iff $A \approx A'$. (Then, and only then, is the algebra Frobenius. See Exercise 14.)

Baer [43b] and Ikeda [51] proved that *any* duality between the lattices of left ideals and right ideals of Artinian rings is given by annihilation, and hence exists only for QF rings. (See Problem 11.)

Another duality by annihilation holds for any injective cogenerator E of mod-R for any ring R: If I is any right ideal, then

$$I = \mathrm{ann}_R \, \mathrm{ann}_E \, I$$

(Morita-Kawada-Tachikawa [57], Morita [58], Tachikawa [58], Azumaya [59], and Rosenberg-Zelinsky [61].) Moreover, any finendo (=finite over endomorphism ring) injective cogenerator is balanced (Dlab and Ringel [72b]). Thus, over Artinian rings, any injective cogenerator is balanced (see 19.16).

Osofsky [66] constructed and characterized rings which are injective cogenerators, and showed they are not necessarily QF (24.32). (Such a ring also occured in Levy [66a] and Klatt-Levy [69]—see Example 24.35.) These rings were precisely the ones over which all faithful modules are generators (Azumaya [66], Utumi [66]. Cf. Kato [68]). Thus the characterization of rings with an R-duality context is accomplished by 24.32 subject to knowing what is meant by selfinjectivity, a problem solved only in special cases (e.g., when R is right or left Noetherian in which case R is QF (24.5), or regular (Utumi [56])).

Selfinjective Artinian rings were characterized as QF rings by Ikeda [52] (where Baer's criterion is called Shoda's condition). Simpler proofs were given by Ikeda-Nakayama [54] and the Artinian condition replaced by the Noetherian. This was sharpened to one-sided conditions by Eilenberg-Nakayama [55] (cf. Faith [66a]), who proved that the global dimension is either 0 or ∞. Moreover group rings of finite groups over a commutative ring K are QF iff K is QF, a result extended to arbitrary K by Connell [63] (Theorem 24.29). Jans [59b] extended the notion of Frobenius algebras (over fields) to infinite dimensional algebras, and proved these were nevertheless selfinjective. Curtis [59] laid the foundations for a Galois theory of quasi-Frobenius rings which however remains incomplete. Nakayama [50b] combined these two topics—Galois theory and (quasi) Frobenius algebras in an inspiring essay, and his bibliography contains other references to these subjects, including some papers of his on the Galois theory of Artinian rings.

Kasch [54, 60, 61] generalized Frobenius rings to Frobenius extensions of rings, and Müller [64, 65] did similarly for quasi-Frobenius rings and extensions. Thus, R is left QF over a subring S provided that R is a finitely generated projective left S-module and that the S-dual module R^* generates R-mod in such a way that R is an (R, S)-direct summand of copies of R^*. (Then, R is a QF ring iff S is. Cf. Morita [62, 67].)

Thrall [48] defined a hierarchy of algebras. QF-1, QF-2, QF-3 generalizing QF algebras. A ring (or algebra) R is right QF-1 if every faithful right R-module M is balanced, that is, the canonical map $R \to \text{Biend}_R M$ is surjective. Thrall and Nesbitt [46; p. 560] proved that QF algebras are QF-1. This was done in two steps: first showing that the basic module $e_0 R$ is a direct summand of any faithful module over a QF-ring. In their paper, $e_0 R$ is called the reduced regular representation, and this result actually proves that any QF-algebra is PF, an implication valid for rings 24.32.

The second step in the Nesbitt-Thrall proof that M is balanced is that any module which has $e_0 R$ as a summand is balanced. In summary: the first step establishes that M is a generator of mod-R, and the second step establishes that generators are balanced, a result (I, 7.1, p. 326) which is valid for arbitrary rings as Morita [58] observed. (See (I, 7.1, p. 326). Cf. also the theorem of Fuller, Chapter 17, Exercise 17.)

More recent papers on QF-1, QF-2, or QF-3 rings include: Camillo [70a], Camillo and Fuller [72] (who, independently with Dlab and Ringel [72], prove for Artin algebras, that is, finitely generated algebras over Artinian centers, that every factor ring is QF-1 iff the ring is primary decomposable serial. Fuller [71] proved the primary-decomposable part. Cf. also Jans [70].), Colby and Rutter [73] (extensive bibliography), Dickson and Fuller [70] (the result of the title is generalized by Ringel [74] to "Noetherian commutative QF-1 rings are QF"; Cf. Camillo [70a], Floyd [68]), Dlab and Ringel [72a, b, c, d], Fuller [68b], Jans [69, 70], Harada [65b, 66], Kato [72], Morita [69] (extending his duality to QF-3 rings), Ringel [74], Ringel and Tachikawa [75/6], and Tachikawa [73] (which is a full length treatise on QF-1 and QF-3 rings).

Following the Goldie theorem were the papers characterizing orders in various kinds of rings, among which are orders in QF rings of: Jans [66], Mewborn and Winton [69], Masaike [71], and Tachikawa [71, p. 251ff.]. (Also see the bibliographies of Elizarov [69] and Faith [71d].)

Returning to the subject of duality in QF rings, yet another aspect of duality was exhibited by Faith and Walker. Thus, an arbitrary ring R is QF iff every projective module is injective (Faith-Walker [67]). The dual theorem was proved by Faith [66a] (see Theorems 24.12 and 24.23). The proofs of these theorems led to a variety of theorems on direct representations of modules, including the concept of $\sum$-injectivity (see Chapter 20).

References

Asano [39], Azumaya [59, 66], Baer [43b], Bass [62], Brauer [63], Brauer and Nesbitt [37], Camillo [70a], Camillo and Fuller [72], Cohn [66a], Connell [63], Colby and Rutter [73], Curtis [59], Curtis and Reiner [62], Dickson and Fuller [70], Dieudonné [58], Dlab and Ringel [72], Eilenberg, Ikeda, and Nakayama [55], Eilenberg, Nagao, and Nakayama [56], Eilenberg and Nakayama [55, 57], Elizarov [69], Faith [66a, 71d, 76b, c, d], Faith and Walker [67], Farkas [73], Floyd [68], Formanek [70], Fuller [68a, b, 69a, 70a, b, c, 71], Gewirtzman [65, 67], Hannula [73], Harada [65, 66], Jacobson [64], Jans [59a, b, 61, 67, 69, 70], Ikeda [51, 52], Ikeda and Nakayama [54], Jennings [41], Kasch [54, 60, 61], Kato [68a, b, 72], Kawada [57], Klatt and Levy [69], Levy [66a], Maisake [71], Mewborn and Winton [69], Morita [58, 62, 66, 67, 69], Morita and Tachikawa [56], Müller [64, 65], Nakayama [39, 40a, b, 42, 50a, b], Onodera [68, 71], Osofsky [66], Pareigis [71], Renault [70], Ringel [74], Ringel and Tachikawa [75/6], Tachikawa [58, 62, 69, 71, 73], Utumi [56, 60, 66, 67], Wagoner [71], Bass [67], Jain and Singh [67, 70].

Chapter 25. Sigma Cyclic and Serial Rings

The first three sections of this chapter present the structure of serial rings of Warfield [75]. The main theorem 25.3.4 characterizes when every finitely presented left module over a ring R is a direct sum of uniserial modules: this happens iff R is itself such a direct sum both as right and left module, that is, iff R is serial. (See Section 0 for definitions.) In this case, then for any finitely generated submodule M of a finitely generated projective module P, there are "stacked" decompositions of P and M into direct sums of uniserial modules (see 25.3.3ff). Moreover, any Noetherian serial ring is decomposable into a finite product of Artinian and (semi)prime rings (25.3.5). This is reminiscent of the theorems of Chatters (20.30) for hereditary rings, and Krull-Asano-Goldie 20.37, for principal ideal rings, and, in fact, Robson's general method (20.35) used to prove these also applies here.

Serial rings characterize rings over which finitely presented modules are direct sums of cyclic uniserial modules (25.3.4). (This generalizes and strengthens some results in Chapter 20, notably 20.41 and 20.43.) Furthermore, any cyclic uniserial module over a semiperfect ring is a local module, in fact, principal cyclic, so over serial rings, any finitely presented module is a direct sum of principal cyclics.

With Warfield's results at hand, in Section 4 we prove the main result on Artinian serial rings of Nakayama [41], adding several contemporary touches: thus, an Artinian ring is serial iff every finitely generated module is a direct sum of principal cyclic modules, and iff every dominant right or left prindec of any factor ring $R/(\operatorname{rad} R)^n$ is quasiinjective (equivalently, injective modulo $(\operatorname{rad} R)^n$), for every n (Fuller [69a], Eisenbud-Griffith [71a]). Over an Artinian serial ring, every module is $\sum$-uniserial 25.4.2. Any Artinian serial ring has finite module type, and these rings all have finite ideal lattices 25.4.4. Primary decomposable serial rings are characterized 25.4.6. The last part of Section 4 is devoted to (injective) σ-cyclic rings. Any FBG ring is similar to a σ-cyclic ring 20.39 and any right Artinian FBG ring has a duality context 25.4.11.

In the short last section, we prove the theorem of Eisenbud-Griffith-Robson to the effect that any proper factor ring R/I of a Noetherian hereditary prime ring R is serial. The proof and, in this generality, theorem, comes from Eisenbud and Griffith [71a], and involves an elegant application of Nakayama's (and Fuller's) characterization of Artinian serial rings (to the effect that dominant prindecs are injective modulo annihilator). (The theorem of Webber sets the stage for this, since it asserts that R/I is right Artinian. See Webber-Chatters 20.29.)

0. Terminology and Examples

A module M is $\sum$-**cyclic** (as defined in Chapter 20) if M is a direct sum of cyclic modules; and σ-**cyclic** if M is a finite direct sum of cyclic modules. Then, the category mod-R of right R-modules is **right** $\sum$-**cyclic** (σ-**cyclic**) if every (finitely generated) right R-module is $\sum$-cyclic (σ-cyclic). It is convenient to say R is right $\sum$-cyclic (σ-cyclic) instead of mod-R, and certainly no confusion should result since R is a cyclic module (and is always $\sum$-cyclic!). Similarly for left $\sum$- or σ-cyclic. We say R is a $\sum$-cyclic (σ-cyclic) ring if R is right and left $\sum$-cyclic (σ-cyclic).

25.0.1 *Example* (Basis theorem for abelian groups, see (I, 27, p. 81)).

25.0.1(a)   The best examples for σ-cyclic are: $R = \mathbb{Z}$, or any almost maximal valuation ring (20.49). For any integer $n \neq 0$, $\mathbb{Z}_n$ is $\sum$-cyclic (see (I, 26, p. 80)).

25.0.1(b)   Any semisimple Artinian ring is not only $\sum$-cyclic, it is $\sum$-simples.

The term **uniserial** (Einreihig) is applied to a module M with linearly ordered submodules:

$$A \subseteq B \quad \text{or} \quad B \subseteq A \qquad\qquad \forall \text{ submodules } A \text{ and } B.$$

A module is said to be $\sum$-**uniserial** (σ-**uniserial**) if it is a direct sum of (finitely many) uniserial modules. We say that R is **right serial** when we mean that R as a right module is σ-uniserial. By symmetry left serial is defined, and **serial** denotes a left and right serial ring.

25.0.2 *Example*

25.0.2(a)   Any ring R with a unique composition series in mod-R is right serial, and then any ring similar to R is also; in particular, the $n \times n$ matrix ring over a right serial ring is right serial. Such a ring need not be left serial as (I, 7.11′.1 and 7.11″.2, p. 337) shows.

25.0.2(b)   For a prime p, the abelian group $\mathbb{Z}_{p^\infty}$ (I, p. 55) is uniserial. Any divisible abelian group is $\sum$-uniserial modules (Exercise). The ring $\mathbb{Z}_{p^n} = \mathbb{Z}/p^n\mathbb{Z}$ is serial, and any torsion module of bounded order over a Dedekind domain, e.g. a principal ideal domain, is $\sum$-uniserial modules. (This also holds for unfaithful torsion modules over Hereditary Noetherian prime rings. See 25.5.1.)

25.0.2(c)   A ring $R = k\langle x\rangle$ of formal power series over a field k is serial since the only ideals besides $J = \operatorname{rad} R$ are the powers J^n of J, $n = 2, 3, \dots$. (Proof?)

25.0.2(d)   A right serial local ring is a right valuation ring (VR) as defined in Chapter 20, *sup.* 20.40.

25.0.3 *Exercise.* Let $\mathbb{Z}_{(p)}$ denote the local ring of $\mathbb{Z}$ at the prime ideal (p); that is,

$$\mathbb{Z}_{(p)} = \{a\,b^{-1} \in \mathbb{Q} \mid (b, p) = 1\}.$$

Then the triangular ring

$$\begin{pmatrix} \mathbb{Z}_{(p)} & \mathbb{Q} \\ 0 & \mathbb{Q} \end{pmatrix}$$

is serial (Proof?), and right but not left Noetherian (see (I, p. 399)).

For a semilocal lift/rad ring, the concept of **right (left) prindec** is defined in 18.23, namely, a right (left) ideal generated by an idempotent e with endomorphism ring eRe a local ring.

In a semiperfect (or semilocal lift/rad) ring an idempotent e generates a prindec iff e is indecomposable in the sense that e is not a sum of two nonzero orthogonal idempotents (see 18.23). Then in a right (left) Artinian ring, a right (left) prindec is **dominant** if it has maximal composition length among all the right (left) prindecs.

25.0.4 *Exercise.* (a) A left prindec Re in a ring R with radical J is uniserial only if Je/J^2e is simple. If, in addition, Re is a left Artinian module, this condition is also sufficient, and in this case,

$$Re \supset Je \supset J^2e \supset \cdots \supset J^ne \supset \cdots$$

is the unique composition series.

(b) In a left Artinian left serial ring R, the only principal cyclic modules are those isomorphic to Re/J^ne, for some left prindec Re, where $J = \mathrm{rad}\, R$.

1. Modules over Semiperfect Rings [1]

If R is a semiperfect ring, then every stable isomorphism class of finitely generated modules contains a unique (up to isomorphism) "minimal" element (1.4 and 1.5). A refinement of the usual notions of the number of generators and relations of a finitely presented module (1.6–1.11) is the main tool used in characterizations of left serial rings (1.13), and of semiperfect rings for which every finitely generated left ideal is principal (1.14).

We repeat a definition (previously defined *sup.* 18.23).

Definition. A module M is **local** if M has a unique maximal proper submodule.

Some of the results stated in this section are restatements of theorems on semiperfect rings, or straight-forward applications of the same. For example:

25.1.1 **Lemma.** *Let R be a semilocal ring. Then following conditions on R are equivalent:*

   (i) *R is a direct sum of local modules in* mod-R.

   (ii) *R is a direct sum of local modules in R-mod.*

   (iii) *R is a lift/rad ring.*

   (iv) *All finitely generated projective modules are direct sums of local cyclic projectives.*

   (v) *R is semiperfect.*

   (vi) *R has an Azumaya diagram.*

Proof. See 18.23 and 22.23. □

---

[1] By numbering sections to conform with the numbering of the paper of Warfield [75] which we are following, we are departing from the style of the other chapters. In order to reduce the decimal notation to the minimum, in this chapter references to results within this chapter will dispense with the prefix. (Thus, 25.1.1 will be referred to as 1.1, etc.)

Actually, in all but (iii), the assumption that R is semilocal is redundant.

25.1.2 Proposition. *Over a semiperfect ring R, any local module is principal cyclic ($=$ an epic image of a right prindec).*

Proof. See 18.23.4. □

25.1.3 Lemma. *If $J = \operatorname{rad} R$, and if M is a finitely generated left R-module, then* (i) *JM is a superfluous submodule of M. Moreover, if R is semilocal, then* (ii) *M/JM is a direct sum of a finite number of simple modules, and* (iii) *then M is local if and only if M/JM is simple.* □

Remark. (i) is just a form of "Nakayama's lemma" 18.4(c), and (ii) and (iii) are trivial.

25.1.4 Theorem. *Let M be a finitely generated module over a semiperfect ring R. Then there is a decomposition $M = N \oplus P$ where P is projective and N has no projective summands. Further, if $M = N' \oplus P'$ is another such decomposition, then $N \approx N'$ and $P \approx P'$.*

Proof. The existence of the indicated decomposition is a triviality—we just start separating off projective summands, if they exist, and the process will eventually stop by the chain condition on M/JM and 1.3. We suppose, then, that we have two such decompositions:

$$M = N \oplus P = N' \oplus P'.$$

By 1.1(iv)–(vi), the endomorphism ring of a finitely generated indecomposable projective module over a semiperfect ring is a local ring, from which it follows that any finitely generated projective module has the exchange property 19.21(n). Applying the exchange property for P, we obtain submodules $P_0' \subseteq P'$ and $N_0' \subseteq N'$, such that $M = P \oplus P_0' \oplus N_0'$ (that, incidentally, is the definition of the exchange property). This clearly implies that $N \approx P_0' \oplus N_0'$, so that $P_0' = 0$ (since N has no nonzero projective summands). Decompose $N' = N_0' \oplus (N' \cap P)$; comparing the two complements of N_0', obtain $(N' \cap P) \oplus P' \approx P$. Since by hypothesis, N' has no nonzero projective summands, it follows that $N' \cap P = 0$, $N' = N_0'$. The above formulas now imply that $N \approx N'$ and $P \approx P'$ as desired. □

Modules X and Y are **stably isomorphic** if there are projectives P and Q such that $X \oplus P \approx Y \oplus Q$. If X and Y are finitely generated, we may assume P and Q are also. The theorem therefore yields the following sharpening of the notion of stable isomorphism for modules over semiperfect rings.

25.1.5 Corollary. *If R is a semiperfect ring then* (i) *every finitely generated module is stably isomorphic to a module with no nonzero projective summands, and* (ii) *two modules with no nonzero projective summands are stably isomorphic if and only if they are isomorphic.* □

This result will allow considerable precision in the next section in applying the Auslander Bridger duality theory, which initially is only a relation between stable isomorphism classes of modules. In order to use that duality theory, we will also need some notions concerning "generators and relations" for modules over semiperfect rings (defined in 1.7). For convenience in reference we restate:

25.1.6 Lemma. *If R is a semiperfect ring with radical J, and M a finitely generated module, there is a projective module P and an epimorphism $f: P \to M$ such that the induced homomorphism $P/JP \to M/JM$ is an isomorphism. If $g: Q \to M$ is another such epimorphism, (with Q projective) then there is an isomorphism $\varphi: P \to Q$, of P onto Q such that $g\varphi = f$.*

This is just a statement of the existence and uniqueness of the projective cover of M. The uniqueness of the projective cover makes it possible to give the:

25.1.7 Definition. Let M be a finitely generated module over a semiperfect ring R, and let S be a simple module. We define **Gen**(M) to be the number of summands in a decomposition of M/JM as a direct sum of simple modules, and **Gen**$(M; S)$ to be the number of such summands isomorphic to S.

25.1.8 Lemma. *A finitely generated module M over a semiperfect ring R is cyclic if and only if for every simple module S, we have $\mathrm{Gen}(M; S) \leq \mathrm{Gen}(_R R; S)$.* $\quad\square$

25.1.9 Definition. Let M be a finitely presented module over a semiperfect ring R, and $f: P \to M$ a projective cover (as in 1.6), and $K = \ker(f)$. Then we define $\mathrm{Rel}(M)$ and $\mathrm{Rel}(M; S)$, for all simple modules S, by $\mathrm{Rel}(M) = \mathrm{Gen}(K)$, $\mathrm{Rel}(M; S) = \mathrm{Gen}(K; S)$. These are well defined by 1.6.

25.1.10 Lemma. *If M is a finitely generated module over a semiperfect ring, and S a simple module, and $M = A \oplus B$, then*

$$\mathrm{Gen}(M; S) = \mathrm{Gen}(A; S) + \mathrm{Gen}(B; S).$$

If M is finitely presented then

$$\mathrm{Rel}(M; S) = \mathrm{Rel}(A; S) + \mathrm{Rel}(B; S).$$

This is easily shown by taking projective covers for A and B and combining them to get one for M.

It is clear from 1.8 that $\mathrm{Gen}(M)$ is only indirectly related to the number of generators required to generate M. However, if we restrict ourselves to certain kinds of elements, the connection becomes closer. If M is an R-module and $x \in M$, we say x is a **local element** if Rx is a local module. The fact that any projective module is generated by local elements (1.1) implies that any module is generated by a set of local elements.

25.1.11 Lemma. *If M is a finitely generated module over a semiperfect ring with radical J, and X is a set of local elements of M, then X is a minimal set of local generators for M if and only if the natural map $M \to M/JM$ takes X bijectively onto a minimal set of local generators of M/JM. The number of elements in any minimal set of local generators is exactly $\mathrm{Gen}(M)$.* $\quad\square$

25.1.12 Definition. A module is **uniserial** if its submodules are linearly ordered with respect to inclusion. A ring R is **left serial** if $_R R$ is a direct sum of uniserial modules. R is **serial** if it is both left and right serial.

Artinian serial rings are traditionally called "generalized uniserial rings", and were introduced by Köthe [35] and Nakayama [40, 41]. Any Artinian principal ideal ring is serial (Köthe [35], Asano [39, 49a]). A ring may be left but not right serial (I, 7.11′.1, p. 337).

25.1.13 Lemma. *If M is a finitely generated module over a left serial ring and N a finitely generated submodule, then*

$$\mathrm{Gen}(N) \leq \mathrm{Gen}(M).$$

Proof. Let L be a local submodule of M such that L is not contained in JM. (You can find such a submodule by taking a projective cover of M (1.6) and looking at the image in M of one of the indecomposable summands of the projective module.) Look at the exact sequence:

$$0 \to N \cap L \to N \to N/N \cap L \to 0.$$

Given any exact sequence $0 \to X \to Y \to Z \to 0$, there is an induced exact sequence $X/JX \to Y/JY \to Z/JZ \to 0$. Note that if X is serial then either $X = JX$ or X/JX is simple. Applying this to the sequence above, and using the fact that $N \cap L$ is serial, we obtain $\mathrm{Gen}(N) \leq 1 + \mathrm{Gen}(N/N \cap L)$. By our choice of L, $\mathrm{Gen}(M/L) + 1 = \mathrm{Gen}(M)$, and since $N/N \cap L$ is a finitely generated submodule of M/L, the result follows by induction on $\mathrm{Gen}(M)$. $\square$

1.13 gives a characterization of left serial rings similar to the well known characterizations of rings whose finitely generated left ideals are principal. (If R is such a ring, M a module which can be generated by n elements, and N a finitely generated submodule, then N can be generated by n elements.) It is fairly easy to characterize semiperfect rings in which every finitely generated left ideal is principal, and this we now do as an application of the previous methods.

First, however, a definition. If M is a uniserial module, we say M is **homogeneously uniserial** if for all pairs of nonzero finitely generated submodules A and B, $A/JA \approx B/JB$. If M satisfies both the ascending and descending chain conditions, this means that all the simple composition factors of M are isomorphic.

25.1.14 Theorem. *The following properties are equivalent for a semiperfect ring R:*

(i) *every finitely generated left ideal is principal,*

(ii) *R is left serial and the indecomposable summands of $_R R$ are homogeneously uniserial,*

(iii) *R is the product of a finite number of full matrix rings over local, left serial rings.*

(iv) *R is similar to a local-decomposable left serial ring.*

Proof. To show that (i) implies (ii), we first show that R must be left serial. If not, $R = P \oplus Q$, where P is an indecomposable projective which is not uniserial. If A is a finitely generated submodule of P such that $\mathrm{Gen}(A) \geq 2$, then

$$\mathrm{Gen}(A \oplus Q) = \mathrm{Gen}(A) + \mathrm{Gen}(Q) > \mathrm{Gen}(_R R)$$

so $A \oplus Q$ is certainly not a cyclic module. We now suppose that R is left serial and has an indecomposable summand P which is not homogeneously uniserial. We will again find a left ideal which is not principal. Again write $R = P \oplus Q$, and let A be a cyclic submodule of P such that $A/JA = S$, where S is a simple module which is not isomorphic to P/JP. In this case, $\text{Gen}(Q; S) = \text{Gen}(_R R; S)$, so

$$\text{Gen}(A \oplus Q; S) = \text{Gen}(_R R; S) + 1.$$

By 1.8, $A \oplus Q$ is not a principal left ideal.

We now show that (ii) implies (iii). Condition (ii) implies that if P and Q are nonisomorphic indecomposable projectives then $\text{Hom}(P, Q) = 0$, since if A were the image of such a homomorphism, $A \neq 0$, then we would have

$$A/JA \approx P/JP \not\approx Q/JQ.$$

Therefore R is a product $\prod_{i=1}^n A_i$ of a finite number of rings A_i satisfying (ii), with the additional property that each of their indecomposable projectives P of $A = A_i$ are isomorphic, so $A \approx P^t$ (P and t depending on i). Then, by (I, 3.33.3, p. 152) and 22.24, A is a full $t \times t$ matrix ring over a local ring $B = \text{End}_A P$. The property of being a left serial ring is preserved under Morita equivalence (since it has the categorical description: projective modules are direct sums of serial modules), so the local rings which arise in each A_i must be left serial rings. This completes the proof of (iii).

To show that (iii) implies (i), we may restrict ourselves to directly indecomposable rings (since the statement of (i) is preserved by finite products of rings). An indecomposable ring satisfying (iii) has only one simple module, up to isomorphism. Combining 1.8 and 1.13 we see that if R is a left serial ring with only one simple module up to isomorphism, then every submodule of a cyclic module is cyclic, which implies (i).

The equivalence of (iii) and (iv) comes from 18.36. $\square$

The semiprimary case of this theorem is 18.38.3. Compare Eisenbud-Griffith [71a], Jacobson [43], and Jategaonkar [70].

2. Auslander-Bridger Duality

In this section, we refine the Auslander-Bridger duality theory, and use it to study rings for which every finitely presented module is a direct sum of local modules. In particular, this property implies that the ring is serial. We conclude by proving a theorem which gives a restriction on the structure of a ring R for which there is an upper bound on the number of generators required for indecomposable finitely presented R-modules.

If R is any ring with identity, and M a finitely presented (FP) left R-module, we define the Auslander-Bridger dual $D(M)$ as follows. Choose an exact sequence $Q \xrightarrow{\varphi} P \to M \to 0$ in which P and Q are finitely generated projective modules. Define $D(M)$ to be the cokernel of the homomorphism $\varphi^*: P^* \to Q^*$ (where $X^* = \text{Hom}(X, _R R)$, and if X is a left module, X^* is a right module). $D(M)$ is well defined up to stable isomorphism (defined *sup.* 1.5), and this defines a duality between the categories of stable isomorphism classes of FP left modules and stable isomorphism classes of FP right modules. (Exercise.)

One should notice that if M is presented by n generators and k relations, by a sequence $R^k \to R^n \to M \to 0$, then $D(M)$ is given by k generators and n relations. For modules over semiperfect rings, we can make this more precise using the invariants defined in the previous section (1.7 and 1.9), and using the fact that over such a ring, each stable isomorphism class of finitely generated modules contains, in some sense, a canonical minimal element (1.4 and 1.5).

25.2.1   Lemma. *If R is a semiperfect ring and P an indecomposable projective left module, then the dual, P^*, is an indecomposable projective right module. Further, if $P/JP \approx S$, then $P^*/P^*J \approx S'$, where S' is the dual of the module S with respect to the ring R/J.*

25.2.2   Lemma. *If R is a semiperfect ring, P a finitely generated projective module and N a submodule not contained in JP, then N contains a nonzero summand of P.*

Proof. Straight-forward generalization of 22.22.   □

If M is a finitely presented module over a semiperfect ring, and $Q \xrightarrow{\varphi} P \to M \to 0$ as stated above, then we call this sequence a **minimal presentation** of M if (i) the induced homomorphism $P/JP \to M/JM$ is an isomorphism, and (ii) if K is the kernel of the homomorphism $P \to M$, then the induced homomorphism $Q/JQ \to K/JK$ is an isomorphism. In this case, the isomorphism types of P/JP and Q/JQ are invariants of M (1.6, 1.7, 1.9). Moreover, a minimal presentation (in fact every such) is obtained by letting $P \to M$ and $Q \to K$ be projective covers. (Then, e.g. $P/JP \approx M/JM$ since $JP = \operatorname{rad} P \supseteq K$.)

25.2.3   Lemma. *If R is a semiperfect ring and M a finitely presented left R-module with no nonzero projective summands, and $Q \to P \to M \to 0$ is a minimal presentation for M, then the induced sequence $P^* \to Q^* \to D \to 0$ is a minimal presentation for D, and D also has no nonzero projective summands.*

Remark. Strictly speaking, the dual $D(M)$ is a stable isomorphism class of which the above module D is a representative. The point of this lemma is that by choosing a suitable resolution, one can assure that the resulting representative is the canonical member of the stable isomorphism class, whose existence is guaranteed by 1.5.

Proof. If $\varphi: Q \to P$ is the homomorphism appearing in the presentation of M, and $\varphi^*: P^* \to Q^*$ is its dual, then we must first show that $\varphi^*(P^*) \subseteq Q^*J$. If not, by 2.2 we could decompose $P^* = A \oplus B$, where φ^* maps B isomorphically onto a nonzero summand of Q^*. Dualizing again, and identifying $P^{**} = P$, we would get a nonzero summand B^* of P in the kernel of the homomorphism from P to M, contradicting the fact the kernel is superfluous.

We now let L be the kernel of the homomorphism $Q^* \to D$, and by our previous argument we know that $L \subseteq Q^*J$. We wish to show that the homomorphism φ^* induces an isomorphism $P^*/P^*J \to L/LJ$. If not, by 2.2 we can decompose $P^* = C \oplus E$, where $C \neq 0$ and C is in the kernel of φ^*. Dualizing again, we obtain $P = C^* \oplus E^*$ and $\varphi(Q) \subseteq E^*$. Hence M would have a projective summand, which it does not.

The final point is that D has no nonzero projective summands. One argues, just as before that if it had such a summand, then $\varphi^*(P^*)$ would lie in a proper summand of Q^*, and dualizing, one would obtain a nonzero summand of Q in the kernel of φ, contradicting the superfluity of the kernel.   □

25.2.4   Theorem. *Let R be a semiperfect ring, $\mathcal{X}$ the class of finitely presented left modules with no nonzero projective summands, and $\mathcal{Y}$ the class of finitely presented right modules with no projective summands. To each $M \in \mathcal{X}$ there is an element $D(M) \in \mathcal{Y}$, uniquely determined up to isomorphism, such that (i) $M \approx N$ if and only if $D(M) \approx D(N)$, and (ii) $D(M \oplus N) \approx D(M) \oplus D(N)$. Furthermore, if to each simple left module S we associate a corresponding simple right module S' as in (2.1), then the following equations hold:*

$$\mathrm{Gen}(M; S) = \mathrm{Rel}(D(M); S'),$$
and
$$\mathrm{Rel}(M; S) = \mathrm{Gen}(D(M); S').$$

This theorem is an immediate consequence of the Auslander-Bridger duality stated at the beginning of this section, and 2.3, 1.5, 1.7, and 1.8. $D(M)$ can be computed explicitly from 2.3. If we start from right modules instead of left modules, we again get such a function D, and, of course, it is the inverse of the one discussed here.

25.2.5   Definition. A module M is **locally presented** (LP) if there is an exact sequence $Q \to P \to M \to 0$ in which P and Q are both local projectives. Equivalently, M is finitely presented and $\mathrm{Gen}(M) = \mathrm{Rel}(M) = 1$.

25.2.6   Theorem. *The following properties of a semiperfect ring R are equivalent:*

(i) *every finitely presented left module is a direct sum of locally presented modules;*

(ii) *every finitely presented left module is a direct sum of uniserial modules;*

(iii) *every finitely presented right module is a direct sum of locally presented modules;*

(iv) *every finitely presented left module and every finitely presented right module is a direct sum of local modules;*

(v) *every finitely presented right or left module is a direct sum of principal cyclic modules;*

(vi) *every finitely presented right module is a direct sum of uniserial modules;*

(vii) *R is serial.*

Proof. Condition (i) says that if M is an idecomposable finitely presented left module which is not projective, then $\mathrm{Gen}(M) = \mathrm{Rel}(M) = 1$. This implies that if P is an indecomposable projective and A a finitely generated submodule $(A \neq 0)$ then $\mathrm{Gen}(A) = 1$ since $\mathrm{Gen}(A) = \mathrm{Rel}(P/A)$ and P/A is indecomposable. (i) therefore implies that R is left serial, and for a left serial ring, (i) and (ii) are trivially equivalent.

Again, since (i) says that if M is an indecomposable FP left module which is not projective, then $\mathrm{Gen}(M) = \mathrm{Rel}(M) = 1$, it follows by 2.4 that this is equivalent to the corresponding condition for right modules, whence (i) and (iii) are equivalent.

Finally, it is clear that (i) and (iii) imply (iv). Conversely, if R satisfies (iv) and M is an indecomposable, non-projective FP module, then $\mathrm{Gen}(M)=1$ by (iv) and $\mathrm{Gen}(D(M))=1$, by (iv) and the fact that $D(M)$ is indecomposable (2.4). Since $\mathrm{Gen}(D(M))=\mathrm{Rel}(M)$ (by 2.4), it follows that M is LP, so (iv) implies (i).

The equivalence of (iv) and (v) in semiperfect rings comes from 18.23.4.

This completes the equivalence of (i)–(v). Since (iv) is right-left symmetric, then the right left symmetry of (ii), namely (vi), is equivalent to the others, that is, (i)–(vi) are equivalent.

Clearly (ii) and (vi) imply (vii), so only the implication (vii) $\Rightarrow$ (ii) is required to complete the proof. This requires a bit more preparation, and is proved in the next section (see 3.4).

3. The Decomposition Theorem

In this section we prove Warfield's theorem stating that a finitely presented module over a serial ring is a direct sum of locally presented modules, a theorem which for noncommutative local rings was first established by Roux [72]. (For the commutative case, consult 20.41 and 20.43.)

We conclude the section with Robson's proof of Warfield's decomposition theorem for Noetherian serial rings: they are direct sums of Artinian and semiprime rings. (Cf. Chatters' theorem 20.30 and the Krull-Asano-Goldie theorem 20.37.)

25.3.1 Lemma. *If M is a finitely presented module over a serial ring, then $\mathrm{Gen}(M)\geq\mathrm{Rel}(M)$. Moreover, M has no nonzero projective summands if and only if $\mathrm{Gen}(M)=\mathrm{Rel}(M)$.*

Proof. The first statement follows by applying 1.13 to a projective cover for M (1.6), and using the definition of $\mathrm{Rel}(M)$ (1.9). If M has no projective summands, then we apply 2.6 to obtain the reverse inequality, $\mathrm{Rel}(M)\geq\mathrm{Gen}(M)$, from the original inequality applied to the dual $D(M)$. Conversely, if $M=N\oplus P$ where P is projective and $\mathrm{Gen}(P)>0$, then $\mathrm{Rel}(M)=\mathrm{Rel}(N)\leq\mathrm{Gen}(N)<\mathrm{Gen}(M)$, so $\mathrm{Rel}(M)<\mathrm{Gen}(M)$ as desired.

25.3.2 Lemma. *Let P be a finitely generated projective module over a serial ring, and $x\in P$ a local element. Then there is an indecomposable summand Q of P such that $x\in Q$.*

Proof. If $x\notin JP$ then Rx is a local summand of P by 2.2. If $x\in JP$ and $M=P/Rx$, then $\mathrm{Gen}(M)=\mathrm{Gen}(P)$ and $\mathrm{Rel}(M)=1$. If we write $M=A\oplus B$, where B is projective and A has no nonzero projective summands (1.4), then by 1.10, $\mathrm{Rel}(A)=1$. By 3.1, this implies $\mathrm{Gen}(A)=1$. If $Q=\{y\in P:\ y+Rx\in A\}$, then Q is clearly a summand of P whose complement is isomorphic to B, and Q is clearly local. $\square$

25.3.3 Theorem. *Let R be a serial ring, P a finitely generated projective module and M a finitely generated submodule of P. Then there is a decomposition*

$P = P_1 \oplus \cdots \oplus P_n$ of P into indecomposable projectives such that if $M_i = M \cap P_i$, then $M = M_1 \oplus \cdots \oplus M_n$.

Proof. We prove the result by induction on $\mathrm{Gen}(M)$ (defined in 1.7). The case $\mathrm{Gen}(M) = 0$ is trivial, and $\mathrm{Gen}(M) = 1$ has actually already been done in 3.2. We will assume, therefore, that the result is known if $\mathrm{Gen}(M) = n$ and prove it under the hypothesis that $\mathrm{Gen}(M) = n + 1$. By 3.1, there is a decomposition $P = A \oplus B$ such that $\mathrm{Gen}(A) = n + 1$, and $M \subseteq A$. We may therefore assume without loss of generality that $\mathrm{Gen}(P) = n + 1$. Assuming this, we remark that to prove the theorem it will suffice to find a decomposition $P = Q \oplus L$, where Q is an indecomposable projective and $M = (M \cap Q) \oplus (M \cap L)$. We will refer to this fact a number of times in the following.

We now recall the basic facts about decompositions of projective modules over a semiperfect ring which we need. Let P be a finitely generated projective module, Q an indecomposable summand of P, and $P = \bigoplus_{i \in I} P_i$ a decomposition of P into indecomposable summands. For any $j \in I$, we let $I(j) = I - \{j\}$, and we say Q replaces P_j if

(*) $\qquad P = Q \oplus (\bigoplus_{i \in I(j)} P_i).$

We let π_i be the projection onto P_i and φ_i the restriction of π_i to Q. In this setting, we draw three conclusions: (i) Q replaces P_j if and only if φ_j is an isomorphism (of Q onto P_j); (ii) if Q replaces P_j, then in the decomposition (*), the projection onto Q is $\varphi_j^{-1} \pi_j$; (iii) there is at least one index j such that Q replaces P_j. The first two of these are essentially trivialities, and the third is just the fact that Q has the exchange property 18.17.

Proof of the theorem. We may suppose that $P = P_1 \oplus \cdots \oplus P_{n+1}$, where each P_i is indecomposable, and that M has a minimal set of local generators $\{x_1, \ldots, x_{n+1}\}$. By the induction hypothesis, applied to the submodule generated by $\{x_1, \ldots, x_n\}$, we may assume that the P_i and x_i have been chosen so that $x_i \in P_i$, $1 \leq i \leq n$. By 3.2, there is an indecomposable projective summand Q of P containing x_{n+1}. In the notation of the previous remarks, we let π_i be the projection onto P_i, and φ_i the restriction of π_i to Q. If φ_{n+1} is an isomorphism, then by (i) above, Q replaces P_{n+1} and we are done. Otherwise, by (iii) above, φ_j is an isomorphism for some j, $1 \leq j \leq n$. Without loss of generality, we may assume that φ_1 is an isomorphism. Since P_1 is a serial module, either $\varphi_1(x_{n+1}) \in R x_1$ or $x_1 \in R \varphi_1(x_{n+1})$. In the first case, $\varphi_1(x_{n+1}) \in R x_1$ implies that $\pi_1(M) \subseteq R x_1 = M \cap P_1$, so $M = (M \cap P_1) \oplus (M \cap (P_2 \oplus \cdots \oplus P_{n+1}))$. This proves the result by induction. In the second case, we write $P = Q \oplus P_2 \oplus \cdots \oplus P_{n+1}$ (as we may by (i)), and, in this decomposition, the projection onto Q is $f = \varphi_1^{-1} \pi_1$. By construction, $f(x_1) \in R x_{n+1}$, from which we conclude that $f(M) \subseteq R x_{n+1} = M \cap Q$. As before, this implies that $M = M \cap Q \oplus M \cap (P_2 \oplus \cdots \oplus P_{n+1})$, and the result is again proved by induction. $\square$

If M is a submodule of a module P, and if there are direct sum decompositions

$$M = M_1 \oplus \cdots \oplus M_k$$

and

$$P = P_1 \oplus \cdots \oplus P_n \qquad (n \geq k)$$

such that $M_i = M \cap P_i$, $i = 1, \ldots, k$, then the decompositions are said to be **simultaneous** (Kaplansky [69]), or **stacked**. If, as in 3.3, P_i is a uniserial module generated by say x_i, $i = 1, \ldots, n$, then there are elements $r_1, \ldots, r_k$ of R, such that $y_i = r_i x_i$ generates M_i, $i = 1, \ldots, k$, and then $\{x\}_{i=1}^n$ and $\{y_i\}_{i=1}^k$ are called **simultaneous** or **stacked** bases of P and M.

In case R is a left valuation ring, then the elements $\{r_i\}_{i=1}^k$ can be chosen such that r_i divides r_{i+1}, that is, $Rr_i \supseteq Rr_{i+1}$, $i = 1, \ldots, k-1$, and are the socalled **elementary divisors**. (Compare Kaplansky [49]. Also see the paper of Cohen and Gluck [70] on stacked bases for modules.)

The next corollary completes the proof of 2.5.

25.3.4 Corollary. *A ring R is serial iff every finitely presented module is a direct sum of cyclic uniserial modules.*

Proof. If R is serial, and if F is a finitely presented module, with presentation $0 \to M \to P \to F \to F \to 0$, where P is finitely generated projective, and M is finitely generated, then the stacked decomposition of 3.3 yields $F \approx P/M \approx \bigoplus_{i=1}^n P_i/M_i$. Inasmuch as every indecomposable finitely generated projective module P_i is principal cyclic in a semiperfect ring (1.1), then P_i is cyclic uniserial, and hence F is a direct sum of cyclic uniserial modules P_i/M_i, $i = 1, \ldots, n$.

The converse is contained implications of 2.5 already proved. $\square$

25.3.5 Decomposition Theorem (Warfield [75]). *A Noetherian serial ring is the direct product of an Artinian ring and a semiprime ring.*

Proof. (Robson [74]) If $c \in \mathscr{C}(N)$, then by 3.3 we may choose a stacked decomposition for R and cR; say $R = \coprod e_i R$, $cR = \coprod (e_i R \cap cR)$, with e_i idempotent, $i = 1, \ldots, n$. It is automatic that $N = \coprod e_i N$. Now the (Goldie) dimension of both R/N and its essential right ideal $cR + N/N$ must be n. Therefore $e_i R \cap cR \nsubseteq e_i N$ for any i. Hence $e_i R \cap cR \supset e_i N$. It follows that $cR \supset N$ and so $cN = N$. Therefore 20.35 applies. $\square$

25.3.6 Corollary (Warfield). *A Noetherian serial ring is a finite product of Artinian rings and hereditary prime rings.*

Proof. By 3.5, it suffices to assume that R is semiprime. In this case, however, R is then a Goldie semiprime ring, and hence is nonsingular by (I, 9.13, p. 397). Now any nonsingular left serial ring is left semihereditary by a theorem of Warfield (loc. cit), a result which we leave for an exercise (or consult Warfield). Thus, R is a hereditary semiprime ring, and hence is a finite product of prime rings by the result of Levy-Chatters (see 20.30 and 20.32). $\square$

25.3.7 Proposition. *If R is a serial ring [resp. right FFM or right FBG], then so is eRe, for any idempotent $e \neq 0$.*

Proof. Let $B = eRe$, and let M be any right B-module, and write

(1) $$M' = M \otimes_B eR = U_1 \oplus \cdots \oplus U_n$$

as a direct sum of uniserial right R-modules. Then, $M'e = M \otimes_B B \approx M$ is a direct sum of the uniserial right B-modules $U_1 e, \ldots, U_n e$. Similarly, if R is right FFM, if M is a finitely generated indecomposable right B-module, and if (1) denotes a decomposition of M' into a direct sum of indecomposable right R-modules, then there exists i such that $M \approx M' e = U_i e$, so that there are only finitely many indecomposable finitely generated right B-modules. [The proof for right FBG is similar.]   □

4. Artinian Serial and Sigma Cyclic Rings

In view of the Decomposition theorem 3.5 for Noetherian serial rings, we may restrict our attention to Artinian or semiprime serial rings, and in this section, we study the former[2].

25.4.1 A   Definition. Let S be a set of modules each of finite length. A module $M \in S$ is said to be **dominant** if M has the maximal possible length. (This assumes that the modules in S have bounded lengths.)

For example, if R is a left Artinian ring, and if Re is a left prindec, then Re is said to be dominant if it is dominant in the set of left prindecs of R.

25.4.1 B *Exercise.* (a) A uniserial module M is cyclic iff M is finitely generated iff M is a local module.

(b) Over a semiperfect ring R, a uniserial module M is principal cyclic iff finitely generated.

25.4.2   Theorem.[3] *For a ring R with radical J the following conditions are equivalent:*

(1) *Every left R-module is a direct sum of finitely generated uniserial modules.*

(1_{bis}) *R is left Artinian, and every finitely generated left R-module is a direct sum of uniserial modules.*

(2) *Every left and right R-module is a direct sum of cyclic uniserial modules.*

(2_{bis}) *R is left Artinian, and every finitely presented left or right module is a direct sum of uniserial cyclic modules.*

(3) *R is left and right Artinian and serial.*

(3_{bis}) *R is left Artinian, every right prindec eR is uniserial modulo eJ^2, and every left prindec Re is uniserial modulo $J^2 e$.*

(4) *R is left Artinian, and every dominant left prindec of any factor ring R/A is quasiinjective in R-mod.*

(5) *R is left Artinian, and every dominant left prindec of R/J^n is injective in R/J^n-mod, for every n.*

(6) *R is left Artinian, and every finitely generated indecomposable right or left module is principal cyclic.*

---

[2] At this juncture, we are departing from the development of serial rings of Warfield [75]. The reader may continue the study of serial rings by reading the subsequent chapters of Warfield's paper on the subject of (semi)prime serial rings.

[3] These rings were first investigated by Köthe [35], Asano [39] (and later [49]), and Nakayama [39, 40]). Köthe contributed the term **uniserial** (*Einreihig*), and Nakayama coined the term ***generalized uniserial*** for what we call Artinian serial rings. Primary-decomposable Artinian serial rings were called uniserial rings. Several parts of this theorem come from Fuller [69a] and Eisenbud and Griffith [71a].

Proof (Schematic).

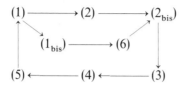

Trivially, $(2) \Rightarrow (1) \Rightarrow (1_{bis})$. Also, $(4) \Rightarrow (5)$ in view of the fact 19.16 that over Artinian rings a module is quasi if and only if it is injective modulo annihilator. $(1) \Rightarrow (5)$. Since every module is a direct sum of finitely generated modules, then 20.17 implies that R is left Artinian. Second, if M is any dominant left prindec of R/J^n, then M is uniserial, hence uniform, and therefore its injective hull $\hat{M}$ in R/J^n-mod is indecomposable, hence uniserial. Since $\hat{M}$ is therefore principal cyclic by 18.23.4 and 18.25, then the length of $\hat{M} \leq$ the length of M. Then $M = \hat{M}$, so M is injective modulo J^n.

$(5) \Rightarrow (1)$ For every left prindec Re, the socle of $Re/J^n e$ is simple for every $n > 0$ such that $J^{n-1} e \neq 0$, and so $J^{n-1} e/J^n e$ is simple for any such n. This proves that any left prindec Re is uniserial. Since every left module M is therefore generated by uniserial principal cyclics, if C is a principal cyclic module of maximal length contained in M, say $C \approx Re/J^k e$ and $J^{k-1} e \neq 0$, then $J^k M = 0$. Since $Re/J^k e$ is a dominant left prindec of R/J^k, then C is injective in R/J^k-mod, and hence a summand of M. By induction, this establishes (1) when M has finite length. In the general case, every submodule P of M is a direct limit of uniserial summands of P which are simultaneously summands of M.

In particular, since the direct limit P of a directed family of pure submodules is again a pure submodule (in the sense that the canonical map $Y \otimes_R P \to Y \otimes_R M$ is monic for every right R-module Y) then there is a maximal pure submodule P of M generated by a maximally independent family $\{C_i\}_{i \in I}$ of uniserial summands of M. If $P \neq M$, then there is a nonzero uniserial summand X/P of M/P, and X is then a pure submodule of M violating the maximality of P. Then $M = P = \sum_{i \in I} \oplus C_i$ is $\sum$-uniserial.

$(6) \Rightarrow (2_{bis})$. The hypothesis (6) implies (v) of 2.6, and therefore (ii) and the right-left symmetry hold, so every finitely presented module is a direct sum of uniserial principal cyclic modules.

$(3) \Rightarrow (3_{bis})$ is trivial.

$(3_{bis}) \Rightarrow (3)$. If eR is any right prindec, then the hypothesis eR/eJ^2 uniserial implies eJ/eJ^2 is simple, hence isomorphic to fR/fJ, for some right prindec fR. Then, eJ^2/eJ^3 is either $=0$, or $\approx fJ/fJ^2$ which is simple. By induction, eJ^n/eJ^{n+1} is simple, or zero, for every n. Since R is left Artinian, $J^n = 0$ for some n, proving that eR is uniserial, and dually, Re is uniserial, so R is a serial ring.

It remains only to prove that R is right Artinian. However, R is semiprimary, and J/J^2 is a finite direct sum of simple modules eJ/eJ^2, so J/J^2 is Noetherian, and so R is right Artinian by 23.19. (Actually, the same argument shows that J^n/J^{n+1} is Noetherian, for any n, so then R has a composition series on the right. Cf. 18.12 and proof.)

$(2_{bis}) \Rightarrow (3)$. Clearly, (2_{bis}) implies that R is serial. Moreover, then R is right Artinian as in the proof of $(3_{bis}) \Rightarrow (3)$.

$(3) \Rightarrow (2_{bis})$ follows from 2.6.

$(3) \Rightarrow (5)$. The proof is similar to $(1) \Rightarrow (5)$, except that we do not know that $\hat{M}$ is uniserial. However, $\hat{M}$ is indecomposable (and uniform), so that every finitely generated submodule is indecomposable, hence uniserial by 2.6, and, by (2_{bis}), isomorphic to a principal cyclic module (see 4.1). This means that the finitely generated submodules of $\hat{M}$ have finite lengths bounded by the length of R, and hence, that they satisfy the a.c.c. Therefore, $\hat{M}$ is a Noetherian module, hence finitely generated, and therefore uniserial, hence principal cyclic of length $\geq$ than the lenght of M. Thus, $M = \hat{M}$ is injective as before.

$(1_{bis}) \Rightarrow (6)$ by (ii) $\Rightarrow$ (v) of 2.6.

$(3) \Rightarrow (4)$ as in the proof of $(3) \Rightarrow (5)$.

$(1) \Rightarrow (2)$. Now $(1) \Rightarrow (3)$ via implications already proved, and ditto for $(3) \Rightarrow (1)$. However, (3) is left-right symmetric, and so the left-right symmetry $(1')$ of (1) holds. Therefore, (1) implies $(2) = (1) + (1')$. $\square$

25.4.3   Corollary. *If R is a semiprimary ring, and if $R/(rad\, R)^2$ is QF, then R is an Artinian serial ring.*

Proof. R is right (and left) Artinian by 23.19. Moreover, every right prindec of $\bar{R} = R/(rad\, R)^2$ is indecomposable and injective, hence has simple socle as well as simple top, proving that it is uniserial. Similarly for left prindecs, so (3_{bis}) of 4.2 holds, and therefore R is serial. $\square$

A theorem of Jans [57], Tachikawa [60] (for algebras) and Colby [66] for Artinian rings states that an infinite lattice of ideals implies that the ring is not FBG in a strong sense. The next result is much weaker. (However, see 25.4.5 following.)

25.4.4   Corollary. *An Artinian serial ring R has finite module type ($=$right and left FFM), and finite lattice of ideals.*

Proof. Let $R = e_1 R \oplus \cdots \oplus e_n R$, for right prindecs $e_i R$, $i = 1, \ldots, n$. Now, by 4.2, every indecomposable cyclic right module is principal cyclic, hence a factor $e_i R/K$, for some submodule K of some $e_i R$. Since $e_i R$ is uniserial, the set of all possible K's in all possible i's is finite, that is, the isomorphism class of indecomposable principal cyclic modules is a finite set, say $C_1, \ldots, C_t$. Moreover, every right R-module M of finite length $\leq d$ is isomorphic to a direct sum

$$M \approx C_1^{e_1} \oplus \cdots \oplus C_m^{e_m}$$

with exponents $e_1, \ldots, e_m$ integers bounded by an integer $f(d)$ depending on d. Thus, by the unique decomposition theorem 18.18, there are just finitely many nonisomorphic modules of length $\leq d$, so R is FFM. Moreover, for ideals A and B, the cyclic modules $R/A \approx R/B$ iff $A = B$, then the lattice of ideals is finite. $\square$

25.4.5   Corollary. *Any right FFM right Artinian ring R has finite ideal lattice.*

Same proof. $\square$

25.4.6 A   Proposition (Nakayama [40], Goursaud [70], Faith [72a]). *If A is a ring which is completely right* PF *in the sense that every factor ring is right* PF, *then A is a primary-decomposable Artinian serial ring.*

Proof. By 24.32, any factor ring of A is a semiperfect ring with finite essential socle, and any nonzero module M is a generator over $A/\text{ann}_A M$, hence has non-zero socle. By 22.29, A is therefore a left perfect ring. Let $J = \text{rad } A$. Since every factor ring has finite right socle, then J/J^2 is finitely generated, and so, by 23.19, A is right Artinian. Since A is right selfinjective, by 24.25, this implies that A is QF. Since every factor ring of A is therefore QF, then A is serial by 4.3. Also, since every onesided ideal is an annulet, 24.36 implies that every ideal is a right and left principal ideal. Therefore, by 19.44, the prime ideals of A commute, and then A is primary-decomposable by 18.37.   □

25.4.6 B   Proposition (Köthe [35], Asano [39, 49], Nakayama [40], Faith [66 b]). *A ring is a primary decomposable serial ring if and only if the following equivalent conditions hold*

(a) *R is right and left Artinian principal ideal ring.*

(b) *R is a serial Artinian right principal ideal ring.*

(c) *R is a right principal ideal QF ring.*

(d) *R is a principal right ideal ring, and every left prindec of R is serial of finite length.*

(e) *Every factor ring of R is QF.*

(f) *R is left or right Artinian, and the injective hull of each cyclic right R-module is cyclic.*

(g) *R is right Noetherian, and the injective hulls of cyclics are cyclic in* mod-R.

Proof. Each ring satisfying one of (a)–(d) is an Artinian principal right ideal ring, and hence, by 19.44, primary-decomposable. Since the hypotheses are preserved by finite products, it suffices to prove the equivalence of (a)–(d) for a matrix ring A_n over a completely primary ring A. We leave this exercise to the reader with two remarks: (1) In general, it is possible for a ring A_n to be a principal right ideal ring, and A not a principal right ideal ring. (Such an example has been given by Swann [61]; cf. 10.21 ff.) However, this cannot happen for Artinian rings; (2) the fact that any ring satisfying one of (a)–(d) is serial follows immediately from the $n = 1$ case for A_n, since A is serial in any case.

Since any factor ring of a serial ring is serial, then any one of (a)–(d) implies (e), whereas (e) implies that R is primary decomposable serial by 4.6 A. Thus, (a)–(e) are equivalent. Furthermore, (f) $\Rightarrow$ (g) is trivial. We complete the proof by (b) $\Rightarrow$ (f), and (g) $\Rightarrow$ (c).

(b) $\Rightarrow$ (f). If $C = R/I$ is cyclic, then $^\perp I = R y$ is principal, and hence $C = R/y^\perp \approx y R$ embeds in R, and hence $\hat{C}$ is a summand of $R = \hat{R}$, hence cyclic.

(g) $\Rightarrow$ (c). Since R is right Artinian, then by (8) of 25.6, R is cyclic only if $R = \hat{R}$. Thus, R is QF 25.3. If $C = R/I$ is any cyclic right module, then $\hat{C}$ is cyclic, and projective by 24.12, hence isomorphic to a summand of R. Since R/I therefore embeds in R, then $^\perp I$ is a cyclic left ideal. Since every left ideal is an annulet, this shows that R is a principal left ideal QF ring, the left-right symmetry of (c). However, since (c) $\Leftrightarrow$ (a), and (a) is right-left symmetric, then we deduce (c). $\square$

25.4.7 *Exercise.* (a) A commutative Artinian ring R is serial if and only if R is QF modulo the square of radical.

(b) A QF ring R is serial if and only if $R/(\text{rad } R)^2$ is QF.

(c) The ring $T_n(F)$ of lower triangular $n \times n$ matrices over a semisimple ring F has the property that it is serial, but not QF. [The injective hull of $T_n(F)$ is the full matrix ring F_n.] If $n > 1$, and if F is a field, then $T_n(F)$ is not primary decomposable.

(d) Defeat the converse of 4.3.

(e) (Eisenbud-Robson [71] and Warfield [75]) Any semiprimary right FBG ring R is right Artinian.

Interlacing

25.4.8 Definition. *Let M_1, and M_2 be right R-modules, and $a: S_1 \to S_2$ a morphism of submodules $S_1 \subseteq M_1$ and $S_2 \subseteq M_2$. Then,*

$$\text{diag } a = \{(s, a\,s) \in M_1 \times M_2 \mid s \in S\}$$

is a submodule of $M_1 \times M_2$, and the factor module

$$M_1 \times M_2/\text{diag } a$$

*is called the **interlacing (module) of** a, and denoted $M_1 \underset{a}{\times} M_2$.*

25.4.9 Proposition. *Let R be a semilocal ring with radical J. If M_1 and M_2 are uniserial modules, each of finite length m, if $M_1 \neq M_2$, and if $f: M_1 \to M_2$ is an isomorphism which induces the identity 1_M on a submodule M of M_1 and M_2 of length $m-1$, then the interlacing of 1_M is an indecomposable module P of length $m^2 - m + 1$, and the top of P has length $m^2 - 2(m-1) > m$. Moreover, the embedding*

$$\begin{cases} M_1 \to P \\ m \mapsto [(m, 0) + \text{diag } 1_M] = [m, 0] \end{cases}$$

is an essential monic.

Proof. Let $[m_1, m_2]$ denote the coset of (m_1, m_2) modulo diag 1_M. Since M_i is uniserial, then by (4) of 18.3, $J = \text{rad } R$ satisfies

$$M = \text{rad } M_i = M_i J$$

$i = 1, 2$. This shows that

$$\text{rad } P = PJ = [M_1 J, M_2 J] = [M, 0] = [0, M]$$

is isomorphic to M under $m \mapsto [m, 0]$. Now,

$$d[P] = m^2 - (m-1) = d[P/PJ] + d[PJ]$$

and $d[PJ] = d[M] = m - 1$, so that

$$d[P/PJ] = (m-1)^2 + 1 > (m-1) + 1 = m.$$

Next let M_1' denote the image of M_1 under the indicated embedding of M in P, and let x be any nonzero element of P. If $x = [m_1, m_2]$ is such that m_2 lies in M, then $x = [m_1, 0] \in M_1'$. Thus, if $x \notin M_1'$, then one of m_1 or m_2 lies outside of M, say $m_1 \notin M$, and then $m_1 j \neq 0$, for some $j \in J$, since M contains socle M_1, the annihilator of J in M_1. But $M = M_1 J$, and hence xj is a nonzero element of M_1' by what has just been proved. This shows that P is an essential extension of M_1'. Therefore, P is indecomposable along with M_1'. $\square$

25.4.10 *Exercise* (Tachikawa [59]). 1. Let R be an Artinian serial ring. If $a: \operatorname{Soc} M_1 \to \operatorname{Soc} M_2$ an isomorphism $i = 1, 2$, then the interlacing of a is decomposable if and only if a is extendable to an embedding of M_1 in M_2, where soc M denotes the socle of M (I, p. 367).

2. Under the same hypotheses, then the interlacing of a has simple socle if and only there is a homomorphism $b: S \to M_2$, where S is a submodule of M_1 of length > 1, and b induces a.

3. Show by example that not every uniserial module over a serial ring is quasinjective.

4. Characterize serial rings for which every uniserial module is quasinjective.

Sigma Cyclic Rings

The terms sigma cyclic, n-gened, BG, and FBG were defined and described in Chapter 20. If R is right Artinian, and FBG, or equivalently, σ-n-gens, for some integer $n > 0$, then R has a duality context 25.11. These rings are characterized by 25.12. If A has a duality context $_B E_A$, and if every indecomposable left B-module of finite length embeds in E, then A is right σ-cyclic 25.13. A sufficient condition for this is that every indecomposable left B-module have square-free socle 25.13(b). Similarly for injective σ-cyclic rings 25.14.

25.4.11 **Proposition.** *If A is right Artinian and right FBG ($= n$-gened, for some $n > 0$), then every indecomposable injective right A-module is finitely generated, and A has a finitely generated injective cogenerator E, hence a duality context $_B E_A$.*

Proof. As in the proof of (3) $\Rightarrow$ (5) of 4.2, any indecomposable injective right A-module is finitely generated. Since A is right Artinian, then $A/\operatorname{rad} A$ is semisimple, so there are just finitely many indecomposable nonisomorphic injective modules $E_1, E_2, \ldots, E_t$ (the injective hulls of the nonisomorphic simple modules). Then $E = E_1 \oplus \cdots \oplus E_t$ is an injective cogenerator which is finitely generated. Then, by a theorem of Morita 23.25, there is a duality context $_B E_A$. $\square$

25.4.12 **Proposition.** *Let A be a right, B a left, Artinian ring, and $_B E_A$ a duality context. The following conditions are equivalent:*

(a) *A is right FBG.*

(b) *For some integer $n > 0$, A is right n-gened, or equivalently, right σ-n-gens.*

(c) *A is similar to a σ-cyclic ring.*

(d) *There exists an integer $m > 0$ such that every finitely generated, indecomposable left B-module can be embedded in E^m.*

(e) *There exists an integer q such that every finitely generated indecomposable left B-module has socle of length $\leq q$.*

(f) *B is left FBG.*

When these conditions hold, and if n (resp. m) is the smallest integer with this property, then $n = m$.

Proof. If $X \leftrightarrow X^*$ under the E-duality, then X is simple if and only if X^* is simple, and by induction, length $X \leq n$ if and only if length $X^* \leq n$. Furthermore, X is indecomposable if and only if X^* is. Thus, (a) $\Leftrightarrow$ (f). Moreover, $A^n \to X \to 0$ is exact in mod-A if and only if

$$(A^n)^* = (A^*)^n = E^n \leftarrow X^* \leftarrow 0$$

is exact in B-mod. Thus, (b) $\Leftrightarrow$ (d). Also (a) $\Leftrightarrow$ (b) is obvious, and (b) $\Leftrightarrow$ (c) is 20.39. Finally, (d) $\Leftrightarrow$ (e) follows, since the injective hull of a B-module over an Artinian ring B is the same as that of its socle, and every semisimple module of length q embeds in E^q (where E can be any injective cogenerator). Since each module M in B-mod has essential socle, an embedding of socle M in E^q extends to an embedding of M. $\square$

Before the complete solution by A. V. Roiter [68], the Brauer-Thrall conjecture for finite dimensional algebras over algebraically closed fields was proved by Curtis and Jans [65] under the hypothesis (b) of the corollary, which shows the algebras are σ-cyclic.

25.4.13 Corollary. *Let $_BE_A$ be a duality context, where A is right, B left, Artinian.*

(a) *A is σ-cyclic if and only if every finitely generated indecomposable left B-module can be embedded in E.*

(b) *A sufficient condition for (a) is for every finitely generated indecomposable left B-module to have square-free socle. If E is a minimal injective cogenerator in B-mod, this condition is also necessary, that is, in this case A is σ-cyclic if and only if every finitely generated indecomposable left B-module has square-free socle.*

Proof. (a) comes from 4.12. (b) comes from the fact that a module M over an Artinian ring embeds in a minimal injective cogenerator if and only if socle M is square-free. $\square$

25.4.14 Corollary. *Let $_BE_A$ be a duality context, where A is right, B left Artinian.*

(a) *A is injective σ-cyclic if and only if every principal indecomposable left B-module Be (where $e = e^2 \in B$), embeds in E.*

(b) *A sufficient condition for (a) is that every principal, indecomposable left B-module Be, $e = e^2 \in B$, has square-free socle. If E is a minimal injective cogenerator for B, this condition is also sufficient.*

Proof. Under the duality $X \leftrightarrow X^*$, X is indecomposable injective if and only if X^* is indecomposable projective. But over an Artinian ring, each such X^* is isomorphic to a principal indecomposable left ideal Be (see 22.23(c)). The rest of the proof is the same as 24.32. $\square$

25.4.15 Corollary. *Let R be QF.*

(a) *Then R is right σ-cyclic if and only if every indecomposable finitely generated left module embeds in R.*

(b) *R is left and right σ-cyclic if and only if every finitely generated module is isomorphic to a direct sum of onesided ideals.*

When (b) *holds, then every right ideal of R contained in a principal indecomposable right ideal eR is cyclic.*

Proof. $_R R_R$ is a duality context. Then (a) of 4.13 implies (a), and (b) is a restatement for the condition (a) applied to both sides. When (b) holds, and if $I \subseteq eR$, then I is indecomposable, hence cyclic. $\square$

25.4.16 Corollary. *Let A be a right Artinian ring. Then there exists a integer $n > 0$ such that every finitely generated indecomposable right A-module is generated by n elements if and only if there exists a duality context $_B E_A$ such that every indecomposable left B-module M embeds in E^n.*

Proof. This is a restatement of (b) $\Leftrightarrow$ (d) of 4.12. $\square$

25.4.17 A   Proposition. *Let R be a ring.*

(a) *If every right R-module embeds in a free module, then R is QF.*

(b) *Let R be right Artinian, or left Noetherian. If every finitely generated right module embeds in a free right R-module, then R is QF. Thus, in order for every finitely indecomposable right module to be isomorphic to a right ideal of R, it is necessary for R to be QF.*

(c) *If R is serial Artinian then every module is isomorphic to a direct sum of cyclic uniserial onesided ideals. In this case every module is isomorphic to a direct sum of cyclic onesided ideals.*

Proof. (a) Assuming (a) then any injective right module E embeds in a free module, hence is a summand of a free module. Since every injective module is thereby projective, this implies R is QF by 24.12.

(b) By 24.1, every right ideal is the right annihilator of a finite subset of R. Thus, the a.c.c. on left ideals implies the d.c.c. on right ideals, so we may suppose R is right Artinian in either case. Now every finitely generated uniform right R-module embeds in R by the argument of 20.15, so these modules have composition series of lengths bounded by the length of R. It follows therefore, as in 4.11, that every indecomposable injective right module is Noetherian, hence also embeds in R. Now any finite direct sum of these is injective, and hence a summand of R, and since R has finite essential right socle, one concludes that R is a finite direct sum of injective hulls of simple modules, that is, R is injective. Then R is QF by 24.5.

(c) If R is serial Artinian, then by 4.2, every right module M is isomorphic to a direct sum of principal cyclic, hence Artinian uniserial modules. Thus, each indecomposable module U is uniform, so by the proof of (b), U embeds in R proving that M is a direct sum of cyclic uniserial right ideals.   □

25.4.17 B   *Exercise.* Show that there exists a (QF) ring R which has the property every finitely generated module is a direct sum of cyclic onesided ideals but R is not serial (see Chapter Notes on Group Rings).

25.4.18 A   **Proposition** (Faith-Walker [67] and Faith [66]). *If R is commutative, then R is QF if and only if each injective module is a direct sum of cyclic modules.*

Proof. One way is established by 24.14. Next let R be a commutative ring having the stated property. Then R is Artinian by 20.18. Therefore R is a direct product of finitely many local Artinian rings, and it suffices to prove that any local Artinian ring with the stated property is QF. Then, if $J = \operatorname{rad} R$, $\widehat{R/J}$ is a cyclic injective cogenerator, hence faithful, hence $\widehat{R/J} \approx R$ is injective. This proves that R is QF 24.3.   □

Right and left Injective Sigma Cyclic Rings

By 24.14, every QF ring R is right and left injective $\sum$-cyclic, and conversely if R is commutative 4.18. The following proposition shows that in general the converse does not hold.

25.4.19   **Proposition.** *Let B be an Artinian commutative ring, and let E denote the injective hull of $B/\operatorname{rad} B$ in B-mod.*

(a) *Then $_B E_A$ is a duality context where $A = \operatorname{End} {}_B E$, and so is $_B E^n_{A_n}$, for any integer $n > 0$ (cf. 25.22).*

(b) *Furthermore, if socle B has length $\leq n$, then the matrix ring A_n is right and left injective σ-cyclic.*

(c) *If B is not quasi-frobenius (QF), then A_n is not. (If B is local then B is QF $\Leftrightarrow n = 1$.)*

Proof. (a) Since every simple B-module appears in $B/\operatorname{rad} B$, then E is an injective cogenerator. Now over a commutative Artinian ring, the injective hull of a module of finite length also has finite length (Morita [58]). Thus, $_B E$ is finitely generated, so $_B E_A$ is a duality context by Morita's theorem. Since $A_n = \operatorname{End} {}_B E^n$, by the same reason, so is $_B E^n_{A_n}$.

(b) If the socle B has length $\leq n$, then so does that of every principal indecomposable left ideal Be, which accordingly can be embedded in E^n. Thus, (b) follows from (a) of the last corollary, and the fact that the conditions are left-right symmetric. (In fact $A = B$, and E is the injective cogenerator of B in B-mod. See Chapter Exercise 16.)

(c) Since B is "dual" to A_n, one of the rings is QF if and only if the other is QF. An Artinian local ring is QF if and only if the right and left socles have length 1.   □

5. Factor Rings of Hereditary Noetherian Prime Rings: The Eisenbud-Griffith-Robson Theorem

We now apply the Webber-Chatters' theorem 20.29, and Nakayama's theorem 4.2 on Artinian serial rings to obtain the theorem of the title of this section, namely:

25.5.1   Theorem (Eisenbud-Griffith [71a], Eisenbud-Robson [70b])[4]. *If R is a Noetherian hereditary prime ring, then R/I is an Artinian serial ring for every ideal $I \neq 0$. In fact, this conclusion holds for any left hereditary ring R with flat injective hull, and any ideal I such that R/I is left Artinian.*

Proof. By the theorem of Goldie and Lesieur-Croisot (I, 9.10, p. 396), R has a simple Artinian classical left quotient ring Q, which by (I, 16.9, p. 529) is a flat right R-module. Now Q is left selfinjective, and the inclusion map Q-mod $\rightsquigarrow R$-mod has exact left adjoint $Q_R \otimes$, so that Q is an injective left R-module by (I, 11.35.3, p. 440). (See also Exercise 19.38.) It follows that Q is the injective hull of R in R-mod. Furthermore, since R is prime, any ideal $I \neq 0$ is an essential left ideal, and hence R/I is left Artinian by 20.29.

Hence, we may suppose that R and I are as stated in the second sentence of the theorem. Since any statement proved about $A = R/I$ also applies to any factor ring of A, in order to prove that A is serial, by 4.2, it suffices to prove that any dominant left prindec of A is injective.

Let E be the injective envelope of R in R-mod. We shall calculate the left A-injective envelope of A in terms of E. Since R is left hereditary, E/I is R-injective. (Use the fact that inj. dim $E/I \leq 1$. See (I, p. 371); also (I, 2.4, p. 535). Let $F = \{x \in E \mid Ix \subseteq I\}$. Clearly F is a left R-submodule of E and is therefore flat. Set $F/I = E'$ and observe that E' is a left A-module containing A. E' is an injective A-module since

$$E' \approx \operatorname{Hom}_R(A, E/I) \approx \{x \in E/I \mid Ix = 0\}.$$

On the other hand, we claim that E' is a projective A-module. By 22.31A, it suffices to show that E' is a flat A-module. Note that $I = IF$ since $1 \in F$; hence $E' = F/IF \approx A \otimes_R F$. But F is a flat R-module, and therefore E' is a flat A-module.

Since E' is injective and projective, by 20.15 we may write $E' = \coprod X_i$, where the X_i are injective indecomposable left prindecs of A. Let X be a dominant left prindec of A. We will show that X is isomorphic to one of the X_i. For every i, we have a map $\varphi_i : X \rightarrow X_i$ given by the composition of maps

$$X \xrightarrow{\text{incl}} A \rightarrow E' = \coprod X_i \xrightarrow{\text{proj}} X_i.$$

Since $X \rightarrow A \rightarrow E'$ is a monomorphism, we can choose an i so that $\ker \varphi_i \not\supseteq \operatorname{soc} X$, and thus Loewy length $(\operatorname{im} \varphi_i) = $ Loewy length(X). Since X is dominant, Loewy length$(X_i) \leq$ Loewy length(X), so φ_i is necessarily onto. Since X_i is projective, φ_i splits. Since X is indecomposable, φ_i is an isomorphism, so X is injective as required. $\square$

---

[4] The prototype of this theorem by Eisenbud and Robson required the additional hypothesis of "enough invertible ideals". Eisenbud and Griffith proved the general case. Singh [75] has sketched a proof of the converse for bounded Noetherian prime rings. (His remark added in proof left off the bounded hypothesis.)

Remark. Any left hereditary ring R is left nonsingular, hence the injective hull Q of R in R-mod is the maximal left quotient ring of R, and Q is Neumann regular and left selfinjective (19.35). By the theorem of Gentile [60] and Sandomierski [68], then Q is a flat right [sic!] R-module (cf. Chapter 11, Exercise 5). However, Q is not necessarily a flat left R-module. For example, assuming that $R \hookrightarrow Q$ is a ring epic (which it is when Q is the classical left quotient ring of R), then Q is left flat only if Q is also the right quotient ring of R (Goodearl [71]). (Cf. Exercise 19.38 (j) and (n).)

Exercises of Chapter 25

1. A local ring B which is right and left injective σ-cyclic is QF (cf. 25.4.19).

2. A ring R is Artinian serial if and only if R is a semiprimary ring with the property that the injective hull of an indecomposable module is indecomposable, equivalently, every indecomposable right or left module has simple socle.

3. (Fuller [69b]) A left Artinian ring is serial if and only if the following equivalent conditions hold: (a) Every indecomposable (right or left) module is QI. (b) Every indecomposable module is QP (=quasi-projective). (c) Every indecomposable QI module is QP. (d) Every indecomposable QP module is QI. (e) Every indecomposable *left* module is both QI and QP. (f) Every left prindec Re, and the injective hull of every simple left module, are uniserial modules. (g) The injective hull of every simple left ideal is uniserial, and every left prindec has simple socle.

4. If R is a finite dimensional algebra over a field k, and if the injective hull of every simple (right or left) module is uniserial, then R is serial.

5. (Warfield [75, 6.8]) If $I_1 \subset I_2 \subset \cdots \subset I_n$ are ideals of a ring, then the direct sum $R/I_1 \oplus \cdots \oplus R/I_n$ is a balanced module. Conclude that any finitely presented module over a discrete valuation ring is balanced. (Also see Warfield's result on balanced modules cited in chapter notes.)

6. (Asano [49a]) Let A be an algebra of finite dimensions over a field. Then, A is uniserial iff A is right uniserial, and hence iff A is left uniserial.

7. (Warfield [75]) Let R be a semiperfect ring. The following conditions are equivalent:

(i) R is left nonsingular and left and right serial.

(ii) R is left nonsingular and left finite dimensional, and all finitely generated nonsingular left modules are projective.

(iii) R is left semihereditary, left finite dimensional, and if Q is the maximal left quotient ring of R, then $_R Q$ is flat.

(iv) R is left and right serial and left and right semihereditary.

8. (Goldie [64]) An indecomposable Artinian serial ring is nonsingular iff isomorphic to the full ring $T_n(D)$ of upper triangular matrices over a division ring.

9. (Murase [63 II, Theorems 17, 18]) Let A be Artinian and serial with radical N. Then, A/N^2 has finite global dimension iff A is a finite product of rings, each of which is similar to a factor ring of the type $T_n(D)$, for various integers n, and fields D (cf. 8).

10. (Murase, Amdal-Ringdal, and Eisenbud-Griffith [71b]) Any Artinian serial ring is a product $A_0 \times A_1 \times A_2 \times A_3$ such that:

(0) A_0 is semisimple, but A_i has no semisimple direct factor, $i \geq 1$.

(1) A_1 is an Artinian PIR.

(2) A_2 has finite global dimension modulo $(\operatorname{rad} A_2)^2$ (compare 9).

(3) A_3 is QF modulo $(\operatorname{rad} A_3)^2$, and has no homogeneous projective modules.

11. (Skornyakov, cited by Eisenbud-Griffith [71b, p. 120]) A ring R is serial Artinian iff every left module is a direct sum of uniserial modules.

12. (Fuller [69b]) If R is left Artinian, and if every module generated by two elements is a direct sum of uniserial modules, then R is serial. (Cf. #3(f); also Ivanov [74].)

Notes for Chapter 25

Uniserial (Einreihig) modules were studied by Köthe [35], who proved that over a primary-decomposable serial ring, every module is $\sum$-uniserial modules, a result that Nakayama [39, 40, 41] extended to Artinian serial rings (see 25.4.2). In the same paper, Köthe characterized commutative Artinian $\sum$-cyclic rings as Artinian principal ideal rings. (Kaplansky and Cohen [51] showed that the Artinian hypothesis was redundant and this led to a number of similar theorems—see Chapter 20.) Asano [39] characterized Artinian rings R with the following two properties: (1) R is $\sum$-cyclic; and (2) every submodule of cyclic module is cyclic. These proved to be the primary-decomposable serial rings. Obviously, the class (2) represents precisely the Artinian principal ideal rings. Actually, Satz 2 of Asano [39] states that an Artinian ring is primary-decomposable serial iff every ideal is a principal right, and a principal left, ideal, so (2) $\Rightarrow$ (1) for Artinian rings. For a primary ring, it suffices for the radical to be right and left principal [Hilfssatz 5]. A more general result of Morita on when the radical is a principal right and left ideal is stated below under *Note on Group Rings*.

Let $e_1, \ldots, e_n$ be a basic set of idempotents of R, that is, $e_0 = e_1 + \cdots + e_n$ is the basic idempotent. Kuppisch [59] showed that if R is serial Artinian, these can be indexed so that if $J = \operatorname{rad} R$, if $\bar{A}$ is the image of $A \subseteq R$ under the canonical map $R \to R/J$, and if $c(M)$ is the length of a module M, then:

(a) $\overline{Re_i} \approx \overline{Ne_{i+1}}$, $i = 1, \ldots, n-1$; and $\overline{Re_n} \approx \overline{Ne_1}$ if $Ne_1 \neq 0$;

(b) $c(Re_i) \geq 2$, $i = 2, \ldots, n$;

(c) $c(Re_{i+1}) \leq c(Re_i) + 1$, $i = 1, \ldots, n-1$;

(d) $c(Re_1) \leq c(Re_n) + 1$.

The sequence $Re_1, \ldots, Re_n$ ordered so that (a)–(d) holds is the **Kuppisch series** for the serial ring R. Kuppisch shows that indecomposable Artinian ring R is QF iff R and Re_i have the same Loewy length, $i = 1, \ldots, n$. For a finite dimensional serial algebra R over an algebraically closed field k, the Kuppisch series for R, the length of R, and the $\dim_k(Re_i/Je_i)$, $i = 1, \ldots, n$, determine a complete set of invariants which determine R up to isomorphism. More generally, Murase [63, 64] classified Artinian serial rings using their Kuppisch series. Also Fuller [68] showed that the global dimension of any Artinian serial ring is completely determined by the Kuppisch series of its indecomposable direct summands. Moreover, the basic ring has the same length on the right as on the left (*loc. cit.*, p. 252, Corollary 2.3). Fuller also studied QF-1 (or balanced) serial Artinian rings, a subject taken up for Noetherian serial algebras by Warfield. (See Warfield's theorem 6.7 quoted below and references given there.)

Murase showed that "many" Artinian serial rings (algebras) may be represented as a "quasi-matrix rings" with elements in a field, and conversely. Moreover, Murase constructed a class of hereditary prime Noetherian rings—infinite quasi-matrix rings—with the property that every quasi-matrix ring is an epic image. Actually, this class consists of "contracted" semigroup rings $K[S]$ with respect to a semigroup with multiplicative zero, over a field K. The Murase quasi-matrix rings have the form $K[S]$, where S is the "Rees factor semigroup" of the semigroup $QM(\infty)$ of infinite upper triangular matrix units modulo an ideal (Clark [68, p. 102, Theorem 1]). The contracted semigroup $K[QM(n)]$ consists of the ring A of all $n \times n$ matrices (a_{ij}) over the (ordinary) polynomial ring $K[x]$ such that $x \mid a_{ij}$ whenever $i > j$, that is,

$$
A = \begin{pmatrix} K[x] & K[x] & \ldots & K[x] \\ (x) & K[x] & \ldots & K[x] \\ \vdots & \vdots & & \vdots \\ (x) & (x) & \ldots & K[x] \end{pmatrix}
$$

and, moreover, every factor ring modulo a nonzero ideal is an Artinian serial ring (Clark, *loc. cit.*, Theorems 2 and 3). (Compare Harada [68, p. 484]. Also see Clark's Theorem 9.)

I lifted the quote "many" in the preceding from Ivanov [74] who has generalized Murase's results to Artinian left serial rings by representing them as rings of matrices over left uniserial rings and modules. The results do not completely determine Artinian left serial rings, only up to determining the left uniserial rings and modules. These results generalize those of Colby and Rutter [68] obtained for Artinian nonsingular left serial rings, and Goldie [64] for twosided Artinian serial nonsingular rings. (Actually, these papers were already generalized by Ivanov [70]). (Also see Chapter Exercises 8–11).

For any finitely generated projective over a serial ring, and any finitely generated submodule, there are stacked decompositions (Theorem of Warfield [75]—see 3.3 ff.). Levy [66 b] proved the converse for Artinian rings.

Primary decomposable Artinian serial rings are characterized by Boyle [73] as rings over which the top and bottom of each finitely generated module are equal. Similarly, for injective hull and projective cover. See Boyle's "pajama" game, Chapter 24, Exercise 30.

Note on Groups Rings

A group ring kG of a finite group G over a field k is semisimple iff the characteristic p of k does not divide $n=|G|$ (theorem of Maschke (I, 13.21, p. 475)). When $p\,|\,n$, then a theorem of Higman [56] states that kG has finite representation type ($=$ FFM) iff the p-subgroups of G are cyclic. Since the p-Sylow subgroups are conjugate, this holds iff G has a cyclic p-Sylow subgroup P. In this case, Higman found a bound $b(n)$ on the number of isomorphism classes of indecomposable finitely generated modules independent of G, and Kasch-Kneser-Kuppisch [57] sharpened this by showing that $b(n)=n$ is an upper bound, and that $b(n)=n$ iff $G \rhd P$ and G/P is abelian of exponent m, and k contains all of the mth roots of unity. Janusz [69] constructed all of the indecomposable modules, and also showed that whenever kG is FFM, then every indecomposable module has squarefree socle.

Prime examples of FFM rings are serial rings. (See 25.4.4.) Janusz [69] characterized when kG is serial assuming that k is a splitting field for G. This happens iff the p-Sylow subgroup of G is cyclic, and every simple kG module F is the tensor $k \otimes F^*$, for some R-module F^*, where R is a complete local domain of characteristic 0 with residue field $\approx k$. (Cf. Srinivasan [60] who showed that every indecomposable module embeds in kG when G is p-solvable with cyclic p-Sylow subgroup. Also see Janusz [70, 72a]. Compare 24.17.A(b).)

Morita [54] determined necessary and sufficient conditions for the radical J of a left Artinian ring R to be a principal left, and a principal right, ideal $J=Ra=bR$. To wit R is serial and quasi-primary-decomposable in the sense that R is a finite product of rings the left prindecs of which have the same "multiplicity". Let G be a finite group with p-Sylow subgroup P, and let H be the largest normal subgroup with order prime to p. Then the group algebra kG over an algebraically closed field k of characteristic p has the stated property iff HP is an normal subgroup of G and P is cyclic.

Note on Warfield's Paper.

Another type of serial ring—the valuation ring—is studied in Chapter 20. Kaplansky showed in [49] that every finitely presented module over a commutative valuation ring is σ-cyclic (see however the footnote, p. 129). Roux [72] generalized this to the noncommutative local serial rings, and obtained a partial converse. Roux's proof is acknowledged by Warfield [75] to be the inspiration for Warfield's results contained in the first three sections of this chapter. I am indebted to Professor Warfield for generously acceding to my request to present them here almost verbatim.

I will close out these notes with additional comments on Warfield's paper, and some problems raised by him.

If a serial ring is nonsingular, it is semihereditary. In particular, a semiperfect, semiprime, left and right Goldie ring is semihereditary if and only if it is serial (cf. Chapter Exercises 8–13). Warfield also gives a fairly complete structure theory for Noetherian serial rings. As we have seen, in 3.5, such a ring is the product of an Artinian serial ring and a finite number of prime rings (necessarily hereditary, by the results just quoted). If R is a prime Noetherian serial ring, which is not semisimple, a theorem of Michler's [69 b] says that there is a discrete valuation ring D (not necessarily commutative) with Jacobson radical J, such that R is a $(D:J)$ block upper triangular matrix ring. (For this terminology, see Robson [72 a]. Any such ring is similar (Morita equivalent) to a matrix ring over D in which the entries below the diagonal are restricted to lie in J.) Warfield gives a new, short proof of Michler's theorem. (Cf. Fuelberth and Kuzmanowitz [74].

In the last section of the paper, he applies his previous results to algebras over commutative Noetherian rings. Assume that R is a commutative Noetherian ring, and A is an R-algebra which is finitely generated as an R-module. For every maximal ideal m of R, let R_m^* be the m-adic completion of R (or, equivalently, of R_m^*). Then the m-adic completion of A is just $A \otimes R_m^* = A_m^*$, and A_m^* is a semi-perfect ring. Warfield's main theorem is:

Theorem (Warfield [75, 6.6]). *The following properties of A are equivalent:* (i) *for every ideal I such that A/I is Artinian, A/I is serial,* (ii) *every finitely generated module is the direct sum of a projective module and a finite number of Artinian serial modules,* (iii) *A is the product of an Artinian serial ring and a finite number of hereditary orders over Dedekind domains, and* (iv) *for every maximal ideal m of R, A_m^* is a serial ring.*

This yields, as a corollary the following structure theorem for a class of hereditary algebras:

Theorem (Warfield [75, 6.7]). *Let R be a commutative Noetherian ring and A an R-algebra which is finitely generated as an R-module. Suppose that A is hereditary. Then the following conditions are equivalent:*

(i) *A is the product of a finite number of block triangular matrix rings over division rings and a finite number of hereditary orders over Dedekind domains.*

(ii) *For every ideal I such that A/I is Artinian, A/I is serial.*

(iii) *If Q is the maximal left quotient ring of A, then $_A Q$ is flat.*

(ii) is related to the Eisenbud-Griffith-Robson theorem 25.5.1.

Furthermore, unlikely as it may seems, these results enable Warfield to characterize those algebras over which every finitely generated module is balanced:

Theorem (Warfield [75, 6.10]). *Let R be a commutative Noetherian ring and A an R-algebra which is finitely generated as an R-module. The following properties of A are equivalent:*

(i) *For every ideal I such that A/I is Artinian, A/I is a principal left and right ideal ring.*

(ii) *every finitely generated module is a direct sum of a projective module with no simple submodules and a finite number of Artinian, homogeneously serial modules (see 1.14),*

(iii) *A is the product of an Artinian ring which is a principal left and right ideal ring, and a finite number of maximal orders over Dedekind domains,*

(iv) *for every maximal ideal m of R, A_m^* (the complete localization) is a principal left and right ideal ring,*

(v) *every finitely generated module is balanced.*

The equivalence of (i), (ii) and (iv) is an immediate consequence of the previous theorem (6.6) and 1.13. The equivalence of (iii) and these is a consequence of (6.6) and Theorem 3.3 of Eisenbud and Griffith [71a]. Actually, (iii) ⇒ (i) was proved by Asano [39, p. 239, Satz 5] for a maximal order in a simple algebra. Also, (iv) ⇒ (i) follows from the theorem of Camillo and Fuller [72] and Dlab and Ringel [72b]. Moreover, that (iii) ⇒ (i) was remarked by Asano [39, p. 239, Satz 5] for a maximal order in a simple algebra.

Warfield's Problems

We next reproduce some problems raised in Warfield [75]:

Question 1. What rings have the property that every finitely presented module is a direct sum of cyclic modules?

Question 2. What rings have the property that every finitely presented module is (a summand) of a direct sum of finitely presented cyclic modules?

Rings with the property of Question 2 are called Köthe rings by Faith [66b]. Cf. Kawada [62, 63, 64] who solved this problem for certain rings. An answer to the parenthetical Question 2 in the commutative case is Theorem 20.45. Descriptions of the rings answering Question 1 have been given by Kaplansky [49] and Lafon [71a], but whether these are "answers" depends on ones taste.[5] In particular, it is still an unsolved problem whether a Bezout domain necessarily has the property described in Theorem 1. If "cyclic" is replaced by "local" in Question 1, then the answer is that the rings in question are exactly the serial rings (Theorems 2.6 and 3.4). One virtue to this question is that the answer is clearly invariant under similarity (= Morita equivalence) of rings. (We have noted in 20.39 necessary and sufficient conditions for a ring A to be similar to a σ-cyclic ring: this happens iff A is σ-n-gens for some integer n.)

Question 4. Under what conditions is a semiperfect ring of bounded module type on its finitely presented indecomposable modules? (Thus, the latter are n-gened for some integer n.)

The theorems of Roiter [68], Auslander [74], and Tachikawa [73] solve this problem, to a certain extent, for Artinian rings and algebras, by showing that such a ring (and FBG ring) has finite module type (= an FFM ring). Also, Eisenbud-Griffith [71a] and Warfield *(loc. cit.)* show that a semiprimary right FBG ring is right Artinian, and therefore by the cited result, is right FFM.

Question 5. Is a semiperfect, semihereditary ring necessarily serial if for some n the indecomposable finitely presented modules are n-gened?

Question 6. When is a direct sum representation of uniserial modules unique?

It is, of course, when the endomorphism rings of the uniserial modules are local rings, for then the direct sum is an Azumaya diagram 21.6. Also, it is when the ring R is commutative (Kaplansky [70b]).

---

[5] For commutative local rings this problem is solved by 20.43 and 20.49.

References

Asano [35, 39, 49 a], Auslander [74], Auslander and Bridger [69], Azamaya [66], Boyle [73], Camillo and Fuller [72], Clark [68], Cohen and Gluck [70], Cohen and Kaplansky [52], Colby and Rutter [68], Curtis and Jans [65], Dickson and Fuller [69], Dlab and Ringel [72a, b], Eisenbud and Griffith 71a, b], Eisenbud and Robson [70b], Faith [66b, 72a], Faith and Walker [67], Fuller [69a, b, 68, 71], Gentile [60], Goldie [64], Goodearl [71], Goursaud [70], Ivanov [70, 74], Jans [57], Janusz [69, 70, 72a], Jategoankar [70], Higman [56], Kaplansky [49, 51], Kasch, Kneser, and Kuppisch [57], Kawada [61, 62, 63], Köthe [35], Kuppisch [59], Levy [66a, b], Michler [69b], Morita [54], Murase [63, 64], Nakayama [39, 40, 41], Robson [74], Roiter [68], Roux [71], Sandomierski [68], Srinivasan [60], Tachikawa [60, 73], Warfield [70, 75], Zaks [74].

Other References

Amdal and Ringdal [68], Brungs [69], Curtis and Jans [65], Dlab and Ringel [72c], Fuller [71, 73], Fuelberth and Kuzmanowitz [74], Ivanov [72, 74], Janusz [72b], Lafon [71a, 73], Morita [58], Shores and Wiegand [74], Tachikawa [59, 60, 61].

Chapter 26. Semiprimitive Rings, Semiprime Rings, and the Nil Radical

A ring R is semiprime (semiprimitive) if and only the intersection of the prime (primitive) ideals is zero. Then, R is a subdirect product of prime (primitive) rings 26.6 and 26.13. The (McCoy) prime radical of a ring is defined to be the intersection of the prime ideals, and is characterized as the set of all strongly nilpotent elements of R (theorem of Levitzki 26.5). When R is commutative, this is just the set of nilpotent elements.

For any separable algebraic field extension P/F then $\mathrm{rad}(A_P)$ is canonically isomorphic to $(\mathrm{rad}\, A)_P$, for any algebra A over F, and scalar extension $A_P = A \otimes P$ (see 26.16). If, on the other hand, P/F is pure transcendental, then $\mathrm{rad}\, A_P = N_P$, where $N = \mathrm{rad}(A_P) \cap A$ is a nil ideal of A (26.17)[1]. These theorems of Amitsur are applied to prove Amitsur's theorems on the semiprimitivity of any group algebra kG over a transcendental field k of characteristic zero 26.20. In particular, kG is semiprimitive whenever k is uncountable of characteristic 0 (see 26.21).

Subdirect Product of Rings and Modules

Let $\{R_i | i \in I\}$ be any family of rings, and let B be any subring of the ring direct product $R = \prod_{i \in I} R_i$. Let π_i denote the projection $R \to R_i$ [where $\forall f \in R$, $\pi_i(f) = f(i)$]. This is a ring epimorphism, and induces a ring homomorphism $\pi_i' : B \to R_i$. If this latter homomorphism is an epimorphism, that is, if $\mathrm{im}\, \pi_i = R_i\ \forall i \in I$, then B is said to be a **subdirect product of the rings** $\{R_i | i \in I\}$.

Example. Any subring B of $\prod_i R_i$ containing $\coprod_i R_i$ is a subdirect product because π_i maps $\coprod_i R_i$, hence B, onto $R_i\ \forall i \in I$.

26.1A Proposition. *A ring B is isomorphic to a subdirect product of rings $\{R_i | i \in I\}$ if and only if the following two conditions are satisfied: (a) To each $i \in I$ there corresponds an ideal Q_i of B such that $B/Q_i \approx R_i$; (b) $\bigcap_{i \in I} Q_i = 0$.*

Proof. Let B be the subdirect sum. Let Q_i be the kernel of the ring epimorphism $B \to R_i$ induced by π_i. Then $B/Q_i \approx R_i$. If $b \in \bigcap_{i \in I} Q_i$, then $\pi_i(b) = 0\ \forall i \in I$, and $b = 0$. Thus (a) and (b) are satisfied. Any ring isomorphic to B also satisfies these conditions.

Conversely, let (a) and (b) be satisfied, and let $\varphi_i : B \to R_i$ be the ring epimorphism having kernel equal to $Q_i\ \forall i$. Then there is a ring homomorphism $f : B \to \pi_i R_i$, where $f(b)(i) = \varphi_i(b)\ \forall b \in B$ and $i \in I$. If $b \in \ker f$, then $\varphi_i(b) = 0\ \forall i$, whence $b \in \bigcap_{i \in I} \ker \varphi_i$. Since $\ker \varphi_i = Q_i$, (2) implies that $b = 0$, and f is an isomor-

---

[1] N is the maximal nil ideal when A is commutative (Snapper [50]).

phism. The projection $\pi_i \colon \prod_i R_i \to R_i$ maps $f(B)$ onto $\varphi_i(B) = R_i \ \forall i \in I$, so im f is a subdirect product of the rings $\{R_i | i \in I\}$, and $B \approx \text{im } f$. $\quad\square$

More generally, a module M is said to be a **subdirect product** of the modules $\{N_i\}_{i \in I}$ iff there is a monomorphism $f \colon M \to \prod_{i \in I} N_i$ such that each of the induced mappings $f_i \colon M \to N_i$ is an epimorphism. (This is equivalent to specifying a collection of submodules $\{M_i\}_{i \in I}$ of M such that $\bigcap_{i \in I} M_i = 0$, and $M/M_i \approx N_i$, since in this case the components $\{f_i\}_{i \in I}$ become the projections $p_i \colon M \to M/M_i$.) M is called **subdirectly irreducible** iff in every representation of M as a subdirect product of modules at least one of the component mappings is an isomorphism. This is equivalent to the statement that the intersection of all nonzero submodules of M is nonzero. This intersection gives a minimal submodule contained in every nonzero submodule, and thus every subdirectly irreducible module is essentially Artinian.

26.1 B Proposition (Beachy [71 b]). *The following conditions on a right R-module M are equivalent:*

(i) *M is essentially Artinian (supra 19.16 B).*

(ii) *M has an essential submodule with d.c.c.*

(iii) *For any collection of submodules $\{M_i\}_{i \in I}$ of M such that $\bigcap_{i \in I} M_i = 0$, there exists a finite subset $F \subseteq I$ such that $\bigcap_{i \in F} M_i = 0$,*

(iv) *If M is represented as a submodule of a direct product $\prod_{i \in I} N_i$, there exists a finite subset $F \subseteq I$ such that M is represented as a submodule of $\prod_{i \in F} N_i$,*

(v) *M is a subdirect product of a finite number of subdirectly irreducible modules.*

(vi) *The socle of M is an essential Artinian submodule.*

(vii) *The injective envelope of M is isomorphic to a direct sum of a finite number of injective envelopes of simple modules.*

Proof. Obviously (i) $\Leftrightarrow$ (ii).

(ii) $\Rightarrow$ (iii). Let N be an essential submodule of M which satisfies the d.c.c. The collection of submodules of N determined by $\{M_i \cap N\}_{i \in I}$ has zero intersection if $\bigcap_{i \in I} M_i = 0$. Since N satisfies the d.c.c., there is a finite subcollection whose intersection is zero.

$$\bigcap_{i \in F} (M_i \cap N) = (\bigcap_{i \in F} M_i) \cap N = 0 \ \Rightarrow \ \bigcap_{i \in F} M_i = 0$$

because N is essential. This proves (iii).

(iii) $\Rightarrow$ (iv). If $f \colon M \to \prod_{i \in I} N_i$ is monic, then $\bigcap_{i \in I} \ker(f_i) = 0$, so $\bigcap_{i \in F} \ker(f_i) = 0$ for a finite subcollection $F \in I$, and (iv) holds.

(iv) $\Rightarrow$ (v). For each nonzero element $m \in M$, let M_m be a submodule of M which is maximal with respect to not containing m. Such a submodule exists by Zorn's lemma, and it follows that $\bigcap_{m \in M} M_m = 0$. Therefore M is a subdirect product of the modules M/M_m. But each of these is subdirectly irreducible, since any submodule properly containing M_m must contain the element m, so the intersection of all submodules of M/M_m contains the element $m \pmod{M_m}$. By assumption, there exists a finite subcollection $F \subseteq M$ such that M is represented as a subdirect product of the modules $\{M/M_m\}_{m \in F}$.

(v) $\Rightarrow$ (vi). Suppose M is a subdirect product of $\{N_i\}_{i \in I}^n$, where each N_i is subdirectly irreducible. Let S_i denote the minimal submodule of N_i. Since S_i is essential in N_i, $\bigoplus_{i=1}^n S_i$ is essential in $\bigoplus_{i=1}^n N_i = \prod_{i=1}^n N_i$, so $M \cap \bigoplus_{i=1}^n S_i \neq 0$. Let $S = M \cap \bigoplus_{i=1}^n S_i$. Then S is completely reducible, Artinian, and is an essential submodule of M. Furthermore, for any minimal submodule $M_0 \subseteq M$, $M_0 \cap S \neq 0$ implies that $M_0 \subseteq S$. This shows that S contains all minimal submodules, and is a sum of simple submodules, so it must be the socle of M.

(vi) $\Rightarrow$ (vii). Let $S = S_1 \oplus \cdots \oplus S_n$ where each S_i is simple. (Because S is Artinian the direct sum must be finite.) Since S is essential in M,

$$E(M) = E(S) \cong E(S_1) \oplus \cdots \oplus E(S_n).$$

(vii) $\Rightarrow$ (i). If the injective envelope of M is isomorphic to a direct sum of finitely many injective envelopes of simple modules, then M is a submodule of a finite direct sum of essentially Artinian modules, and so is essentially Artinian. $\square$

Part (iv) points up a duality, since a module is finitely generated iff whenever it can be represented as a quotient of a direct sum of a collection of modules, it can be represented as a quotient of a finite subcollection of these modules (Proof?). A module satisfies the a.c.c. iff every submodule is finitely generated, and the dual statement is that a module satisfies the d.c.c. iff every quotient module is essentially Artinian (See 19.16 B.).

26.2 *Exercises.* (a) A subdirect of rings $\{R_i | i \in I\}$ is a commutative ring if and only if each R_i is.

(b) (1) $\mathbb{Z}$ is a subdirect product of finite fields $\{\mathbb{Z}/p\}$, one for each prime p.

(2) $\mathbb{Z}$ is a subdirect product of rings $\mathbb{Z}/p^{e_i}$, where p is a prime, and e_i is any integer ≥ 1.

(c) A ring R is **subdirectly irreducible** provided that for any representation of R as a subdirect product of rings $\{R_i | i \in I\}$, where $R_i \approx R/Q_i$, for ideals Q_i of R, there is always one ideal $Q_i = 0$.

(1) A ring R is subdirectly irreducible if and only if the intersection of the nonzero ideals of R is nonzero.

(2)* (Birkhoff) Every ring is isomorphic to a subdirect product of subdirectly irreducible rings.

(c)* (Ginn and Moss [75]). A right and left Noetherian ring R is right Artinian iff essentially right Artinian.

Prime Radical

An ideal P of R is **prime** provided that R/P is a prime ring. If K and Q are ideals of R, and if $K \supseteq Q$, then

$$0 \to K/Q \to R/Q \to R/K \to$$

is exact so that there is a ring isomorphism $R/K \approx (R/Q)/(K/Q)$. This shows the following proposition.

26.3 Proposition. *If $K \supseteq Q$ are ideals of R, then K is a prime ideal of R if and only if K/Q is a prime ideal of R/Q.* □

The **prime radical** of R is defined to be the intersection of the prime ideals of R. In view of 26.3, we have the following corollary.

26.4 Corollary. Prime rad$(R/$prime rad $R)=0$. □

An element $a \in R$ is **strongly nilpotent** if, for each infinite sequence $\{a_n | n \geq 0\}$ such that $a_0 = a$, and $a_{n+1} \in a_n R a_n$, $n = 0, 1, \ldots$, there exists an integer k such that $a_n = 0 \; \forall n \geq k$. If a is strongly nilpotent, and if $\{a_n | n = 0, 1, \ldots\}$ is the sequence $a_0 = a$, $a_1 = a^2, \ldots, a_n = a^{2^n}$, then $a_{n+1} = a^{2^{n+1}} = a^{2^n} \cdot a^{2^n} = a_n^2 \in a_n R a_n \; \forall n$. Thus, $a_k = a^{2^k} = 0$ for some k, so each strongly nilpotent element is nilpotent. *If R is commutative, then conversely, each nilpotent element is strongly nilpotent.*

26.5 Proposition (Levitzki [51]). *The prime radical is the set of all strongly nilpotent elements of R.*

Proof. Let a be an element of R not in prime rad R. Then a lies outside of some prime ideal P and $a R a \not\subseteq P$, so there is an $a_1 \in a R a$ and $a_1 \notin P$. Assuming $a_n \notin P$, then $a_n R a_n \not\subseteq P$, so there is an $a_{n+1} \in a_n R a_n$ and $a_{n+1} \notin P$. Since $a_n \notin P \; \forall n$, $a_n \neq 0 \; \forall n$, so a is not strongly nilpotent.

Conversely, assume a is not strongly nilpotent, and let $\{a_n | n = 0, 1, \ldots\}$ be a sequence of elements in R such that $a_0 = a$, and $a_{n+1} \in a_n R a_n \; \forall n$. Let

$$T = \{a_n | n = 0, 1, \ldots\}.$$

Then $0 \notin T$, and by Zorn's lemma, there exists an ideal P that is maximal in the set of ideals not containing an element of T.

Next let A, B be right ideals of R such that $A \not\subseteq P$, $B \not\subseteq P$. Since $A + P \neq P$, $B + P \neq P$, both $A + P$ and $B + P$ meet T, say $a_i \in A + P$, $a_j \in B + P$. If $m = \max \{i, j\}$, then

$$a_{m+1} \in a_m R a_m \subseteq (A + P)(B + P) \subseteq AB + P.$$

But $a_{m+1} \notin P$, hence $AB \not\subseteq P$. Thus, P is prime, and $a_0 = a \notin P$. Thus $a \notin$ prime rad R. □

26.6 Exercise. (a) Show that the (Baer [43a]) lower nil radical (as defined by 24.3(e)) is the prime radical.

(b) For a ring R, the following are equivalent:

(1) R is semiprime.

(2) Prime rad $R = 0$.

(3) R is a subdirect product of prime rings.

(4) For any pair A, B of ideals, $AB = 0$ if and only if $A \cap B = 0$.

Nil Radicals

An ideal A of a ring R is **nil** provided that every element of A is nilpotent. A chain of nil ideals of R is again a nil ideal, hence by Zorn's lemma there exists a maximal nil ideal N of R. If A is any nil ideal of R, then $A + N$ is a nil ideal containing N, hence $A + N = N$, and $N \supseteq A$. Thus R has a largest nil ideal N, called the **nil radical** of R. By 18.8 and 26.5, we have the inclusions rad $R \supseteq$ nil rad $R \supseteq$ prime rad R, which may be proper in each case; however, if R is commutative, each nilpotent element is strongly nilpotent, so 26.5 implies the next proposition.

26.7 Proposition. *If R is commutative then* nil rad $R =$ prime rad R. $\square$

We now define the **nil radical of an ideal** A of R to be $\eta^{-1}(\text{nil rad } R/A)$, where $\eta: R \to R/A$; that is, nil rad A is the largest ideal of R which is nil modulo A. Prime rad A is defined similarly.

26.8 Corollary. *If R is commutative, and A is an ideal, then* nil rad A *is the intersection of the prime ideals containing A.* $\square$

The prime ideals containing A are called the **prime ideals of** (belonging to) A. A **minimal prime ideal** of A is just one that is minimal in the set of prime ideals of A ordered by inclusion.

26.9 Proposition (McCoy [49]). *If A is an ideal of a ring R, then every prime ideal belonging to A contains a minimal prime ideal belonging to A, and prime rad A is the intersection of the minimal prime ideals of A.*

Proof. Immediate from the proof of 26.5, since the ideal P constructed in the second paragraph is actually a minimal prime ideal. $\square$

An ideal I of a ring R is **semiprime** in case $\forall$ right ideals K, if $K^n \subseteq I$ for some n, then $K \subseteq I$. Expressed otherwise, I is semiprime if and only if the factor ring R/I is semiprime. A combination of 26.3 and 26.6 establishes the next proposition.

26.10 Proposition. *An ideal I of R is semiprime if and only if I is the intersection of the prime ideals of R containing it. Therefore every semiprime ideal of R contains prime rad R.* $\square$

26.11 Corollary (Levitzki [51]). *Let $N(\alpha)$ be the ideal of R defined inductively for any ordinal α by setting:*

$N(0) =$ *the sum of all nilpotent ideals of R;*
$N(\alpha + 1) =$ *the inverse image in R of the ideal $N(0)$ defined for $R/N(\alpha)$;*
$N(\alpha) = \bigcup_{\beta < \alpha} N(\beta)$ *when α is a limit ordinal.*

Then, there is a least ordinal α such that $N(\alpha) = N(\alpha + 1)$, and $N(\alpha)$ is then called **the Baer lower nil radical.** *Moreover, $N(\alpha) =$ prime rad R.*

Proof. Clearly $M =$ prime rad $R \supseteq N(0)$, and, moreover, assuming there exists an ordinal α_0 so that $M \supseteq N(\beta) \ \forall \beta \leq \alpha_0$, one sees that $M \supseteq N(\alpha_0)$. (This follows since M contains any ideal which is nilpotent modulo $M(\beta)$.) Thus, by transfinite induction, $M \supseteq N(\alpha)$. However, $N(\alpha) \supseteq M$ by 26.10. $\square$

Semiprimitive Rings

An ideal I of R is **primitive** provided that R/I is a primitive ring. Since any primitive ring is a prime ring, every primitive ideal is a prime ideal.

If V is a faithful simple R/I-module, then V is a simple R-module whose annihilator ideal is I. Conversely, if M is any (simple) R-module with annihilator ideal I, then M is a faithful (simple) R/I-module. These two statements prove the next proposition.

26.12 Proposition. *An ideal I of R is primitive if and only if I is the annihilator ideal of a simple right R-module.* $\square$

A ring R is **semiprimitive** provided that the intersection of its primitive ideals is zero. **Left semiprimitive** rings are defined symmetrically.

26.13 Theorem (Jacobson [45]). *The following statements about a ring R are equivalent:*

(a) R *is semiprimitive.*

(b) rad $R=0$.

(c) R *is a subdirect product of primitive rings.*

(d) R *is left semiprimitive.*

Proof. (a) $\Leftrightarrow$ (c) is a consequence of 26.1. Since (b) is left-right symmetric (rad $R=$ left rad R), (b) $\Leftrightarrow$ (c) is a consequence of (a) $\Leftrightarrow$ (b) established by 18.0. $\square$

26.14 Corollary. rad R *is the intersection of all the primitive ideals of R.* $\square$

Let A be an algebra over a field F. Then, for any extension field P, let $A_P = A \otimes_F P$. If f is an automorphism of P leaving the elements of F fixed, then $1 \otimes f$ is an automorphism of A_P fixing the elements of $A \otimes_F F \approx A$. If P/F is Galois, with Galois group $\mathrm{Gal}(P/F) = G$, then the

$$G_P = \{1 \otimes f \mid f \in G\}$$

is a group of automorphism of A_P, and the subring fix G_P of elements of A_P left fixed by every element of G_P is the image B of the canonical embedding $A \to A \otimes_F P$. For notational simplicity, let $A = B$.

26.15 Lemma. *If A is a semiprimitive algebra over a field F, and P/F is a finite Galois extension, then $A_P = A \otimes_F P$ is semiprimitive.*

Proof. For any ring B, let $J(B) = \mathrm{rad}\, B$, and $z \in A \cap J(A_P)$. If z' is the quasinverse of z in A_P, $(1+z)(1+z')=1$ then the fact that $f(z)=z \ \forall\, f \in G$ implies that $f(z')=z'$. This proves that $A \cap J(A_P) = J(A) = 0$. Now the G-trace is defined by $\mathrm{Tr}_G(x) = \sum_{f \in G} f(x)$, for any $x \in P$. Since P/F is Galois, if $p_1, \dots, p_n$ is a free basis of P over F, then the discriminant of P/F is defined to be $\det(\mathrm{Tr}_G(p_i\, p_j))$, and is $\neq 0$. Now suppose $z = \sum_{i=1}^{n} a_i \otimes p_i \in J(A_P)$, where $a_1, \dots, a_n \in A$. For each j,

$$t_j = \sum_{i=1}^{n} a_i \otimes \mathrm{Tr}_G(p_i\, p_j) \in J(A_P)$$

is an element of $J(A_P)$ since $J(A_P)$ is mapped into itself by the elements of G_P. But $t_j \in J(A_P) \cap A$, since $\mathrm{Tr}_G(p_i\, p_j) \in F$, $i, j = 1, \dots, n$. Since $J(A_P) \cap A \subseteq J(A)$, then $t_j = 0 \ \forall\, j$. The fact that the discriminant of P/F is nonzero now implies that $a_i = 0$, $i = 1, \dots, n$, and then $z = 0$. $\square$

26.16 Corollary (Amitsur [57/58]). *If P/F is a separable and algebraic field extension, then $\mathrm{rad}(A_P) = (\mathrm{rad}\, A)_P$ canonically.*

Proof. This follows from the proposition for P/F finite dimension. The general case is an exercise. $\square$

26.17A Proposition (Amitsur [56, 57/58]). *Let P/F be a pure transcendental field extension. Then $N = \text{rad}(A_P) \cap A$ is a nil ideal such that $N_P = \text{rad}(A_P)$ canonically.*

Proof. The proof, like that of 26.15 and 16, follows Jacobson [64, p. 252 ff.].

(1) $J(A_P) \neq 0 \Rightarrow J(A_P) \cap A \neq 0.$

To prove (1), let $B = \{x_i\}_{i \in I}$ be a transcendence basis of the polynomial algebra $Q = F[B]$ such that $P = F(B)$ is the quotient field. Thus, for any finite subset $x_{i_1}, \ldots, x_{i_k}$, the set of all products

$$\{x_{i_1}^{n_1} x_{i_2}^{n_2} \cdots x_{i_k}^{n_k} \mid n_i \in \mathbb{Z}^+, \ i = 1, \ldots, k\}$$

is a linearly independent set over F. Since we identify A (resp. P) with its canonical image under $A \to A_P$ (under $P \to A_P$), then this set, considered as a subset of A_P, is linearly independent over A as well. If z is a nonzero element of $J(A_P)$, then there is a polynomial $q(x) \in Q$ such that $q(x)z$ is a nonzero polynomial in the x_i with coefficients in A. Hence, there exists an nonzero element $z \in J(A_P)$ of least total degree t in $A[B] = A \otimes_F Q$. If $t = 0$, then z is a nonzero element of $A \cap J(A_P)$, as desired. Otherwise, there exists $j \in I$ such that if B' denotes $B - \{x_j\}$, then $z = \sum_{i=0}^{s} z_i x_j^i$ is a polynomial of degree $s > 0$ in x_j with coefficients in the polynomial ring $A[B']$. For notational simplicity let $x = x_j$, and let f be the automorphism of P such that $f(x) = x + 1$, and $f(b) = b \ \forall b \in F[B']$. Then, $g = 1 \otimes f$ is an element of G_P such that $z - g(z)$ is an element of $A[B'] \cap J(A_P)$ of degree $< s$. Hence, by the induction assumption, $z = g(z)$ which is impossible unless F has characteristic $p > 0$. Then, f is an automorphism of finite order p, and $E = F(B')(x^p - x)$, the subfield of P generated by $F(B')$ and $x^p - x$, is the set of elements left fixed by f. Thus, $z = f(z)$ is an element of $A_E \cap J(A_P)$. Now P/E is a Galois extension of degree p, and $A_P = (A_E)_P$, so 26.15 implies $J(A_E) = J(A_P) \cap A_E$, that is, $z \in J(A_E)$. Let $F' = F(B')$. Then, $P = F'(x)$, and $E = F'(x^p - x)$. Let $h: P \to E$ be the isomorphism such that $h(x) = x^p - x$, and $h|F' = 1_{F'}$, and let $k = 1 \otimes h$ be the extended isomorphism $A_P \to A_E$. Then, $w = k^{-1}(z)$ is an element of $J(A_P)$ of total degree $< t$. This contradicts the choice of z.

(2) $J(A_P) = N_P,$

where $N = J(A_P) \cap A$ is a nil ideal.

Clearly (1) implies that $J(A_P) = N_P$, and it remains only to prove that N is a nil ideal. Assume that $N \neq 0$, and assume the notation in the proof of (1). Thus, x is one of the x_i, and $F' = F(B')$ is the subfield generated by remaining x_j's. If $z \neq 0$ is an element of N, then $xz \in J(A_P)$, and has quasinverse $g(x)^{-1}w$, where $g(x) = \sum_{i=0}^{m} g_i x^i \in F'[x]$ and $w \in A_{F'}[x]$. Thus,

(3) $w + g(x) x z + w x z = 0.$

Since $z \neq 0$, then $w \neq 0$. Write $w = \sum_{i=0}^{n} w_i x^i$, with $w_i \in A_{F'}$, $i = 0, \ldots, n$, and $w_n \neq 0$. Since $P = F'(x)$, then $J(A_P) = (A_{F'} \cap J(A_P))_P$. Thus, $w \in J(A_P)$ implies that $w_i \in J(A_P)$,

$i=0, \ldots, n$. The identity (3), and the fact that $z \in A \subseteq A_{F'}$ implies that $g(x) x z = -w - w x z$ has degree $m+1$ as a polynomial in x over $A_{F'}$, where $m = \deg g(x)$. Moreover, $m+1 = \deg(w + w x z) \leq n+1$, hence $m \leq n$. Now $m = n$ would imply by (3) that $g_m z + w_m z = 0$, and $z + g_m^{-1} w_m z = 0$. Since $g_m^{-1} w_m \in J(A_P)$ is quasi-regular, this would imply $z = 0$, contrary to the hypothesis. Hence assume that $m < n$. Then comparision of the coefficients of $x^{n+1}, x^n, \ldots, x^{m+1}$ in (3) yields identities

(4) $\qquad w_n z = 0, \; w_n + w_{n-1} z = 0, \; \ldots, \; w_{n+2} + w_{n+1} z = 0$

and

$$g_m z + w_{m+1} + w_m z = 0.$$

Therefore,

$$z^{n-m+1} - g_m^{-1} w_m z^{n-m+1} = 0,$$

which by quasiregularity of $g_m^{-1} w_m$ implies $z^{n-m+1} = 0$. Thus, z is nilpotent, and therefore, N is a nil ideal. $\quad \Box$

26.17 B   Corollary. *If A has no nil ideals $\neq 0$, then A_P is semiprimitive for every pure transcendental extension field P of F.* $\quad \Box$

If field extension P/F is **separably generated** in case there is a pure transcenion E/F such that P/E is separably algebraic 13.13. The next result follows from 26.16 and 26.17 B.

26.18 A   Corollary. *If P/F is a separably generated (transcendental) field extension, then $J(A) = 0$ implies $J(A_P) = 0$.* $\quad \Box$

26.18 B   Exercise. (a) Prove for any ring A that $\mathrm{rad}\, A[x] = N[x]$, where $A[x]$ is the polynomial ring and N is a nil ideal.

(b) Conclude that $A[x]$ is semiprimitive when A has no nil ideals $\neq 0$, and hence that $A[x]$ may be semiprimitive even if A is not.

Group Algebras over Formally Real Fields

A field F is **formally real** provided that

$$x_1^2 + x_2^2 + \cdots + x_n^2 = 0 \;\Rightarrow\; x_1 = x_2 = \cdots = x_n = 0$$

for any elements $x_1, \ldots, x_n \in F$.

26.19   Lemma. *If G is a group, and if F is a formally real field, then the group algebra FG has no nil ideals $\neq 0$.*

Proof. The involution of G sending $g \mapsto g^{-1}$ extends to an involution of FG: namely if $a = \sum_{g \in G} a_g g \in FG$, where $a_g \in F \; \forall g \in G$, let $a^* = \sum_{g \in G} a_g g^{-1}$. The mapping $t: FG \to F$ such that $t(a) = a_1$, where 1 is the group identity, is a linear transformation over F, and $t(a b) = t(b a) \; \forall a, b \in F(G)$ (cf. 13.22). Moreover, $t(a a^*) = \sum_{g \in G} a_g^2$. Thus,

(5) $\qquad a a^* = 0 \;\Rightarrow\; a = 0$

since F is formally real. Now if b is an element in a nil ideal of FG, and $b \neq 0$, then $a = bb^* \neq 0$ is a nilpotent element of FG such that $a^* = a$. Suppose that $a^t \neq 0$, and $a^{t+1} = 0$. Then $c = a^t$ satisfies $c^2 = 0$. But, $c^* = c$, and so $cc^* = 0$, whence $c = 0$ by (5). This contradicts $c = a^t \neq 0$. □

A field F is **absolutely algebraic** if F is algebraic over the prime subfield.

26.20 Proposition (Amitsur [59]). *If F is a field of characteristic 0, and if F is not absolutely algebraic, then every group algebra FG is semiprimitive, for any group G.*

Proof. The prime subfield is $\mathbb{Q}$ and is formally real. Let P be a subfield of F such that $P/\mathbb{Q}$ is pure transcendental, and F/P is separably algebraic. Then, by 26.17A if $A = \mathbb{Q}G$, then $J(A_P) = N_P$, where N is a nil ideal of $\mathbb{Q}G$. Thus, $J(A_P) = 0$ by the last lemma, and $J(A_F) = 0$ by 26.8. □

26.21 Corollary. *Every group algebra FG over an uncountable field F of characteristic 0 is semiprimitive.*

Proof. An absolutely algebraic field is countable. □

Exercises of Chapter 26

A ring R is **radical** over a subring B provided that to every $a \in R$ there is an integer $n = n(a)$ such that $a^n \in B$.

1. (Wedderburn [05]) Any finite field is commutative, and a radical extension of every subring.

2. (Jacobson [45]) Any algebraic division algebra over a finite field is commutative. (Lemma. Every algebraic division algebra contains an element separable over its center.)

3. (Jacobson) If every element of a ring R satisfies $a^{n(a)} = a$, then R is commutative.

4. (Herstein [53]) If every element a of a ring R satisfies $a^{n(a)} - a \in \text{center } R$, then R has (commutator) nil ideal $\neq 0$, or else R is commutative.

5. (Kaplansky [51]) If P is a commutative field, and then P is radical over a commutative subfield $F \neq P$ if and only if P has characteristic $p \neq 0$, and either P/F is purely inseparable, or else P is algebraic over $GF(p)$.

6. (Kaplansky [51]) A field R is radical over its center only if R is commutative.

7. (Faith [60]) A field R is radical over a subring $\neq R$ only if R is commutative.

8. (Faith [61]) If R/A is a radical extension, then (a) If R is (semi)primitive, so is A. (b) If R is a field, so is A. (c) If $\text{rad } R = 0$, and if A is commutative, R is also commutative. (d) (Armendariz [67]) If A/B is radical, then $\text{rad } B = B \cap \text{rad } A$.

9. A subdirectly irreducible primitive ring is left primitive. (Not every primitive ring is left primitive. See Bergman [64] and Jategaonkar [68].)

10. (Snapper [50]) If A is a commutative ring, and x an indeterminate over A, then rad $A[x] = N[x]$, where N is the maximal nil ideal of A.

11. (Goldman [51]-Krull [51]) If A is a finitely generated commutative ring, then rad A is nil.

12*. (Amitsur [56]) Over an uncountable field F. (a) The radical of a finitely generated algebra is nil. (b) The maximal nil ideal contains every nil right or left ideal. (c) If A is algebraic over F, then A_P is algebraic over P, for any extension field, and the total matrix algebra A_n is algebraic over F. (d) (Amitsur [56b]) For any algebra R over F we have rad $R[x] = N[x]$, where $R[x]$ is the polynomial ring, and N is the maximal nil ideal. (Cf. 26.18B.) For the polynomial ring $R[x_\alpha]$ in infinitely many indeterminates (cardinality α), this result holds without assuming F is uncountable.

13. (Patterson [61]) For any ring R, if R_ω is the ring of row-finite $\omega \times \omega$ matrices over R, then rad $R_\omega = (\text{rad } R)_\omega$ canonically if and only if rad R is right vanishing.

14. A prime ideal P is right **inessential** if P is not an essential right ideal. Any right inessential prime ideal is a right annulet. Any maximal annihilating ideal is prime.

15*. (Lenagen [73]) If R has "*Krull dimension*", then nil rad R is nilpotent. (This happens e.g. if R is right Noetherian by Levitzki's theorem (I, 9.15, p. 398). Also see Chapter 17.)

16*. (Formanek [72b, 73a]) The free algebra of rank > 1 over a field k is primitive. (Similarly for group algebras kG for $G = a$ free product $\neq \mathbb{Z}_2 * \mathbb{Z}_2$.) The group algebra kS_ω of the infinite symmetric group is semiprimitive.

Notes for Chapter 26

As I indicated in the Preface, Chapter 18 and the present chapter are introductory to the main themes of Jacobson's Colloquium volume [55, 64]. (In it, and elsewhere, Jacobson employs the term "semisimple" for what we call "semiprimitive" (after Nagata's suggestion) in parallel with the term "semiprime".)

The semiprimitivity of a group ring kG over a transcendental field k of characteristic zero (of an arbitrary group G), has been shown by Amitsur [57c, 59]. Similarly, Amitsur (l.c.) showed that $D[x]$ is not primitive for an algebraic central division algebra over a non-denumberable field k. (Also see Amisur [56b].) In case $D[x] = D \otimes_k k[x]$ is primitive, then one can show that $D(x) = D \otimes_k k(x)$ is a simple right and left principal ideal ring not a field. (This is always the case when D is a transcendental algebra over its center k, that is, in order for $D[x]$ not to be primitive, then it is necessary for D to be algebraic over k. See Exercises 13.11.6–13.11.13 of Volume I, p. 468.) Any commutative group algebra of characteristic 0 is semiprimitive (Amitsur [59].)

Bergman [64] gave the first right-but-not-left primitive ring, and Jategaonkar [68] gave such an example which was a principal ideal domain.

Jacobson conjectured that the radical of an algebra A which is a factor algebra of a free algebra on finitely many letters (=a finitely generated algebra) must be nil. Amitsur [56a] verified this for algebras over a nondenumerable field, and Goldman [51] and Krull [51] verified this for commutative A. Amitsur generalized the latter to PI (polynomial identity) algebras (see his 1967 Proc. Amer. Math. Soc. paper). Also see Amitsur-Procesi [66] and Procesi [67b].

As discussed in Notes for Chapter 18, Köthe's radical, when it exists, is a nil ideal K containing every onesided nil ideal of the ring R (Köthe [30]). It is not known if the Köthe radical exists for an arbitrary ring, but again, Amitsur [56a] showed it does for an algebra over a nondenumerable field. (Also consult Amitsur [73].)

Regarding Exercise 13 and Patterson's theorem, the radical of R_ω has been explicitly determined (see Slover [69]).

The Krull intersection theorem for Noetherian commutative R states that $\bigcap_{n<\omega}(\operatorname{rad} R)^n = 0$, and Jacobson's conjecture that this also holds for Noetherian rings has been verified by Jategaonkar [74a] and Cauchon [76] for fully bounded Noetherian rings.

Invariant Wedderburn Factors

Assume that an algebra A of finite dimension over a field k is separable modulo radical (I, p. 471). If S_1 and S_2 are two semisimple factors of A (13.18), then there exists an element $x \in \operatorname{rad} R$ such that $S_1 = (1-x)^{-1} S_2 (1-x)$ (Mal'cev [42]). (This remains true, assuming only that $A/\operatorname{rad} A$ is finite dimensional (Eckstein [69]; also see Curtis [54], as cited by Eckstein).)

A semisimple factor S (also called a **Wedderburn factor**) is said to be G-**invariant** relative to a finite group G of automorphisms and anti-automorphisms of A provided that $g(S) \subseteq S \ \forall g \in G$. If A is finite dimensional, and G is finite, then A has G-invariant Wedderburn factors provided that the characteristic of k does not divide $|G|$ (Taft [57]), or when G is a completely reducible group acting on A and k has characteristic 0 (Mostow [56]). (Cf. Taft [68].) Taft [64] established a strong uniqueness of G-invariant Wedderburn factors for a completely reducible group G.

Notes on Prime Ideals

Going beyond the classical correspondence between prime ideals in polynomial rings over algebraically closed fields and algebraic varieties, the importance of prime ideals in general commutative rings no doubt stems from the information about the ring obtained "locally" via the local ring R_p at a prime ideal P. (See Notes for Chapter 18 on the chronological development of local rings.) However, this cannot guarantee the importance of prime ideals in noncommutative rings inasmuch as the local ring may not exist. For example, a prime ring R may not have a right or left (classical) quotient ring $Q(R)$. Although the point of the Goldie-Lesieur-Croisot theorem (I, 9.9, p. 394) is that $Q(R)$ exists if R is, say, right Noetherian, nevertheless, there may not exist a "local" ring R_p at a prime ideal P^2. This difficulty is partially surmounted if we restrict our attention to polynomial identity (PI) rings, since, by a theorem of Posner [60] and Small [71], one may construct R_p (in the canonical way) provided that R and R/P satisfy exactly the same polynomial identities.

Jategaonkar [74c] extended the rank theorem for prime ideals for Noetherian commutative rings to Noetherian PI algebras. Kaplansky has given a historical and lucid account of Krull's Principal Ideal Theorem (in Kaplansky [68]); and Jategaonkar [75] has generalized this to PI-rings.

Krull [28] defined such concepts the highest prime ideal dividing an ideal A, the prime ideals belonging to A, and isolated components, for noncommutative rings satisfying "finiteness conditions weaker than the Noetherian chain conditions".

Fitting [35] defined the prime and primary ideals of an ideal in a noncommutative ring "without finiteness conditions". However, assuming the ascending chain conditions each coirreducible ideal, that is, an ideal not the intersection of two larger ideals is primary, and, of course, every ideal is the intersection of primary ideals. Further properties of the prime ideals are developed, and the radical of an ideal A is defined as the set $\sqrt{A}$ of properly nilpotent elements modulo A; that is, elements c which generate a nilpotent ideal modulo A. Fitting applies his results to characterize when an order R is a simple algebra (of finite dimensions) is a product of primary ideals. This happens iff the proper prime ideals are comaximal and commutative. (Compare the Chinese remainder theorem 18.30 and also 18.32.)

Notes on the Prime Radical

Mathematicians nevertheless failed to get at the characterization in general rings of the (McCoy) prime radical, namely the intersection of the prime ideals, until McCoy [49] characterized it as the set of all elements r such that any "m-system" which contains r contains 0. Here, an m-**system** M (generalizing the concept of a multiplicatively closed system) is a set of elements of R such that a and $b \in M \Rightarrow \exists x \in R$ & $axb \in M$. McCoy also proved that the prime radical is the intersection of the minimal primes.

Notes on the Baer Lower Nil Radical

In [43] Baer defined the lower nil radical $L(R)$ as the intersection of all radical ideals, that is, of all nil ideals I such that R/I contains no nilpotent ideals. (Baer also constructed $L(R)$ by transfinite induction from below.) McCoy [49] shows that the prime radical $P(R)$ is contained in $L(R)$, and Levitzki [51] proved that $P(R)=L(R)$, namely *the prime radical coincides with the Baer Lower nil radical.*

---

[2] See, however, the local "envelope" at P as defined in Goldie [67], and developed in Lambek-Michler [73, 74] and Jategaonkar [74b]. (Jategaonkar characterizes when this exists by a very simple condition on the injective hull of R/P. See his Theorem 4.5].)

Levitzki also gave another characterization of $P(R)$ (the one in 26.5) as follows. An *m*-**sequence** $\{a_n\}_{n=1}^{\infty}$ is a sequence such that for every integer m there exists b_m so that $a_{m+1} = a_m b_m a_m$. An element a is **associated with an *m*-sequence** $\{a_n\}_{n=1}^{\infty}$ if $a_0 = a$. Then a is **strongly** nilpotent iff there is associated *m*-sequence containing 0 (in the sequence). Levitzki's characterization: $P(R)$ is the set of all strongly nilpotent elements. (The terminology is that of Lambek [66].)

For other results on the prime radicals, see McCoy [64] and Jacobson [64, pp. 193–202].

For some related results on radicals, consult Notes for Chapter 17, 18, and 22. For general theories of radicals, consult Brown and McCoy [47, 48], Amitsur [54], and Kurosch [53] (cf. Jacobson [64]). Also, see Amitsur [73a].

For the primitive (maximal) ideals I of the universal enveloping algebra U of a finite dimensional complex nilpotent Lie algebra g, consult Dixmier [68]. (The quotients U/I are the Weyl algebras A_n, for the same n for "almost all I".) See J. C. McConnell and Borho-Gabriel-Rentschler [73] for more general Lie algebras, and generalizations.

Other Notes on Group Rings

Formanek and Snider [72] constructed the first primitive group rings. (See Exercises for Chapter 19, #11–19.) Lawrence [74a, 75a] determined the coefficient rings of primitive group rings. Moreover, the free algebra of rank >1 over a field is primitive (Formanek [73a]).

The book and report of Passman [71,74] on group rings (soon to be superceded by Passman's new book on group rings) are source materials for results on semiprimitivity and primitivity of RG, and numerous other group ring questions. For example, Kaplansky's theorem on the Dedekind finiteness of KG in fields of characteristic 0 (called von **Neumann finite** *l.c.*), is obtained in the section of Passman [74] on algebraic elements, including theorems on the trace of idempotents. (For characteristic 0, the trace $\operatorname{Tr} e$ of an idempotent $e \ne 1$ is a rational number between 0 and 1 (strict), by a theorem of Kaplansky and Zalesskii.)

Another example: KG satisfies a polymomial identity iff G has an abelian subgroup of finite index. (Theorem of S. A. Amitsur, I. M. Isaacs, D. S. Passman, and M. Smith.) This is related to group algebras KG all irreducible representations of which are of bounded degree. (See Kaplansky [49b] and Amitsur [61]; also Snider [74] for a generalization.)

Note on Polycyclic by Finite Group Rings

One of the most famous problems on group rings is the **zero divisor question**: if G is torsionfree, is KG an integral domain? Yes, if G is supersolvable (Formanek), a result generalized by D. Farkas and R. Snider to G polycyclic by finite.

A ring R is a **Hilbert ring** if R is a Noetherian ring and if rad R/A is nilpotent for every ideal A. A **capital** of a ring R is a factor ring R/A where A is a maximal ideal. (If R is commutative then R/A is then a field.) A field K is **absolute** is K has prime characteristic and K is absolutely algebraic. Henceforth, assume that G is polycyclic by finite.

Theorem (Rosenblade [73]). *If K is any absolute field, then any simple KG module is finite dimensional over K.*

This generalized a theorem of P. Hall for finitely generated nilpotent G. Moreover, by the work of Hall cited by Rosenblade (*l.c.*), this implies that the simple modules of the integral group ring $\mathbb{Z}G$ are all finite. (Cf. Jategaonkar [74a] who proves residual nilpotence of the augmentation ideal.) Furthermore:

Corollary (Rosenblade). *If R is any commutative Hilbert ring all of whose capitals are absolute, then any simple RG module is finite dimensional over a capital of R.*

Rosenblade also proved, for any ring R, that RG is a Hilbert ring iff R is, generalizing various forms of the Hilbert Nullstellensatz. For polynomial rings over fields, this is a theorem of Goldman [51] and Krull [51] (See Rosenblade, *l.c.*, p. 309).

Mitchell [72] placed the Jacobson radical and ring theory in the setting of additive categories with finitely many objects.

References

Armendariz [67], Amitsur [56a, b, 57/58, 59, 73], Baer [43a], Bergman [64], Brenner [7], Brown and McCoy [47, 48], Curtis [54], Eckstein [69], Faith [60, 62], Fitting [35], Goldman [51], Hampton and Passman [72], Herstein [53], Jacobson [43, 45a, b, c, 55, 64], Jategaonkar [68, 71], Köthe [30a, b], Krull [28, 48, 51], Lambek [66], Lenagen [73], Levitzki [51], Mal'cev [42], McCoy [49, 55, 56, 57a, 64], Mostow [56], Patterson [61], Shock [72], Snapper [50], Taft [57, 64, 68], Wedderburn [05].

Additional References

Amitsur [54, 61, 67, 73a], Amitsur and Procesi [66], Beachy [71b], Borho, Gabriel, and Rentschler [73], Cauchon [76], Dixmier [68], Formanek [72b, 73a], Formanek-Snider [72], Fluch [65], Jategaonkar [74a, 75], Kaplansky [48, 49, 68], Kurosch [53], Lawrence [74a, 75a], Mitchell [72], Passman [71, 74], Rosenblade [73], Small [71, 73], Smith [71], Snider [74], Vámos [68, 75].

Bibliography

[72]   Abhyankar, S.S., Heinzer, W., Eakin, P.: On the uniqueness of the coefficient ring in a poly-
        nomial ring. J. Algebra 23, 310–342 (1972).
[73]   Ahsan, J.: Rings all of whose modules are quasi-injective. Proc. Lond. Math. Soc. (3) 27,
        425–439 (1973).
[39]   Albert, A.A.: Structures of Algebras. Coll. Pub. Amer. Math. Soc. 24, Providence, R.I., 1939.
[61]   Albrecht, F.: On projective modules over semi-hereditary rings. Proc. Amer. Math. Soc. 12,
        638–639 (1861).
[68]   Amdal, I.K., Ringdal, F.: Catégories unisériales. C. R. Acad. Sci. Paris, Sér. A 267, 85–87,
        247–249 (1968).
[54a] Amitsur, S.A.: A general theory of radicals, II. Radicals in rings and bicategories. Amer. J.
        Math. 76, 100–125 (1954).
[54b] Amitsur, S.A.: Differential polynomials and division algebras. Ann. of Math. 59, 245–278
        (1954), (7–56).
[55]   Amitsur, S.A.: Finite subgroups of division rings. Trans. Amer. Math. Soc. 80, 361–386 (1955).
[56a] Amitsur, S.A.: Algebras over infinite fields. Proc. Amer. Math. Soc. 7, 35–48 (1956).
[56b] Amitsur, S.A.: Radical of polynomial rings. Canad. J. Math. 8, 355–361 (1956).
[57a] Amitsur, S.A.: Derivations in simple rings. Proc. Lond. Math. Soc. (3) 7, 87–112 (1957).
[57b] Amitsur, S.A.: A generalization of Hilbert's Nullstellensatz. Proc. Amer. Math. Soc. 8,
        649–656 (1957).
[57c] Amitsur, S.A.: The radical of field extensions. Bull. Res. Council of Israel (Israel J. Math.)
        7F, 1–10 (1957).
[59]   Amitsur, S.A.: On the semi-simplicity of group algebras. Mich. Math. J. 6, 251–253 (1956).
[60]   Amitsur, S.A.: Finite dimensional central division algebras. Proc. Amer. Math. Soc. 11, 28–31
        (1960).
[61]   Amitsur, S.A.: Groups with representations of bounded degree II. Illinois J. Math. 5, 198–205
        (1961).
[63]   Amitsur, S.A.: Remarks on principal ideal rings. Osaka Math. J. 15, 59–69 (1963).
[67]   Amitsur, S.A.: A generalization of Hilbert's Nullstellensatz. Proc. Amer. Math. Soc. 8,
        649–656 (1967).
[68]   Amitsur, S.A.: Rings with involution. Israel J. Math. 6, 99–106 (1968).
[70]   Amitsur, S.A.: A noncommutative Hilbert basis theorem and subrings of matrices. Trans.
        Amer. Math. Soc. 149, 133–142 (1970).
[71a] Amitsur, S.A.: Rings of quotients and Morita context. J. Algebra 17, 273–298 (1971).
[71b] Amitsur, S.A.: Embeddings in matrix rings. Pac. J. Math. 36, 21–29 (1971).
[72a] Amitsur, S.A.: On central division algebras. Israel J. Math. 12, 408–420 (1972).
[72b] Amitsur, S.A.: On rings of quotients. Istitutio Nazionale di Alta Matematica Symposia VIII,
        149–164 (1972).
[73a] Amitsur, S.A.: Nil radicals. Historical notes and some new results, published in the book
        edited by Kertész [73].
[73b] Amitsur, S.A.: Polynomial identities and Azumaya algebras. J. Algebra 27, 117–125 (1973).
[75]   Amitsur, S.A.: Central embeddings in semi-simple rings. Pac. J. Math. 56, 1–6 (1975).
[66]   Amitsur, S.A., Procesi, C.: Jacobson rings and Hilbert algebras. Annali di Mat. 71, 61–72
        (1966).
[72]   Anderson, F.W., Fuller, K.R.: Modules with decompositions that complement direct sum-
        mands. J. Algebra 22, 241–253 (1972).

[74] Anderson, F. W., Fuller, K. R.: Rings and Categories of Modules. Graduate Text in Mathe-
      mathics. Berlin-Heidelberg-New York: Springer 1974.
[48] Arens, R. F., Kaplansky, I.: Topological representations of algebras. Trans. Amer. Math. Soc.
      63, 457–481 (1948).
[67] Armendariz, E. P.: On radical extensions of rings. J. Austral. Math. Soc. 7, 552–554 (1967).
[73] Armendariz, E. P.: A note on semiprime rings with torsionless injective envelopes. Canad.
      Math. Bull. 16, 429–431 (1973).
[27a] Artin, E.: Zur Theorie der hyperkomplexen Zahlen. Abh. Math. Sem. U. Hamburg 5, 251–260
      (1927).
[27b] Artin, E.: Zur Arithmetik hyperkomplexer Zahlen. Abh. Math. Sem. U. Hamburg 5, 261–289
      (1927).
[50] Artin, E.: The influence of J. H. M. Wedderburn on the development of modern algebra. Bull.
      Amer. Math. Soc. 56, 65–72 (1950).
[55] Artin, E.: Galois Theory. Notre Dame Univ., South Bend 1955.
[57] Artin, E.: Geometric Algebra. Interscience, N. Y. 1957.
[59] Artin, E.: Theory of Algebraic Numbers. Mathematical Institute, Göttingen 1959.
[43] Artin, E., Whaples, G.: The theory of simple rings. Amer. J. Math. 65, 87–107 (1943).
[44] Artin, E., Nesbitt, E., Thrall, R.: Rings with Minimum Condition. U. of Mich. Press, Ann
      Arbor 1944.
[69] Artin, M.: On Azumaya algebras and finite dimensional representations of rings. J. Algebra
      11, 532–563 (1969).
[38] Asano, K.: Nichtkommutative Hauptidealringe I. Act. Sci. Ind. 696, Paris 1938.
[39] Asano, K.: Über verallgemeinerte Abelsche Gruppen mit hyperkomplexen Operatorenring
      und ihre Anwendungen. Japan J. Math. Soc. Japan 15, 231–253 (1939).
[49a] Asano, K.: Über Hauptidealringe mit Kettensatz. Osaka Math. J. 1, 52–61 (1949).
[49b] Asano, K.: Über die Quotientenbildung von Schiefringen. J. Math. 2, 73–79 (1949).
[50] Asano, K.: Zur Arithmetik in Schiefringen II. J. Polytech. Osaka City U. Ser. A. Math. 1,
      1–27 (1950)
[61] Asano, S.: On the radical of quasi-Frobenius algebras; Remarks concerning two quasi-
      Frobenius rings with isomorphic radicals; Note on some generalizations of quasi-Frobenius
      rings. Kodai Math. Sem. Reps. 13, 135–151, 224–226, 227–334 (1961).
[55] Auslander, M.: On the dimension of modules and algebras III. Nagoya Math. J. 9, 67–77
      (1955).
[57] Auslander, M.: On regular group rings. Proc. Amer. Math. Soc. 8, 658–664 (1957).
[66a] Auslander, M.: Remarks on a theorem of Bourbaki. Nagoya Math. J. 27, 361–369 (1966).
[66b] Auslander, M.: Coherent functors. Proceedings of the Conference on Categorical Algebra
      (La Jolla, 1965). Springer-Verlag, No. 9 189–231 (1966).
[74] Auslander, M.: Representation theory of Artin algebras, I, II. Comm. Algebra 1, 177–268,
      177–310 (1974).
[69] Auslander, M., Bridger, M.: Stable Module Theory. Memoirs of the Amer. Math. Soc. No. 94,
      Providence 1969.
[57] Auslander, M., Buchsbaum, D. A.: Homological dimension in local rings. Trans. Amer. Math.
      Soc. 85, 390–405 (1957).
[59] Auslander, M., Buchsbaum, D. A.: Unique factorization in regular local rings. Proc. Nat.
      Acad. Sci. 45, 733–734 (1959).
[68] Auslander, M., Brumer, A.: Brauer groups of discrete valuation rings. Nederl. Akad. Wetensch.
      Proc. Ser. A. Indag. Math. 30, 286–296 (1968).
[60a] Auslander, M., Goldman, O.: Maximal orders. Trans. Amer. Math. Soc. 97, 1–24 (1960).
[60b] Auslander, M., Goldman, O.: The Brauer group of a commutative ring. Trans. Amer. Math.
      Soc. 97, 367–409 (1960).
[74] Auslander, M., Reiten, I.: Stable equivalence of dualizing R-varieties, I–IV. Advances in
      Math. 12, 306–366 (1974).
[75] Auslander, M., Reiten, I.: Representation theory of Artin algebras, III. Almost split sequences.
      Comm. Algebra 3, 239–294 (1975).
[48] Azumaya, G.: On generalized semiprimary rings and Krull-Remak-Schmidt's theorem. Japan
      J. Math. 19, 525–647 (1948).
[50] Azumaya, G.: Corrections and supplementaries to my paper concerning Krull-Remak-
      Schmidt's theorem. Nagoya Math. J. 1, 117–124 (1950).

[51]   Azumaya, G.: On maximally central algebras. Nagoya Math. J. **2**, 119–150 (1951).

[59]   Azumaya, G.: A duality theory for injective modules (Theory of quasi-Frobenius modules). Amer. J. Math. **81**, 249–278 (1959).

[66]   Azumaya, G.: Completely faithful modules and self-injective rings, Nagoya Math. J. **27**, 697–708 (1966).

[74]   Azumaya, G.: Characterizations of semi-perfect and perfect modules. Math. Z. **140**, 95–103 (1974).

[40]   Baer, R.: Abelian groups that are direct summands of every containing abelian group. Bull. Amer. Math. Soc. **46**, 800–806 (1940).

[42]   Baer, R.: Inverses and zero-divisors. Bull. Amer. Math. Soc. **48**, 630–638 (1942).

[43a] Baer, R.: Radical ideals. Amer. J. Math. **65**, 537–568 (1943).

[43b] Baer, R.: Rings with duals. Amer. J. Math. **65**, 569–584 (1943).

[48]   Baer, R.: Direct decompositions into infinitely many summands. Trans. Amer. Math. Soc. **64**, 551–519 (1948).

[52]   Baer, R.: Kriterien für die Existenz eines Einselementes in Ringen. Math. Z. **56**, 1–17 (1952).

[68]   Baer, R.: Dualisierbare Moduln und Praemoduln. In: Studies on abelian groups, pp. 37–68. Springer 1968.

[66]   Balcerzyk, S.: On projective dimension of direct limits of modules. Bull. acad. Polon. Sci., Sér. Sci. Math. Astron. Phys. **14**, 241–244 (1966).

[70]   Bang, C. M.: Countably generated modules over complete discrete valuation rings. J. Algebra **14**, 552–560 (1970).

[60]   Bass, H.: Finitistic dimension and a homological generalization of semiprimary rings. Trans. Amer. Math. Soc. **95**, 466–488 (1960).

[61]   Bass, H.: Projective modules over algebras. Ann. of Math. **73**, 532–542 (1961).

[62a] Bass, H.: The Morita Theorems. Math. Dept., U. of Oreg., Eugene 1962.

[62b] Bass, H.: Injective dimension in Noetherian rings. Trans. Amer. Math. Soc. **102**, 18–29 (1962).

[62c] Bass, H.: Torsion free and projective modules. Trans. Math. Soc. **102**, 319–327 (1962).

[63a] Bass, H.: Big projective modules are free. Ill. J. Math. **7**, 24–31 (1963).

[63b] Bass, H.: On the ubiquity of Gorenstein rings. Math. Z. **82**, 8–28 (1963).

[64a] Bass, H.: Projective modules over free groups are free. J. Algebra **1**, 367–373 (1964).

[64b] Bass, H.: K-theory and stable algebra. Publications I. H. E. S. No. 22, 5–60 (1964).

[67]   Bass, H.: Lectures on topics in algebraic K-theory. Tata Institute for Advanced Study, Colaba 1967.

[68]   Bass, H.: Algebraic K-theory. Benjamin, N. Y., and Amsterdam 1968.

[73]   Bass, H.: Introduction to some methods of algebraic K-theory. Conference Board of the Math. Sciences Regional Conference Series in Math., vol. 20, Amer. Math. Soc., Providence 1973.

[71a] Beachy, J. A.: Bicommutators of cofaithful, fully divisible modules. Canad. J. Math. **23**, 202–213 (1971); Corrigendum **26**, 256 (1974).

[71b] Beachy, J. A.: On quasi-Artinian rings. J. Lond. Math. Soc. (2) **3**, 449–452 (1971).

[71c] Beachy, J. A.: Generating and Cogenerating structures. Trans. Amer. Math. Soc. **158**, 75–92 (1971).

[75]   Beachy, J. A., Blair, W. D.: Rings when faithful left ideals are cofaithful. Pac. J. Math. (1975).

[61]   Beaumont, R., Pierce, R.: Torsion-free rings. Ill. J. Math. **5**, 61–98 (1961).

[71]   Beck, I.: Injective modules over a Krull domain. J. Algebra **17**, 116–131 (1971).

[72a] Beck, I.: $\sum$-injective modules. J. Algebra **21**, 232–249 (1972).

[72b] Beck, I.: Projective and free modules. Math. Z. **129**, 231–234 (1972).

[60]   Behrens, E.-A.: Distributiv darstellbare Ringe. Math. Z. **73**, 409–432 (1960).

[61a] Behrens, E.-A.: Distributiv darstellbare Ringe, II. Math. Z. **76**, 367–384 (1961).

[61b] Behrens, E.-A.: Einreihige Ringe, Math. Z. **77**, 207–218 (1961).

[37]   Bell, E. T.: Men of Mathematics. Simon and Shuster, New York 1937.

[64]   Bergman, G. M.: A ring primitive on the right but not the left. Proc. Amer. Math. Soc. **15**, 473–475 (1964).

[69a] Bergman, G. M.: Centralizers in free Associative algebras. Trans. Amer. Math. Soc. **137**, 327–344 (1969).

[69b] Bergman, G. M.: Ranks of tensors and change of base field. J. Algebra **11**, 613–621 (1969).

[71a] Bergman, G. M.: Groups acting on hereditary rings. Proc. Lond. Math. Soc. **3** (23), 70–80 (1971).

[71b] Bergman, G. M.: Commutative rings and centres of hereditary rings. Proc. Lond. Math. Soc. **23**, 214-236 (1971).

[72] Bergman, G. M.: Boolean rings of projection maps. J. Lond. Math. Soc. (2) **4**, 593-598 (1972).

[74] Bergman, G. M.: Some examples in P. I. ring theory. Israel J. Math. **18**, 257-277 (1974).

[69] Bergman, G. M., Cohn, P. M.: Symmetric elements in free powers of rings. J. Lond. Math. Soc. (2) **1**, 525-534 (1969).

[71] Bergman, G. M., Cohn, P. M.: The centres of 2-firs and hereditary rings. Proc. Lond. Math. Soc. **23**, 83-98 (1971).

[73] Bergman, G. M., Clark, W. E.: The automorphism class group of the category of rings. J. Algebra **24**, 80-99 (1973).

[73] Bergman, G. M., Isaacs, I. M.: Rings with fixed-point-free. Proc. Lond. Math. Soc. (3) **27**, 69-87 (1973).

[67] Birkhoff, G.: Lattice Theory (revised). Coll. Pub. Vol. **25**, Amer. Math. Soc., Providence 1967.

[69] Björk, J. E.: Rings satisfying a minimum condition on principal ideals. J. reine u. angew. Math. **236**, 466-488 (1969).

[70a] Björk, J. E.: On subrings of matrix rings over fields. Proc. Cam. Phil. Soc. **68**, 275-284 (1970).

[70b] Björk, J. E.: Rings satisfying certain chain conditions. J. reine u. angew. Math. **237** (1970).

[71] Björk, J. E.: Conditions which imply that subrings of semiprimary rings are semiprimary. J. Algebra **19**, 384-395 (1971).

[72a] Björk, J. E.: Radical properties of perfect modules. J. reine u. angew. Math. **253**, 78-86 (1972).

[72b] Björk, J. E.: The global dimension of some algebras of operators. Inv. Math. **17**, 67-78 (1972).

[74] Björk, J. E.: Noetherian and Artinian chain conditions of associative rings. Arch. Math. **24**, 366-378 (1974).

[73] Blair, W. D.: Right Noetherian rings integral over center. J. Algebra **24** (1973).

Blair, W. D. (see Beachy).

[73] Boisen, M. B., Larsen, M. D.: On Prüfer rings as images of Prüfer rings. Proc. Amer. Math. **40**, 87-90 (1973).

[67] Bokut', L.: The embedding of rings in fields. Soviet Math. Dokl. **8**, 901-904 (1967).

[68] Bongale, P. R.: Filtered Frobenius algebras II. J. Algebra **9**, 79-93 (1968).

[69] Borel, A.: Linear Algebraic Groups. Benjamin, New York 1969.

[73] Borho, W., Gabriel, P., Rentschler, R.: Primideale in einhüllenden auslösbaren Lie-Algebren. Lecture Notes in Mathematics, vol. 357. New York-Heidelberg-Berlin: Springer 1973.

[58-64] Bourbaki, N.: Eléments de Mathématique, Algébre, Chapitres 1-8, (A. S. I. Nᵒˢ: 1144 (§1), 1032-1236 (§2), 1044 (§3), 1102 (§§4, 5), 1179 (§§ 6, 7), 1236 (§8)). Hermann, Paris 1958, 1959, 1964.

[61a] Bourbaki, N.: Eléments de Mathématique (A. S. I. N° 1290), Algèbre Commutative, Chapitre 1 (Modules Plat) et 2 (Localisation). Hermann, Paris 1961.

[61b] Bourbaki, N.: Eléments de Mathématique (A. S. I. N° 1293), Algèbre Commutative, Chapitre 3 (Graduations, Filtrations, et Topologies) et 4 (Idéaux Premiers Associés et Décomposition Primaire). Hermann, Paris 1961.

[64] Bourbaki, N.: Eléments de Mathématique (A. S. I. N° 1308), Algèbre Commutative, Chapitre 5 (Entires) et 6 (Valuations). Hermann, Paris 1964.

[65] Bourbaki, N.: Eléments de Mathématique (A. S. I. N° 1314), Algèbre Commutative, Chapitre 7 (Diviseurs). Hermann, Paris 1965.

[73] Boyle, A. K.: When projective covers and injective hulls are isomorphic. Bull. Austral. Math. Soc. **8**, 471-476 (1973).

[74] Boyle, A. K.: Hereditary QI-rings. Trans. Amer. Math. Soc. **192**, 115-120 (1974).

[75] Boyle, A. K., Goodearl, K. R.: Rings over which certain modules are injective. Pac. J. Math. **58**, 43-53 (1975).

[67] Bowtell, A. J.: On a question of Mal'cev. J. Algebra **7**, 126-139 (1967).

[29] Brauer, R.: Über Systeme hyperkomplexer Zahlen. Math. Z. **39**, 79-107 (1929).

[63] Brauer, R.: Representations of finite groups. Lectures on Modern Mathematics, Vol. I, pp. 133-175 (T. L. Saaty, Ed.). John Wiley and Sons, Inc., New York 1953.

[37] Brauer, R., Nesbitt, C.: On the regular representations of algebras. Proc. Nat. Acad. Sci. **23** (1937).

Bridger, M. (see Auslander).

[51] Brown, B.: An extension of the Jacobson radical. Proc. Amer. Math. Soc. **2**, 114-117 (1951).

[47] Brown, B., McCoy, N.: Radicals and subdirect sums. Amer. J. Math. **69**, 46-58 (1947).

[48] Brown, B., McCoy, N.: The radical of a ring. Duke Math. J. **15**, 495–499 (1948).

[50] Brown, B., McCoy, N.: The maximal regular ideal of a ring. Proc. Amer. Math. Soc. **1**, 165–171 (1950).

[63] Brumer, A.: Structure of hereditary orders. Bull. Amer. Math. Soc. **69** (1963); addendum, *ibid.* **70**, 185 (1964).

Brumer, A. (see Auslander).

[69] Brungs, H.H.: Generalized discrete valuation rings. Canad. J. Math. **21**, 1404–1408 (1969).

[70] Brungs, H.H.: Idealtheorie für eine Klasse noetherscher Ringe. Math. Z. **118**, 86–92 (1970).

[73 a] Brungs, H.H.: Non commutative Krull domains. J. reine u. angew. Math. **264**, 161–171 (1973).

[73 b] Brungs, H.H.: Left Euclidian rings. Pac. J. Math. **45**, 27–33 (1973).

[55] Buchsbaum, D.A.: Exact categories and duality. Trans. Amer. Math. Soc. **80**, 1–34 (1955).

[61] Buchsbaum, D.A.: Some remarks on factorization in power series rings. J. of Math. and Mech. (now Indiana J. Math.) **10**, 749–754 (1961).

Buchsbaum, D. (see Auslander).

[65] Bumby, R.T.: Modules which are isomorphic to submodules of each other. Arch. Math. **16**, 184–185 (1965).

[69 a] Burgess, W.D.: On semiperfect group rings. Canad. Math. Bull. **12**, 645–652 (1969).

[69 b] Burgess, W.D.: Rings of quotients of group rings. Canad. J. Math. **21**, 865–875 (1969).

[63] Bush, G.G.: The embedding theorems of Mal'cev and Lambek. Canad. J. Math. **15**, 49–58 (1963).

[64] Butts, H.S.: Unique factorization of ideals into nonfactorable ideals. Proc. Amer. Math. Soc. **15** (1964).

[65] Butts, H.S.: Quasi-invertible prime ideals. Proc. Amer. Math. Soc. **16**, 291–292 (1965).

[66] Butts, H.S., Gilmer, R.W., Jr.: Primary ideals and prime power ideals. Canad. J. Math. **18**, 1183–1195 (1966).

[69] Cailleau, A.: Une caractérisation des modules $\sum$-injectifs. C. R. Acad. Sci. Paris **269**, 997–999 (1969).

[70] Cailleau, A., Renault, G.: Etude des modules $\sum$-injective. C. R. Acad. Sci. Paris **270**, 1391–1394 (1970).

[68] Caldwell, W.: Hypercyclic rings. Pac. J. Math. **24**, 29–44 (1967).

[70 a] Camillo, V.P.: Balanced rings and a problem of Thrall. Trans. Amer. Math. Soc. **149**, 143–153 (1970).

[70 b] Camillo, V.P.: A note on commutative injective rings. Pac. J. Math. **35**, 59–64 (1970).

[73] Camillo, V.P.: A note on hereditary rings. Pac. J. Math. **45**, 35–41 (1973).

[75 a] Camillo, V.P.: Commutative rings whose quotients are Goldie. Glasgow Math. J. **16**, 32–33 (1975).

[75 b] Camillo, V.P.: Distributive modules. J. Algebra **36**, 16–25 (1975).

[73] Camillo, V.P., Cozzens, J.H.: A theorem on hereditary rings. Pac. J. Math. **45**, 35–41 (1973).

[72] Camillo, V.P., Fuller, K.R.: Balanced and QF-1 algebras. Proc. Amer. Math. Soc. **34**, 373–378 (1972).

[74] Camillo, V.P., Fuller, K.R.: On Loewy length of rings. Pac. J. Math. **53**, 347–354 (1974).

[72] Carson, A.B.: Coherence of polynomial rings over semisimple algebraic algebras. Proc. Amer. Math. Soc. **34**, 20–24 (1972).

[56] Cartan, H., Eilenberg, S.: Homological Algebra. Princeton U. Press, Princeton 1956.

[69 a] Cateforis, V.C.: Flat regular quotient rings. Trans. Amer. Math. Soc. **138**, 241–249 (1969).

[69 b] Cateforis, V.C.: On regular selfinjective rings. Pac. J. Math. **30**, 39–45 (1969).

[70] Cateforis, V.C.: Two-sided semisimple maximal quotient rings. Trans. Amer. Math. Soc. **149**, 339–349 (1970).

[76] Cauchon, G.: Les *T*-anneaux, la condition (*H*) de Gabriel et ses conséquences. Comm. Algebra **4**, 1–10 (1976).

[60] Chase, S.U.: Direct products of modules. Trans. Amer. Math. Soc. **97**, 457–473 (1960).

[61] Chase, S.U.: A generalization of the ring of triangular matrices. Nagoya Math. J. **18**, 13–25 (1961).

[62 a] Chase, S.U.: On direct products and sums of modules. Pac. J. Math. **12**, 847–854 (1962).

[62 b] Chase, S.U.: A remark on direct products of modules. Proc. Amer. Math. Soc. **13**, 214–216 (1962).

[65] Chase, S.U., Faith, C.: Quotient rings and direct products of full linear rings. Math. Z. **88**, 250–264 (1965).

[65]   Chase, S. U., Harrison, D., Rosenberg, A.. Galois theory and Galois cohomology of commutative rings. Memoirs of the Amer. Math. Soc. **52**, 15–33 (1965).

[71]   Chatters, A. W.: The restricted minimum condition in Noetherian hereditary rings. J. London Math. Soc. **4**, 83–87 (1971).

[72]   Chatters, A. W.: A decomposition theorem for Noetherian hereditary rings. Bull. Lond. Math. Soc. **4**, 125–126 (1972).

[51]   Chevalley, C.: Algebraic Functions of a Single Variable. Surveys of the Amer. Math. Soc., Providence 1951.

[67]   Clark, W. E.: Algebras of global dimension one with a finite ideal lattice. Pac. J. Math. (23) **3**, 463–471 (1967).

[68]   Clark, W. E.: Murase's quasi-matrix rings and generalizations. Sci. Pap. Coll. Gen. Educ., U. Tokyo **18**, 99–109 (1968).

       Clark, W. E. (see Bergman).

[61]   Clifford, A. H., Preston, G. B.: Algebraic theory of semigroups, Vol. I. Surveys of the Amer. Math. Soc., Vol. 7. Providence 1961.

[50]   Cohen, I. S.: Commutative rings with restricted minimum condition. Duke Math. J. **17**, 27–42 (1950).

[51]   Cohen, I. S., Kaplansky, I.: Rings for which every module is a direct sum of cyclic modules. Math. Z. **54**, 97–101 (1951).

[70]   Cohen, J. M., Gluck, H.: Stacked bases for modules over principal ideal rings domains. J. Algebra **14**, 493–505 (1970).

[58]   Cohn, P. M.: On a class of simple rings. Mathematika **5**, 103–117 (1958).

[61]   Cohn, P. M.: On the embedding of rings in skew fields. Proc. Lond. Math. Soc. **11**, 511–530 (1961).

[65]   Cohn, P. M.: Universal Algebra. Harper and Row, New York 1965.

[66]   Cohn, P. M.: Morita Equivalence and Duality. University of London, Queen Mary College, Mile End Road, London, (Bookstore) 1966.

[67]   Cohn, P. M.: Torsion modules over free ideal rings. Proc. Lond. Math. Soc. **17**, 577–599 (1967).

[71]   Cohn, P. M.: Free ideal rings. Academic Press, New York 1971.

[67]   Cohn, P. M., Saciada, E.: An example of a simple redical ring. J. Algebra **5**, 373–377 (1967).

       Cohn, P. M. (see Bergman).

[66]   Colby, R. R.: On indecomposable modules over rings with minimum condition. Pac. J. Math. **19**, 23–33 (1966).

[68]   Colby, R. R., Rutter, E. A., Jr.: The structure of certain Artinian rings with zero singular ideal. J. Algebra **8**, 156–164 (1968).

[71]   Colby, R. R., Rutter, E. A., Jr.: $\prod$-Flat and $\prod$-Projective modules. Arch. Math. **22**, 246–251 (1971).

[73]   Colby, R. R., Rutter, E. A., Jr.: Generalizations of QF-3 algebras. Trans. Amer. Math. Soc. **153**, 371–385 (1973).

[70]   Coleman, D. B.: On group rings. Canad. J. Math. **22**, 249–254 (1970).

[71]   Coleman, D. B., Enochs, E. E.: Polynomial invariance of rings. Proc. Amer. Math. Soc. **27**, 247–252 (1971).

[63]   Connell, I.: On the group ring. Canad. J. Math. **15**, 650–685 (1963).

[63]   Corner, A. L. S.: Every countable reduced torsion-free ring is an endomorphism ring. Proc. Lond. Math. Soc. (3) **13**, 687–710 (1963).

[69]   Corner, A. L. S.: Additive categories and a theorem of W. G. Leavitt. Bull. Amer. Math. Soc. **75**, 78–92 (1969).

[65]   Courter, R. C.: The dimension of a maximal commutative subalgebra of K_n. Duke Math. J. **32**, 225–232 (1965).

[69]   Courter, R. C.: Finite direct sums of complete matrix rings over perfect completely primary rings. Canad. J. Math. **21**, 430–446 (1969).

[70]   Cozzens, J. H.: Homological properties of the ring of differential polynomials. Bull. Amer. Math. Soc. **76**, 75–79 (1970).

[72]   Cozzens, J. H.: Simple principal left ideal rings. J. Algebra **23**, 66–75 (1972).

[73]   Cozzens, J. H.: Twisted group rings and a problem of Faith. Bull. Austral. Math. Soc. **9**, 11–19 (1973).

[75]   Cozzens, J. H., Faith, C.: Simple Noetherian Rings. Cambridge Tracts in Mathematics and Physics, Cambridge U. Press, Cambridge 1975.

[72] Cozzens, J. H., Johnson, J.: Some applications of differential algebra to ring theory. Proc. Amer. Math. Soc. **31**, 354–356 (1972).

Cozzens, J. H. (see Camillo).

[63] Crawley, P., Jonsson, B.: Direct decomposition of algebraic systems. Bull. Amer. Math. Soc. **69**, 541–547 (1963).

[64] Crawley, P., Jonsson, B.: Refinements for infinite direct decompositions of algebraic systems. Pac. J. Math. **14**, 797–855 (1964).

Croisot, R. (see Lesieur).

[59] Curtis, C. W.: Quasi-Frobenius rings and Galois theory. Ill. J. Math. **3**, 134–144 (1959).

[62] Curtis, C. W., Reiner, I.: Representation Theory of Finite Groups and Associative Algebras. Interscience, New York 1962.

[65] Curtis, C. W., Jans, J. P.: On algebras with a finite number of indecomposable modules. Trans. Amer. Math. Soc. **114**, 122–132 (1965).

[71] Dade, E. C.: Deux groupes finis distincts ayant la même algebre de groupe sur tout corps. Math. Z. **119**, 345–348 (1971).

[62] Davis, E. D.: Overrings of commutative rings, I. Trans. Amer. Math. Soc. **104**, 52–61 (1962).

[64] Davis, E. D.: Overrings of commutative rings, II. Intergrally closed overrings. Trans. Amer. Math. Soc. **110**, 196–212 (1964).

[1897, 1900] Dedekind, R.: Über Zerlegungen von Zahlen, durch ihre größten gemeinsamen Teiler. Festschrift Techn. Hoch. Braunschweig 1897, and Ges. Werke, vol. 2, 103–148 (1900).

[1900] Dedekind, R.: Über die von drei Moduln erzeugte Dualgruppe. Math. Ann. **53**, 371–403 (1900), and Ges. Werke, vol. 2, 236–271 (1900).

Deleanu, A., Bucur, I. (see Bucur and Deleanu).

[71] DeMeyer, F., Ingraham, E.: Separable Algebras Over Commutative Rings. Lecture Notes in Mathematics, vol. 81. Springer, Berlin-Heidelberg-New York 1971.

[71] Deshpande, M. G.: Structure of right subdirectly irreducible rings, I. J. Algebra **17**, 317–325 (1971).

[35, 68] Deuring, M.: Algebren. Ergebnisse der Math. und ihrer Grenzgebiete, Bd. 74, Springer, Berlin-Heidelberg-New York 1935, 1968.

[23] Dickson, L. E.: Algebras and their Arithmetics. U. of Chicago (1923) and Stechert & Co. (reprint) 1938.

[66] Dickson, S. E.: A torsion theory for Abelian categories. Trans. Amer. Math. Soc. **121**, 223–235 (1966).

[69] Dickson, S. E.: Algebras of finite representation type. Trans. Amer. Math. Soc. **135**, 127–141 (1969).

[69] Dickson, S. E., Fuller, K. R.: Algebras for which every indecomposable right module is invariant in its injective envelope. Pac. J. Math. **31**, 655–658 (1969).

[70] Dickson, S. E., Fuller, K. R.: Commutative QF-1 Artinian rings are QF. Proc. Amer. Math. Soc. **24**, 667–670 (1970).

[70] Dickson, S. E., Kelly, G. M.: Interlacing methods and large indecomposables. Bull. Austral. Math. Soc. **3**, 337–348 (1970).

[42] Dieudonné, J.: Les déterminants sur un corps non-commutatif. Bull. Soc. Math. France **71**, 27–45 (1943).

[58] Dieudonné, J. A.: Remarks on quasi-Frobenius rings. Ill. J. Math. **2**, 346–354 (1958).

[68] Dixmier, J.: Sur les algèbres de Weyl. Bull. Soc. Math. France **96**, 209–242 (1968).

[69] Dlab, V.: Rank theory of modules. Fund. Math. **64**, 313–324 (1969).

[71] Dlab, V., Ringel, C. M.: Exceptional rings. Colloquia Math. Soc. János Bolyai **6**, 167–171 (1971).

[72a] Dlab, V., Ringel, C. M.: Rings with the double centralizer property. J. Algebra **22**, 480–501 (1972).

[72b] Dlab, V., Ringel, C. M.: Balanced rings. Proceedings. Lecture Notes in Mathematics Vol. 246, pp. 74–143. Springer, Berlin-Heidelberg-New York 1972.

[72c] Dlab, V., Ringel, C. M.: Decompositions of modules over right uniserial rings. Math. Z. **129**, 207–230 (1972).

[72d] Dlab, V., Ringel, C. M.: A class of balanced non-uniserial rings. Math. Ann. **195**, 279–291 (1972).

[73a] Dlab, V., Ringel, C. M.: Représentations indécomposables des algèbres. C. R. Acad. Sci. Paris **276**, 1393–1396 (1973).

[73b] Dlab, V., Ringel, C. M.: Sur la conjecture de Brauer-Thrall. C. R. Acad. Sci. Paris **276**, 1441-1442 (1973).

[66] Dubois, D.: Modules of sequences of elements of a ring. J. Lond. Math. Soc. **41**, 177-180 (1966).

[72] Dyson, F. J.: Quaternion determinants. Helv. Phys. Acta **45**, 289-302 (1972).

[68] Eakin, P. M.: The converse to a well-known theorem on Noetherian rings. Math. Ann. **177**, 278-282 (1968).

[72] Eakin, P. M.: A note on finite dimensional subrings of polynomial rings. Proc. Amer. Math. Soc. **31**, 75-80 (1972).

[70] Eakin, P., Heinzer, W.: Non finiteness in finite dimensional Krull domains. J. Algebra **14**, 333-340 (1970).

[72] Eakin, P., Heinzer, W.: A cancellation problem for rings. Lecture Notes in Mathematics, Vol. 311, 333-341. Berlin-Heidelberg-New York: Springer 1972.

Eakin, P. (see Abhyankar).

[53] Eckmann, B., Schopf, A.: Über injective Moduln. Arch. Math. **4**, 75-78 (1953).

[69] Eckstein, F.: On the Mal'cev theorem. J. Algebra **12**, 372-385 (1969).

[56] Eilenberg, S.: Homological dimension and syzygies. Ann. of Math. **64**, 328-336 (1956).

Eilenberg, S. (see Cartan).

[55] Eilenberg, S., Ikeda, M., Nakayama, T.: On the dimension of modules and algebras, I. Nagoya Math. J. **8**, 49-57 (1955).

[56] Eilenberg, S., Nagao, H., Nakayama, T.: On the dimension of modules and algebras, IV. Dimension of residue rings of hereditary rings. Nagoya Math. J. **10**, 87-95 (1956).

[55, 57] Eilenberg, S., Nakayama, T.: On the dimension of modules and algebras, II (Frobenius algebras and quasi-Frobenius rings). Nagoya Math. J. **9**, 1-16 (1955); IV (dimension of residue rings) loc. cit. **11**, 9-12 (1957).

[57] Eilenberg, S., Rosenberg, A., Zelinsky, D.: On the dimension of modules and algebras. VIII. Dimension of tensor products. Nagoya Math. J. **12**, 71-93 (1957).

[70] Eisenbud, D.: Subrings of Artinian and Noetherian rings. Math. Ann. **185**, 247-249 (1970).

[70a] Eisenbud, D., Robson, J. C.: Modules over Dedekind prime rings. J. Algebra **16**, 67-85 (1970).

[70b] Eisenbud, D., Robson, J. C.: Hereditary Noetherian prime rings. J. Algebra **16**, 86-104 (1970).

[71a] Eisenbud, D., Griffith, P.: Serial rings. J. Algebra **17**, 389-400 (1971).

[71b] Eisenbud, D., Griffith, P.: The structure of serial rings. Pac. J. Math. **36**, 109-121 (1971).

[69] Elizarov, V. P.: Quotient rings. Algebra and Logic **8**, 219-243 (1969).

[60] Endo, S.: Note on PP-rings (a supplement to Hattori's paper). Nagoya Math. J. **17**, 167-170 (1960).

[62] Endo, S.: On flat modules over commutative rings. J. Math. Soc. Japan **14**, 284-291 (1962).

[63] Endo, S.: Projective modules over polynomial rings. J. Math. Soc. Japan **15**, 339-352 (1963).

[67] Endo, S.: Completely faithful modules and quasi-Frobenius algebras. J. Math. Soc. Japan **19**, 437-456 (1967).

[67a] Endo, S., Watanabe, Y.: On separable algebras over a commutative ring. Osaka J. Math. **4**, 233-242 (1967).

[67b] Endo, S., Watanabe, Y.: The centers of semisimple algebras over a commutative ring, I, II. Nagoya Math. J. **30**, 285-293 (1967); **39**, 1-6 (1970).

Enochs, E. E. (see Coleman).

[59, 61] Faith, C.: Rings with minimum condition on principal ideals, I, II. Arch. Math. **10**, 327-330 (1959); **12**, 179-181 (1961).

[60] Faith, C.: Algebraic division ring extensions. Proc. Amer. Math. Soc. **11**, 43-53 (1960).

[61] Faith, C.: Radical extensions of rings. Proc. Amer. Math. Soc. **12**, 274-283 (1961).

[62] Faith, C.: Strongly regular extensions of rings. Nagoya Math. J. **20**, 169-183 (1962).

[64] Faith, C.: Noetherian simple rings. Bull. Amer. Math. Soc. **70**, 730-731 (1964).

[65] Faith, C.: Orders in simple Artinian rings. Trans. Amer. Math. Soc. **114**, 61-64 (1965).

[66a] Faith, C.: Rings with ascending condition on annihilators. Nagoya Math. J. **27**, 179-191 (1966).

[66b] Faith, C.: On Köthe rings. Math. Ann. **164**, 207-212 (1966).

[67a] Faith, C.: Lectures on injective modules and quotient rings, Lecture Notes in Mathematics vol. 49. Springer, New York-Heidelberg-Berlin 1967.

[67b] Faith, C.: A general Wedderburn theorem. Bull. Amer. Math. Soc. **73**, 65-67 (1967).

[71a] Faith, C.: The correspondence theorem and the structure of Noetherian simple rings. Bull. Amer. Math. Soc. **77**, 338-342 (1971).

[71b] Faith, C.: Orders in semilocal rings. Bull. Amer. Math. Soc. **77**, 960-962 (1971).

[71 c]   Faith, C.: Big decompositions of modules. Notices of the Amer. Math. Soc. **18**, 400 (1971).

[72 a]   Faith, C.: A correspondence theorem for projective modules, and the structure of simple Noetherian rings. Symposium Matematica **8**, 309–345 (1972).

[72 b]   Faith, C.: Modules finite over endomorphism ring. Lecture Notes in Mathematics, Vol. 246. Springer, Berlin-Heidelberg-New York 1972.

[72 c]   Faith, C.:. Galois subrings of Ore domains are Ore domains. Bull. Amer. Math. Soc. **78**, 1077–1082 (1972).

[73]   Faith, C.: When are proper cyclics injective? Pac. J. Math. **45**, 97–112 (1973).

[74]   Faith, C.: On the structure of indecomposable injective modules. Comm. Algebra **2**, 559–574 (1974).

[75 a]   Faith, C.: On a theorem of Chatters. Comm. Algebra **3**, 169–184 (1975).

[75 b]   Faith, C.: Projective ideals in Cohen rings. Archiv Math. **26**, 588–594 (1975).

[75 c]   Faith, C.: Embedding modules in projectives. Preprint, Rutgers, U., New Brunswick, N.J. 08903, 1975.

[76 a]   Faith, C.: On hereditary rings and Boyle's conjecture. Archiv Math. (1976).

[76 b]   Faith, C.: Injective cogenerator rings, and Tachikawa's theorem. Proc. Amer. Math. Soc. (1976).

[76 c]   Faith, C.: Semiperfect Prüfer rings and FPF rings. Israel J. Math. **27**, 113–119 (1976).

[76 d]   Faith, C.: Characterizations of rings by faithful modules. Lecture Notes, Math. Dept., Israel Institute of Technology (TECHNION), Haifa (1976).

Faith, C. (see Chase; also Cozzens).

[64 a]   Faith, C., Utumi, Y.: Quasi-injective modules and their endomorphism rings. Arch. Math. **15**, 166–174 (1964).

[64 b]   Faith, C., Utumi, Y.: Intrinsic extensions of rings. Pac. J. Math. **14**, 505–512 (1964).

[65 a]   Faith, C., Utumi, Y.: On Noetherian prime rings. Trans. Amer. Math. Soc. **114**, 53–60 (1965).

[65 b]   Faith, C., Utumi, Y.: Maximal quotient rings. Proc. Amer. Math. Soc. **16**, 1084–1089 (1965).

[67]   Faith, C., Walker, E. A.: Direct sum representations of injective modules. J. Algebra **5**, 203–221 (1967).

[73]   Farkas, D.: Self-injective group rings. J. Algebra **25**, 313–315 (1973).

[61 a]   Feller, E. H., Swokowski, E. W.: Reflective N-prime rings with the ascending condition. Trans. Amer. Math. Soc. **99**, 264–271 (1961); corrections, ibid. p. 555.

[61 b]   Feller, E. H., Swokowski, E. W.: Reflective rings with the ascending chain condition. Proc. Amer. Math. Soc. **12**, 651–653 (1961).

[70]   Fields, K. L.: On the global dimension of skew polynomial rings. J. Algebra **14**, 528–530 (1970).

[58]   Findlay, G. D., Lambek, J.: A generalized ring of quotients, I, II. Canad. Math. Bull. **1**, 77–85, 155–167 (1958).

[71]   Fisher, J. L.: Embedding free algebras in skew fields. Proc. Amer. Math. Soc. **30**, 453–458 (1971).

[70]   Fisher, J. W.: On the nilpotency of nil subrings. Canad. J. Math. **22**, 1211–1216 (1970).

[72]   Fisher, J. W.: Nil subrings of endomorphism rings of modules. Proc. Amer. Math. Soc. **34**, 75–78 (1972).

[74]   Fisher, J. W., Snider, R. L.: Prime von Neumann regular rings and primitive group algebras. Proc. Amer. Math. Soc. **44**, 244–250 (1974).

[33]   Fitting, H.: Die Theorie der Automorphismenringe Abelscher Gruppen und ihr Analogon bei nicht kommutativen Gruppen. Math. Ann. **107**, 514–542 (1933).

[35 a]   Fitting, H.: Über die direkten Produktzerlegungen einer Gruppe in direkt unzerlegbare Faktoren. Math. Z. **39**, 19–41 (1935).

[35 b]   Fitting, H.: Primärkomponentenzerlegung in nichtkommutativen Ringen. Math. Ann. **111**, 19–41 (1935).

[68]   Floyd, D. R.: On QF-1 algebras. Pac. J. Math. **27**, 81–94 (1968).

[65]   Fluch, W.: Gruppen ohne endlich-dimensionale Darstellungen. Math. Scand. **16**, 164–168 (1965).

[70]   Formanek, E.: A short proof of a theorem of Jennings. Proc. Amer. Math. Soc. **26**, 406–407 (1970).

[72 b]   Formanek, E.: A problem of Passman on semisimplicity. Bull. Lond. Math. Soc. **4**, 375–376 (1972).

[72 c]   Formanek, E.: Central polynomials for matrix rings. J. Algebra **23**, 129–132 (1972).

[73 a]   Formanek, E.: Group rings of free products are primitive. J. Algebra **26**, 508–511 (1973).

[73b] Formanek, E.: Idempotents in Noetherian group rings. Canad. J. Math. **25**, 366–369 (1973).
[74]   Formanek, E., Jategaonkar, A. V.: Subrings of Noetherian rings. Proc. Amer. Math. Soc. **46**, 181–186 (1974).
[72]   Formanek, E., Snider, R. L.: Primitive group rings. Proc. Amer. Math. Soc. **36**, 375–376 (1972).
[71]   Fossum, R. M.: Injective modules over Krull orders. Math. Scand. **28**, 233–246 (1971).
[73]   Fossum, R. M.: The Divisor Class Group of a Krull Domain. Ergebnisse der Math. und ihrer Grenzgebiete, Bd. 74. Springer, Berlin-Heidelberg-New York 1973.
[58]   Fraenkel, A. A., Bar-Hillel, Y.: Foundations of Set Theory. North Holland, Amsterdam 1958.
[60]   Freyd, P.: Functor Theory (Dissertation), Ph. D. Thesis. Princeton U. 1960.
[64]   Freyd, P.: Abelian Categories. Harper and Row, New York 1964.
[67a] Fuchs, L.: Algebraically compact modules over Noetherian rings. Indian J. Math. **9**, 357–374 (1967).
[67b] Fuchs, L.: Note on purity and algebraic compactness for modules. Studies on Abelian Groups. Dunod, Paris 1967.
[69]   Fuchs, L.: On quasi-injective modules. Scuola Norm. Sup. Pisa **23**, 541–546 (1968).
[70a] Fuchs, L.: Abelian Groups (Second Edition), Vol. I. Pergamon, New York 1970.
[70b] Fuchs, L.: Torsion preradicals and ascending Loewy series of modules. J. reine u. angew. Math. **239**, 169–179 (1970).
[71]   Fuchs, L.: On the substitution property of modules. Monatshefte für Math. **75**, 198–204 (1971).
[72]   Fuchs, L.: The cancellation problem for modules. Lectures on Rings and Modules. Lecture Notes in Mathematics. Springer, Berlin-Heidelberg-New York 1972.
[74]   Fuchs, L.: On torsion abelian groups quasi-projective over their endomorphism rings. Proc. Amer. Math. Soc. **42**, 13–15 (1974).
[71]   Fuchs, L., Loonstra, F.: On the cancellation property of modules in direct sums over Dedekind domains. Indagationes Math. **33**, 163–169 (1971).
[70]   Fuchs, L., Rangaswamy, K. M.: Quasi-projective abelian groups. Bull. Soc. Math. France **98**, 5–8 (1970).
[72]   Fuelberth, J. D., Teply, M. L.: A splitting ring of global dimension two. Proc. Amer. Math. Soc. **35**, 317–324 (1972).
[75a] Fuelberth, J. D., Kuzmanowitz, J.: On the structure of splitting rings. Comm. Algebra **3**, 913–949 (1975).
[75b] Fuelberth, J. D., Kuzmanowitz, J.: The structure of semiprimary and Noetherian hereditary rings. Trans. Amer. Math. Soc. **212**, 83–111 (1975).
[68a] Fuller, K. R.: Generalized uniserial rings and their Kuppisch series. Math. Z. **106**, 248–260 (1968).
[68b] Fuller, K. R.: Structure of QF-3 rings. Trans. Amer. Math. Soc. **134**, 343–354 (1968).
[69a] Fuller, K. R.: On indecomposable injectives over Artinian rings. Pac. J. Math. **29**, 115–135 (1969).
[69b] Fuller, K. R.: On direct representations of quasi-injectives and quasi-projectives. Arch. Math. **20**, 495–502 (1969); (Corrections, **21**, 478 (1970)).
[70a] Fuller, K. R.: Double centralizers of injectives and projectives over Artinian rings. Ill. J. Math. **14**, 658–664 (1970).
[70b] Fuller, K. R.: Relative projectivity and injectivity classes determined by simple modules. J. Lond. Math. Soc. **5**, 423–431 (1972).
[70c] Fuller, K. R.: Primary rings and double centralizers. Pac. J. Math. **34**, 379–383 (1970).
[73]   Fuller, K. R.: On generalized uniserial rings and decompositions that complement direct summands. Math. Ann. **200**, 175–178 (1973).
[74]   Fuller, K. R.: Density and equivalence. J. Algebra **29**, 528–580 (1974).
[76]   Fuller, K. R.: On rings whose [*sic*] left modules are direct sums of finitely generated modules. Proc. Amer. Math. Soc. **54**, 39–44 (1976).
       Fuller, K. R. (see Camillo).
       Fuller, K. R. (see Anderson; also Camillo; also Dickson).
[70]   Fuller, K. R., Hill, D. A.: On quasi-projective modules via relative projectivity. Arch. Math. **21**, 369–373 (1970).
[75]   Fuller, K. R., Reiten, I.: Note on rings of finite representation type. Proc. Amer. Math. Soc. **50**, 92–94 (1975).
[75]   Fuller, K. R., Shutters, W. A.: Projective modules over non-commutative semilocal rings. Tôhoku Math. J. **27**, 303–311 (1975).

[62]   Gabriel, P.: Des catégories abeliennes. Bull. Soc. Math. France **90**, 323–448 (1962).
[72]   Gabriel, P.: Unzerlegbare Darstellungen I. Manuscripta Math. **6**, 71–103 (1972).
       Gabriel, P. (see Borho).
[60]   Gentile, E.: On rings with one-sided fields of quotients. Proc. Amer. Math. Soc. **11**, 380–384 (1960).
[67]   Gewirtzman, L.: Anti-isomorphisms of the endomorphism rings of torsion-free modules. Math. Z. **98**, 391–400 (1967).
[71]   Gill, D. T.: Almost maximal valuation rings. J. Lond. Math. Soc. (2) **4**, 140–146 (1971).
[68]   Gilmer, R. W.: Multiplicative Ideal Theory, Part I and II. Queens Papers on Pure and Applied Math. vol. 12. Queen's U., Kingston, Ontario 1968.
[73]   Gilmer, R. W.: Semigroup rings as Prüfer rings. Duke Math. J. **41**, 219–230 (1973).
[75]   Gilmer, R. W.: Polynomial rings over a commutative regular ring. Proc. Amer. Math. Soc. **49**, 294–296 (1975).
       Gilmer, R. W. (see Butts).
[75]   Ginn, S. M., Moss, P. B.: Finitely embedded modules over Noetherian rings. Bull. Amer. Math. Soc. **81**, 709–710 (1975).
[71]   Golan, J.S.: Quasiperfect modules. Quart. J. Math. Oxford (2) **22**, 173–182 (1971).
[72]   Golan, J.S.: Characterization of rings using quasiprojective modules, III. Proc. Amer. Math. Soc. **31** (1972).
[75]   Golan, J.S.: Localization of Noncommutative Rings. New York: Dekker 1975.
[58]   Goldie, A. W.: The structure of prime rings under ascending chain conditions. Proc. Lond. Math. Soc. VIII, 589–608 (1958).
[60]   Goldie, A. W.: Semi-prime rings with maximum condition. Proc. Lond. Math. Soc. X, 201–220 (1960).
[61]   Goldie, A. W.: Rings with maximum condition (Lecture Notes). Yale U. Math. Dept. 1961.
[62]   Goldie, A. W.: Non-commutative principal ideal rings. Arch. Math. **13**, 213–221 (1962).
[64]   Goldie, A. W.: Torsionfree modules and rings. J. Algebra , 268–287 (1964).
[67]   Goldie, A. W.: Localization in non-commutative Noetherian rings. J. Algebra **5**, 89–105 (1967).
[69]   Goldie, A. W.: Some aspects of ring theory. Bull. Lond. Math. Soc. **1**, 129–154 (1969).
[74]   Goldie, A. W.: Lectures on quotient rings and rings with polynomial identity. Math. Inst. Gießen 1974.
[73]   Goldie, A. W., Small, L. W.: A note on rings of endomorphisms. J. Algebra **24**, 392–395 (1973).
[46]   Goldman, O.: Semisimple extensions of rings. Bull. Amer. Math. Soc. **51**, 1028–1032 (1946).
[51]   Goldman, O.: Hilbert rings, and the Hilbert Nullstellensatz. Math. Z. **54**, 136–140 (1951).
[69]   Goldman, O.: Rings and modules of quotients. J. Algebra **13**, 10–47 (1969).
       Goldman, O. (see Auslander).
[70]   Goodearl, K. R.: Some representation theorems for involution rings. J. Algebra **14**, 299–311 (1970).
[71]   Goodearl, K. R.: Embedding non-singular modules in free modules. J. Pure and Applied Algebra **1**, 275–279 (1971).
[72]   Goodearl, K. R.: Singular Torsion and the Splitting Properties. Memoirs of the Amer. Math. No. 124, Providence 1972.
[73a] Goodearl, K. R.: Triangular representations of splitting rings. Trans. Amer. Math. Soc. **185**, 271–285 (1973).
[73b] Goodearl, K. R.: Prime ideals in regular selfinjective rings. Canad. J. Math. **25**, 829–839 (1973).
[73c] Goodearl, K. R.: Idealizers and nonsingular rings. Pac. J. Math. **48**, 395–402 (1973).
[73d] Goodearl, K. R.: Prime ideals in regular self-injective rings, II. J. Pure and Applied Algebra **3**, 357–373 (1973).
[74a] Goodearl, K. R.: Simple self-injective rings need not be Artinian. Comm. Algebra **2**, 83–89 (1974).
[74b] Goodearl, K. R.: Localization and splitting in hereditary Noetherian prime rings. Pac. J. Math. **53**, 137–151 (1974).
[75]   Goodearl, K. R.: Simple regular rings and rank functions. Math. Ann. **214**, 267–287 (1975).
[76]   Goodearl, K. R., Boyle, A. K.: Dimension theory for nonsingular injective modules. Memoirs of the Amer. Math. Soc. Providence 1972.
       Goodearl, K. R. (see Boyle).
[75]   Goodearl, K. R., Handelman, D.: Simple self-injective rings. Comm. Algebra **3**, 797–834 (1975).
[73]   Gordon, R., Robson, J.C.: Krull dimension. Memoirs Amer. Math. Soc. **133** (1973).
[68]   Gorenstein, D.: Finite Groups. Harper and Row, New York 1968.

[70] Goursaud, J.M.: Une caractérisation des anneaux uniseriels. C. R. Acad. Sci. Paris **270**, 364–367 (1970).

[73] Goursaud, J.M.: Sur les anneaux de groupes semi-parfaits. Canad. J. Math. **25**, 922–928 (1973).

[75] Goursaud, J.M., Jeremy, L.: L'enveloppe injective des anneaux réguliers. Comm. Algebra **3**, 763–779 (1975).

[75] Goursaud, J.M., Valette, J.: Sur l'enveloppe injective des anneaux des groupes réguliers. Bull. Math. Soc. France **103**, 91–102 (1975).

[74] Greenberg, B.: Global dimension of cartesian squares. J. Algebra **32**, 31–43 (1974).

[70] Griffith, P.: On the decomposition of modules and generalized left uniserial rings. Math. Ann. **184**, 300–308 (1970).

[70] Griffith, P., Robson, J.C.: A theorem of Asano and Michler. Proc. Amer. Math. Soc. **24**, 837–838 (1970).

[57] Grothendieck, A.: Sur quelques points d'algèbre homologique. Tohoku Math. J. **9**, 119–221 (1957).

[65] Grothendieck, A.: Le groupe de Brauer I. Séminaire Bourbaki, Exposé 290 (1965).

[68] Gupta, R.N.: Characterization of rings whose classical quotient rings are perfect rings.

[68] Gupta, R.N.: Self-injective quotient rings and injective quotient modules. Osaka J. Math. **5**, 69–87 (1968).

[67] Haines, J.S.: A note on direct product of free modules. Amer. Math. Monthly **74**, 1079–1080 (1967).

[39] Hall, M.H., Jr.: A type of algebraic closure. Ann. of Math. **40**, 360–369 (1939).

[40] Hall, M.H., Jr.: The position of the radical in an algebra. Trans. Amer. Math. Soc. **48**, 391–404 (1940).

[72] Hampton, C.R., Passman, D.S.: On the semisimplicity of group rings of solvable groups. Trans. Amer. Math. Soc. **173**, 289–301 (1972).

[75] Handelman, D.: When is the maximal ring of quotients projective? Proc. Amer. Math. Soc. **52**, 125–130 (1975).

[75] Handelman, D., Lawrence, J.: Strongly prime rings. Trans. Amer. Math. Soc. **211**, 209–223 (1975).

Handelman, D. (see Goodearl).

[73] Hannula, T.A.: On the construction of QF-rings. J. Algebra **25**, 403–414 (1973).

[56] Harada, M.: Note on the dimension of modules and algebras. J. Inst. Polytechnics, Osaka City U. **7**, 17–28 (1956).

[63a] Harada, M.: Hereditary orders. Trans. Amer. Math. Soc. **107**, 273–290 (1963).

[63b] Harada, M.: Multiplicative ideal theory in hereditary orders. J. Math. Osaka City U. **14**, 83–106 (1963).

[64] Harada, M.: Hereditary semiprimary rings and triangular matrix rings. Nagoya J. Math. **27**, 463–484 (1966).

[65a] Harada, M.: Note on quasi-injective modules. Osaka J. Math. **2**, 351–356 (1965).

[65b] Harada, M.: QF-3 and semiprimary PP-rings. Osaka J. Math. **2**, 357–368 (1965).

[66] Harada, M.: QF-3 and semiprimary PP-rings. Osaka J. Math. **2**, 21–27 (1966).

[71] Harada, M.: On categories of indecomposable modules II. Osaka J. Math. **8**, 309–321 (1971).

[72] Harada, M.: On quasi-injective modules with a chain condition over a commutative ring. Osaka J. Math. **9**, 421–426 (1972).

[73] Harada, M.: Perfect categories, I–IV. Osaka J. Math. **10**, 329–367 (1973).

[72] Harada, M., Ishii, T.: On endomorphism rings of Noetherian quasi-injective modules. Osaka J. Math. **9**, 217–223 (1972).

[58] Harada, M., Kanzaki, T.: On Kronecker products of primitive algebras. J. Inst. Polytech. Osaka City U. **9**, 19–28 (1958).

[70] Harada, M., Sai, Y.: On categories of indecomposable modules, I. Osaka J. Math. **7**, 323–344 (1970).

[58] Harris, B.: Commutators in division rings. Proc. Amer. Math. Soc. **9**, 628–630 (1958).

Harrison, D. (see Chase).

[63] Hattori, A.: Semisimple algebras over a commutative ring. J. Math. Soc. Japan **15**, 404–419 (1963).

Heinzer, W. (see Abyankar; also Eakin).

[53] Herstein, I.N.: Finite multiplicative subgroups in division rings. Pac. J. Math. **1**, 121–126 (1953).

[68] Herstein, I. N.: Noncommutative Rings. The Carus Mathematical Monographs, No. 15. Amer. Math. Ass., and Wiley, New York 1968.

[70] Herstein, I. N.: On the Lie structure of an associative ring. J. Algebra **14**, 561–571 (1970).

[71] Herstein, I. N.: Notes from a ring theory conference. Conference Board of the Mathematical Sciences, Regional Conference Board in Math. No. 9. Amer. Math. Soc. Providence, R. I., (1971), iv + 38 pp.

[62] Herstein, I. N., Small, L.: Nil rings satisfying certain chain conditions. Canad. J. Math. **11**, 180–184 (1962), addendum, *ibid.* **14**, 300–302 (1965).

[64] Herstein, I. N., Small, L.: Nil rings satisfying certain chain conditions. Canad. J. Math. **16**, 771–776 (1964).

[54] Higman, D. G.: Indecomposable representations at characteristic *p*. Duke Math. J. **21**, 377–381 (1954).

[56] Higman, G.: On a conjecture of Nagata. Proc. Cambridge Phil. Soc. **52**, 1–4 (1956).

      Hill, D. A. (see Fuller).

[63] Hinohara, Yu: Projective modules over weakly noetherian rings. J. Math. Soc. Japan **15**, 75–88 (1963), supplement 474–475.

[56] Hochschild, G.: Relative homological algebra. Trans. Amer. Math. Soc. **82**, 246–269 (1956).

[58] Hochschild, G.: Note on relative homological algebra. Nagoya Math. J. **13**, 89–94 (1958).

[65] Hochschild, G.: The structure of Lie groups. San Francisco, London, Amsterdam, Holden Day 1965.

[70] Hoffman, K. H.: The duality of compact semigroups and *C**-bigebras. Lecture Notes in Mathematics, vol. 129, Springer, Berlin-Heidelberg-New York 1970.

[39] Hopkins, C.: Rings with minimal condition for left ideals. Ann. of Math. **40**, 712–730 (18–18) (1939).

[62] Hsu, C. S.: Theorems on direct sums of modules. Proc. Amer. Math. Soc. **13**, 540–542 (1962).

[67] Huppert, B.: Endliche Gruppen I. Die Grundlehren der math. Wiss. in Einzeldarstellungen, Bd. 134. Springer, Berlin-Heidelberg-New York 1967.

[69] Hutchinson, J. J.: Intrinsic extensions of rings. Pac. J. Math. **30**, 669–677 (1969).

      Ingraham, E. (see DeMeyer).

[51] Ikeda, M.: Some generalizations of quasi-Frobenius rings. Osaka J. Math. **3**, 227–239 (1951).

[52] Ikeda, M.: A characterization of quasi-Frobenius rings. Osaka J. Math. **4**, 203–210 (1952).

[66] Ikeda, M.: Über die einstufigen nichtkommutativen Ringe. Nagoya Math. J. **27**, 371–379 (1966).

[54] Ikeda, M., Nakayama, T.: On some characteristic properties of quasi-Frobenius and regular rings. Proc. Amer. Math. Soc. **5**, 15–19 (1954).

      Ikeda, M. (see Eilenberg).

      Isaacs, I. M. (see Bergman).

      Ishii, T. (see Harada).

[70] Ivanov, G.: Rings with zero singular ideal. J. Algebra **16**, 340–346 (1970).

[72] Ivanov, G.: Ph. D. Thesis, Australian National University, Canberra 1972.

[74] Ivanov, G.: Left generalized uniserial rings. J. Algebra **31**, 166–181 (1974).

[75] Ivanov, G.: Decomposition of modules over uniserial rings. Comm. Algebra **3**, 1031–1036 (1975).

[71] Jacobinski, H.: Two remarks about hereditary orders. Proc. Amer. Math. Soc. **28**, 1–8 (1971).

[43] Jacobson, N.: The theory of rings, Surveys of the Amer. Math. Soc. vol. 2, Providence 1942.

[45a] Jacobson, N.: The radical and semisimplicity for arbitrary rings. Amer. J. Math. **67**, 300–342 (1945).

[45b] Jacobson, N.: The structure of simple rings without finiteness assumptions. Trans. Amer. Math. Soc. **57**, 228–245 (1945).

[45c] Jacobson, N.: Structure theory for algebraic algebras of bounded degree. Ann. of Math. **46**, 695–707 (1945).

[50] Jacobson, N.: Some remarks on one-sided inverses. Proc. Amer. Math. Soc. **1**, 352–355 (1950).

[51, 53, 64] Jacobson, N.: Lectures in Abstract Algebra, vol. I–III. Van Nostrand, New York and Princeton 1965, 1953, 1964.

[55, 64] Jacobson, N.: Structure of Rings, Revised. Colloquium Publication, vol. 37. Amer. Math. Soc., Providence 1955, 1964.

[75] Jacobson, N.: PI-Algebras. Lecture Notes in Mathematics, Vol. 441. New York-Heidelberg-Berlin: Springer 1975.

[60] Jaffard, P.: Les systèmes d'Ideaux. Dunod, Paris 1960.

[73a] Jain, S.: Flat and FP-injectivity. Proc. Amer. Math. Soc. (1973).

[73b] Jain, S.: Flat injective modules and FP-injectivity. Rutgers Ph. D. Thesis. Rutgers, The State University, New Brunswick, N. J. 1973.

Jain, S. K., Singh, S. (see Singh and Jain).

[69]   Jain, S. K., Mohamed, S. H., Singh, S.: Rings in which every right ideal is quasi-injective. Pac. J. Math. **31**, 73–79 (1969).

[56]   Jans, J. P.: On the indecomposable representations of an algebra. Ann. of Math. **66**, 418–429 (1967).

[59a] Jans, J. P.: Projective injective modules. Pac. J. Math. **9**, 1103–1108 (1959).

[59b] Jans, J. P.: On Frobenius algebras. Ann. of Math. **69**, 392–407 (1959).

[59c] Jans, J. P.: Some remarks on symmetric and Frobenius algebras. Nagoya Math. J. **16**, 65–71 (1960).

[61]   Jans, J. P.: Duality in Noetherian rings. Proc. Amer. Math. Soc. **12**, 829–835 (1961).

[63a] Jans, J. P.: On finitely generated modules over Noetherian rings. Trans. Amer. Math. Soc. **106**, 330–340 (1963).

[63b] Jans, J. P.: Module classes of finite type. Pac. J. Math. **13**, 603–609 (1963).

[67]   Jans, J. P.: On orders in quasi-Frobenius rings. J. Algebra **7**, 35–43 (1967).

[68]   Jans, J. P.: A note on injectives. Math. Ann. **175**, 239–242 (1968).

[70]   Jans, J. P.: On the double centralizer property. Math. Ann. **188**, 85–89 (1970).

[56]   Jans, J. P., Nakayama, T.: On the dimension of modules and algebras, VIII. Nagoya Math. J. **11**, 67–76 (1956).

[67]   Jans, J., Wu, L.: On quasi-projectives. Ill. J. Math. **11**, 439–448 (1967).

[69]   Janusz, G. J.: Indecomposable modules for finite groups. Ann. of Math. **89**, 209–241 (1969).

[70]   Janusz, G. J.: Faithful representations of p-groups at characteristic p, I. J. Algebra **15**, 335–351 (1970).

[72a] Janusz, G. J.: Faithful representations of p-groups at characteristic p, II. J. Algebra **22**, 137–160 (1972).

[72b] Janusz, G. J.: Some left serial algebras of finite type. J. Algebra **23**, 404–411 (1972).

[70a] Jategaonkar, A. V.: Left principal ideal rings. Lecture Notes in Mathematics. Springer, Berlin-Heidelberg-New York 1970.

[70b] Jategaonkar, A. V.: Orders in artinian rings. Bull. Amer. Math. Soc. **75**, 1258–1259 (1970).

[72]   Jategaonkar, A. V.: Structure and classification of hereditary noetherian prime rings, pp. 171–229. (Proceedings) Ring Theory. Academic Press, New York 1972.

[74a] Jategaonkar, A. V.: Jacobson's conjecture and modules over fully bounded noetherian rings. J. Algebra **30**, 103–121 (1974).

[74b] Jategaonkar, A. V.: Injective modules and localization in non-commutative noetherian rings. Trans. Amer. Math. Soc. **190**, 109–123 (1974).

[74c] Jategaonkar, A. V.: Relative Krull dimension and prime ideals in right Noetherian rings. Comm. Algebra **2**, 429–468 (1974).

[74d] Jategaonkar, A. V.: Integral group rings of polycyclic-by-finite groups. J. Pure and Appl. Algebra **4**, 337–343 (1974).

[75]   Jategaonkar, A. V.: Principal ideal theorem for Noetherian P. I. rings.

[74]   Jategaonkar, V. A. [sic!]: Global dimension of tiled orders over commutative Noetherian domains. Trans. Amer. Math. Soc. **190**, 357–374 (1974).

[41]   Jennings, S. A.: The structure of the group ring of a p-group over a modular field. Trans. Amer. Math. Soc. **50**, 175–185 (1941).

[63]   Jenson, C. U.: On characterization of Prüfer rings. Math. Scand. **13**, 90–98 (1963).

[66a] Jenson, C. U.: A remark on flat and projective modules. Canad. J. Math. **18**, 943–949 (1966).

[66b] Jenson, C. U.: A remark on semi-hereditary local rings. J. Lond. Math. Soc. **41**, 479–482 (1966).

[66c] Jenson, C. U.: Arithmetical rings. Acta Math. Acad. Sci. Hungar. **17**, 115–123 (1966).

Jeremy, L. (see Goursaud).

Johnson, J. J. (see Cozzens).

[51a] Johnson, R. E.: The extended centralizer of a ring over a module. Proc. Amer. Math. Soc. **2**, 891–895 (1951).

[51b] Johnson, R. E.: Prime rings. Duke Math. J. **18**, 799–809 (1951).

[53]   Johnson, R. E.: Representations of prime rings. Trans. Amer. Math. Soc. **74**, 351–357 (1953).

[61]   Johnson, R. E.: Quotient rings with zero singular ideal. Pac. J. Math. **11**, 1385–1392 (1961).

[63]   Johnson, R. E.: Principal right ideal rings. Canad. J. Math. **15**, 297–301 (1963).

[65a] Johnson, R.E.: Rings with zero right and left singular ideals. Trans. Amer. Math. Soc. **118**, 150–157 (1965).

[65b] Johnson, R.E.: Unique factorization in principal right ideal domains. Proc. Amer. Math. Soc. **16**, 526–528 (1965).

[69] Johnson, R.E.: Extended Mal'cev domains. Proc. Amer. Math. Soc. **21**, 211–213 (1969).

[61] Johnson, R.E., Wong, E.T.: Quasi-injective modules and irreducible rings. J. Lond. Math. Soc. **36**, 260–268 (1961), (see also Wong and Johnson).

[70] Jonah, D.: Rings with minimum condition for principal right ideals have the maximum condition for principal left ideals. Math. Z. **113**, 106–112 (1970).

Jonnson, B. (see Crawley).

[66] Kanzaki, T.: On Galois extensions of rings. Nagoya Math. J. **27**, 43–49 (1966).

Kanzaki, T. (see Harada).

[42] Kaplansky, I.: Maximal fields with valuation. Duke Math. J. **9**, 303–321 (1942).

[46] Kaplansky, I.: On a problem of Kurosch and Jacobson. Bull. Amer. Soc. **52**, 496–500 (1946).

[47] Kaplansky, I.: Semi-automorphisms of rings. Duke Math. J. **14** (1947).

[48] Kaplansky, I.: Rings with polynomial identity. Bull. Amer. Math. Soc. **54**, 575–580 (1948).

[49] Kaplansky, I.: Elementary divisors and modules. Trans. Amer. Math. Soc. **66**, 464–491 (1949).

[50] Kaplansky, I.: Topological representations of algebras, II. Trans. Amer. Math. Soc. **68**, 62–75 (1950)

[51] Kaplansky, I.: A theorem on division rings. Canad. J. Math. **3**, 290–292 (1951).

[52] Kaplansky, I.: Modules over Dedekind rings and valuation rings. Trans. Amer. Math. Soc. **72**, 327–340 (1952).

[53] Kaplansky, I.: Dual modules over a valuation ring. Proc. Amer. Math. Soc. **4**, 213–219 (1953).

[58a] Kaplansky, I.: Projective modules. Ann. of Math. **68**, 372–377 (1958).

[58b] Kaplansky, I.: On the dimension of modules and algebras, X. A right hereditary ring which is not left hereditary. Nagoya Math. J. **13**, 85–88 (1958).

[59] Kaplansky, I.: Homological dimension of rings and modules. Mimeographed notes. U. of Chicago 1969.

[60] Kaplansky, I.: A characterization of Prüfer rings. J. Indian Math. Soc. **24**, 279–281 (1960).

[62] Kaplansky, I.: The splitting of modules over integral domains. Arch. Math. **13**, 341–343 (1962).

[66] Kaplansky, I.: The homological dimension of a quotient field. Nagoya Math. J. **27**, 139–142 (1966).

[68] Kaplansky, I.: Commutative rings. Proceedings of the Canadian Mathematical Congress, Manitoba 1968.

[69a] Kaplansky, I.: Fields and Rings, Chicago Lectures in Math. U. of Chicago Press, Chicago and London 1969.

[69b] Kaplansky, I.: Infinite Abelian Groups (Second Edition). U. of Michigan Press, Ann Arbor 1969.

[70a] Kaplansky, I.: Problems in the theory of rings. Amer. Math. Monthly **77**, 445–454 (1970).

[70b] Kaplansky, I.: Commutative Rings. Allyn and Bacon, Inc., Boston 1970.

Kaplansky, I. (see Arens; also Cohen).

[54] Kasch, F.: Grundlagen einer Theorie der Frobenius-Erweiterungen. Math. Ann. **127**, 453–474 (1954).

[60/61] Kasch, F.: Projektive Frobenius-Erweiterungen. Sitzungsber. Heidelberger Akad. **4**, 89–109 (1960/61).

[61] Kasch, F.: Dualitätseigenschaften von Frobenius-Erweiterungen. Math. Z. **77**, 229–237 (1961).

[57] Kasch, F., Kneser, M., Kuppisch, H.: Unzerlegbare modulare Darstellungen endlicher Gruppen mit zyklischer p-Sylow Gruppe. Arch. Math. **8**, 320–321 (1957).

[66] Kasch, F., Mares, E.A.: Eine Kennzeichnung semi-perfekter Moduln. Nagoya Math. J. **27**, 525–529 (1966).

[67] Kato, T.: Self-injective rings. Tohoku Math. J. **19**, 485–494 (1967).

[68a] Kato, T.: Some generalizations of QF-rings. Proc. Japan Acad. **44**, 114–119 (1968).

[68b] Kato, T.: Torsionless modules. Tohoku Math. J. **20**, 234–243 (1968).

[72] Kato, T.: Structure of left QF-3 rings. Proc. Japan Acad. **48**, 479–483 (1972).

[57] Kawada, Y.: On similarities and isomorphisms of ideals in a ring. J. Math. Soc. Japan **9**, 374–380 (1957).

[62, 63, 64] Kawada, Y.: On Köthe's problem concerning algebras for which every indecomposable module is cyclic I, II, and III. Sci. Reps. Tokyo Kyoiku Daigaku vols. 7, 8, and 9; 154–230, 1–62, and 166–250 (1962, 1963, and 1964).

Kawada, Y. (see Morita).

[73] Kertész, A. (Editor): Associative Rings, Modules, and Radicals (Proceedings of the Colloquium at Keszthely, 1971). János Bolyai Mathematical Society and North-Holland Publishing Company, Amsterdam, London, and Budapest 1973.

[69] Klatt, G. B., Levy, L. S.: Pre-self-injective rings. Trans. Amer. Math. Soc. **122**, 407–419 (1969).

[67] Klein, A. A.: Rings nonembeddable in fields with multiplicative semigroups embeddable in groups. J. Algebra **7**, 100–125 (1967).

[69] Klein, A. A.: Necessary conditions for embedding rings into fields. Trans. Amer. Math. Soc. **137**, 141–151 (1969).

[72] Klein, A. A.: A remark concerning embeddability of rings in fields. J. Algebra **21**, 271–274 (1972).

Kneser, M. (see Kasch).

[70] Knus, M. A.: Algèbres d'Azumaya et modules projectifs. Comment. Math. Helv. **45**, 372–383 (1970).

[70] Koehler, A.: Quasi-projective covers and direct sums. Proc. Amer. Math. Soc. **24**, 655–658 (1970).

[65] Koh, K.: A note on a self-injective ring. Canad. Math. Bull. **8**, 29–32 (1965).

[70] Koh, K., Luh, J.: On a finite dimensional quasi-simple module. Proc. Amer. Math. Soc. **25**, 801–807 (1970).

[48a] Kolchin, E.: Algebraic matric groups and the Picard-Vessiot theory of homogeneous linear differential equations. Ann. of Math. (2) **49**, 1–42 (1948).

[48b] Kolchin, E. R.: On certain subgroups in the theory of algebraic matric groups. Ann. of Math. **49**, 774–789 (1948).

[30a] Köthe, G.: Die Struktur der Ringe deren Restklassenring nach dem Radical vollständig reduzibel ist. Math. Z. **32**, 161–186 (1930).

[30b] Köthe, G.: Über maximale nilpotente Unterringe und Nilringe. Math. Ann. **103**, 359–363 (1938).

[31] Köthe, G.: Schiefkörper unendlichen Ranges über dem Zentrum. Math. Ann. **105**, 15–39 (1931).

[35] Köthe, G.: Verallgemeinerte Abelsche Gruppen mit Hyperkomplexem Operatorenring. Math. Z. **39**, 31–44 (1935).

[1870] Kronecker, L.: Auseinandersetzung einiger Eigenschaften der Klassenzahl idealer complexer Zahlen. Monatsber. Berl. Akad. 881–889 (1870) (see also, *Werke*, Teubner, Leipzig 273–282 (1895).

[24] Krull, W.: Die verschiedenen Arten der Hauptidealringe. Sitzungsber. Heidelberger Akad. **6** (1924).

[25] Krull, W.: Über verallgemeinerte endliche Abelsche Gruppen. Math. Z. **23**, 161–196 (1925).

[26] Krull, W.: Theorie und Anwendung der verallgemeinerten Abelschen Gruppen. Sitzungsber. Heidelberger Akad. **7**, 1–32 (1926).

[28] Krull, W.: Zur Theorie der Allgemeinen Zahlringe. Math. Ann. **99**, 51–70 (1928).

[29] Krull, W.: Zur Theorie der zweiseitigen Ideale in nichtkommutativen Bereichen. Math. Z. **28**, 481–503 (1928).

[38] Krull, W.: Dimensionstheorie in Stellenringen. J. reine u. angew. Math. **179**, 204–226 (1938).

[48] Krull, W.: Idealtheorie. Chelsea, New York 1948.

[50] Krull, W.: Jacobsonscher Radical und Hilbertscher Nullstellensatz. International Congress of Mathematicians, Cambridge (1950, Conference in Algebra, pp. 56–64).

[51] Krull, W.: Jacobsonsche Ringe, Hilbertscher Nullstellensatz, Dimensionstheorie. Math. Z. **54**, 354–387 (1951).

[69] Kruse, R. L., Price, D. T.: Nilpotent rings. Gordon and Breach, New York 1969.

[59] Kuppisch, H.: Beiträge zur Theorie nichthalbeinfacher Ringe mit Minimalbedingung. J. reine u. angew. Math. **201**, 100–112 (1959).

[75] Kuppisch, H.: Quasi-frobenius algebras of finite representation type. Lecture Notes in Math., Vol. 488, 184–199 (1975).

[35] Kurosch, A.: Durchschnittsdarstellungen mit irreduziblen Komponenten in Ringen und sogenannten Dualgruppen. Mat. Sbornik **42**, 613–616 (1935).

[43] Kurosch, A.: Isomorphisms of direct decompositions. Izvestia Akad. Nauk SSSR **7**, 185–199 (1943); (English Summary) 199–202.

[53]   Kurosch, A.: Radicals of rings and algebras. Mat. Sbornik **33**, 13–26 (1953).

[70]   Kurshan, R. P.: Rings whose cyclic modules have finitely generated socles. J. Algebra **15**, 376–386 (1970).

[72]   Kuzmanovitch, J.: Localizations of Dedekind prime rings. J. Algebra **21**, 378–393 (1972).

[71 a] Lafon, J. P.: Anneaux locaux commutatifs sur lesquels tout module de type fini est somme directe de modules momogenes. J. Algebra **17**, 575–591 (1971); Corrections

[71b] Lafon, J. P.: Sur les anneaux commutatifs d'Hermite et a diviseurs elementaires. C. R. Acad. Sci. Paris **273**, 964–966 (1971).

[73]   Lafon, J. P.: Modules de présentation finie et de type fini sur un anneau arithmétique. Symposia Math. **11**, 121–141 (1973).

[51]   Lambek, J.: The immersibility of a semigroup into a group. Canad. J. Math. **3**, 34–43 (1951).

[63]   Lambek, J.: On Utumi's ring of quotients. Canad. J. Math. **15**, 363–370 (1963).

[65]   Lambek, J.: On the ring of quotients of a noetherian ring. Canad. Math. Bull. **8**, 279–290 (1965).

[66]   Lambek, J.: Rings and Modules. Blaisdell, New York 1966.

[71]   Lambek, J.: Torsion Theories, Additive Semantics, and Rings of Quotients. Lecture Notes in Mathematics vol. 177, Springer, Berlin-Heidelberg-New York 1971.

[72]   Lambek, J.: Localization and completion. J. Pure and Applied Algebra **1**, 343370 (1972).

[76]   Lambek, J.: Localization at epimorphisms and quasi-injectives. J. Algebra (1976).

       Lambek, J. (see Findlay).

[73]   Lambek, J., Michler, G. O.: The torsion theory at a prime ideal of a right Noetherian ring. J. Algebra **25**, 364–389 (1973).

[74]   Lambek, J., Michler, G. O.: Localization of right Noetherian rings at semiprime ideals. Can. J. Math. **26**, 1069–1084 (1974).

       Lambek, J., (see Findlay).

[76]   Lambek, J., Michler, G. O.: On products of full linear rings. Pub. Math. Debrecen (1976).

[73]   Lambek, J., Rattray, B.: Localization at injectives in complete categories. Proc. Amer. Math. Soc. **41**, 1–9 (1973).

[75]   Lambek, J., Rattray, B.: Localizations and duality in additive categories. Houston J. Math. **1**, 87–100 (1975).

[65]   Lang, S.: Algebra. Addison-Wesley, Cambridge 1965.

[69]   Lanski, C.: Nil subrings of Goldie rings are nilpotent. Canad. J. Math. **21**, 904–907 (1969).

[67]   Larsen, M. D.: Equivalent conditions for a ring to be a P-ring and a note on flat overrings. Duke Math. J. **34**, 273–280 (1967).

[71]   Larsen, M. D., McCarthy, P. J.: Multiplicative Theory of Ideals. Academic Press, New York 1971.

       Larsen, M. D. (see Boisen).

[74]   Lawrence, J.: A singula primitive ring. Proc. Amer. Math. Soc. **45**, 59–62 (1974).

[75a] Lawrence, J.: The coefficient ring of a primitive group ring. Canad. J. Math. **27**, 489–494 (1975).

[75b] Lawrence, J.: Infinite group rings. Ph. D. Thesis. Carleton U., Ottawa, 1975.

       Lawrence, J. (see Goodearl).

       Lawrence, J. (see Handelman).

[64]   Lazard, D.: Sur les modules plats. C. R. Acad. Sci. Paris **258**, 273–280 (1964).

[71]   Lenagan, T. H.: Bounded Asano orders are hereditary. Bull. Lond. Math. Soc. **3**, 67–69 (1971).

[73]   Lenagan, T. H.: Nil radical of rings with Krull dimension. Bull. Lond. Math. Soc. **5**, 307–311 (1973).

[74]   Lenagan, T. H.: Nil ideals in rings with finite Krull dimension. J. Algebra **29**, 77–87 (1974).

[69]   Lenzing, H.: Endlich Präsentierbare Moduln. Arch. Math. **20**, 262–266 (1969).

[59]   Lesieur, L., Croisot, R.: Sur les anneaux premiers noethériens à gauche. Ann. Sci. École Norm. Sup. **76**, 161–183 (1959).

[71]   Levine, J.: On the injective hulls of semisimple modules. Trans. Amer. Math. Soc. **155**, **115**–126 (1971).

[31]   Levitzki, J.: Über nilpotente Unterringe. Math. Ann. **105**, 620–627 (1931).

[38]   Levitzki, J.: The equivalence of nilpotent elements of a semisimple ring. Compositio Math. **5**, 392–402 (1938).

[39]   Levitzki, J.: On rings which satisfy the minimum condition for right-hand ideals. Compositio Math. **7**, 214–222 (1939).

[43]   Levitzki, J.: On the radical of a general ring. Bull. Amer. Math. Soc. **49**, 462–466 (1943).

[44]   Levitzki, J.: A characteristic condition for semiprimary rings. Duke Math. J. **11**, 367–368 (1944).

[45a] Levitzki, J.: Solution of a problem of G. Koethe. Amer. J. Math. **67**, 437–442 (1945).

[45b] Levitzki, J.: On three problems concerning nil-rings. Bull. Amer. Math. Soc. **51**, 913–919 (1945).

[46] Levitzki, J.: On a problem of A. Kurosch. Bull. Amer. Math. Soc. **51**, 1033–1035 (1946).

[51] Levitzki, J.: Prime ideals and the lower radical. Amer. J. Math. **73**, 25–29 (1951).

[63] Levitzki, J.: On nil subrings (Posthumous paper edited by S. A. Amitsur). Israel J. Math. **1**, 215–216 (1963).

[63a] Levy, L. S.: Unique direct sums of prime rings. Trans. Amer. Math. Soc. **106**, 64–76 (1963).

[63b] Levy, L. S.: Torsionsfree and divisible modules over non-integral domains. Canad. J. Math. **15**, 132–151 (1963).

[66a] Levy, L. S.: Commutative rings whose homomorphic images are self-injective. Pac. J. Math. **18**, 149–153 (1966).

[66b] Levy, L. S.: Decomposing pairs of modules. Trans. Amer. Math. Soc. **122**, 64–80 (1966).

[72] Levy, L. S.: Almost diagonal matrices over Dedekind domains. Math. Z. **124**, 89–99 (1972).

Levy, L. S. (see Klatt).

[33] Littlewood, D. E.: On the classification of algebras. Proc. Lond. Math. Soc. (2) **35**, 200–240 (1933).

[61] Lissner, D.: Matrices over polynomial rings. Trans. Amer. Math. Soc. **98**, 285–305 (1961).

[05] Loewy, A.: Über die vollständig reduciblen Gruppen, die zu einer Gruppe linearer homogener Substitutionen gehören. Trans. Amer. Math. Soc. **6**, 504–533 (1905).

[17] Loewy, A.: Über Matrizen und Differentialkomplexe, I, II, and III. Math. Ann. **78**, 1–51, 343–368 (1917).

Loonstra, F. (see Fuchs).

[60] Lubkin, S.: Imbedding of abelian categories. Trans. Amer. Math. Soc. **97**, 410–417 (1960).

Luh, J. (see Koh).

[50] MacLane, S.: Duality in groups. Bull. Amer. Math. Soc. **56**, 485–516 (1950).

[63] MacLane, S.: Homology. Academic, New York 1963.

[36] Mal'cev, A. I.: On the immersion of an algebraic ring into a field. Math. Ann. **113**, 686–691 (1936).

[42] Mal'cev, A. I.: On the representation of an algebra as a direct sum of its radical and a semi-simple algebra. Dokl. Akad. Nauk. SSSR **36**, 42–45 (1942).

[49] Mal'cev, A. I.: On infinite soluble groups. Dokl. Akad. Nauk. SSSR (N.S.) **67**, 23–25 (1949).

[51] Mal'cev, A. I.: On some classes of infinite soluble groups. Mat. Sbornik (N.S.) **28**, 567–588 (1951) (Russian).

[63] Mares, E. A.: Semiperfect modules. Math. Z. **82**, 347–360 (1963).

Mares, E. A. (see Kasch).

[70a] Masaike, K.: Endomorphism rings of modules over orders in Artinian rings. Proc. Japan Acad. **46**, 89–93 (1970).

[70b] Masaike, K.: Remarks on results of Zelmanowitz. Sci. Rpts. Tokyo Kyoiku Daigaku, Sec. A **10**, 47–48 (1970).

[71a] Masaike, K.: On quotient rings and torsionless modules. Sci. Rpts. Tokyo Kyoiku Daigaku, Sec. A **11**, 26–31 (1971).

[71b] Masaike, K.: Quasi-Frobenius maximal quotient rings. Sci. Rpts. Tokyo Kyoiku Daigaku, **11**, 1–5 (1971).

[58] Matlis, E.: Injective modules over noetherian rings. Pac. J. Math. **8**, 511+528 (1958).

[59] Matlis, E.: Injective modules over Prüfer rings. Nagoya Math. J. **15**, 57–69 (1959).

[60a] Matlis, E.: Divisible modules. Proc. Amer. Math. Soc. **11**, 385–392 (1960).

[60b] Matlis, E.: Modules with descending chain condition. Trans. Amer. Math. Soc. **97**, 495–508 (1960).

[66] Matlis, E.: Decomposable modules. Trans. Amer. Math. Soc. **125**, 147–179 (1966).

[73] McCarthy, P. J.: The ring of polynomials over a von Neumann regular ring. Proc. Amer. Math. Soc. **39**, 253–254 (1973).

McCarthy, P. J. (see Larsen).

[69] McConnell, J. C.: The Noetherian property in complete rings and modules. J. Algebra **12**, 143–153 (1969).

[74] McConnell, J. C.: Representations of solvable Lie algebras and the Gelfand-Kirillov conjecture. Proc. London Math. Soc. **29**, 453–484 (1974).

[75] McConnell, J. C.: Representations of solvable Lie Algebras: Twisted group rings. Ann. Sci. École Norm. Sup. (4) **2**, 157–178 (1975).

[49]   McCoy, N. H.: Prime ideals in general rings. Amer. J. Math. **71**, 823–833 (1949).

[55]   McCoy, N. H.: Subdirect sum representations of prime rings. Duke Math. J. **22**, 357–364 (1955).

[56]   McCoy, N. H.: The prime radical of a polanomial ring. Publ. Math. Debrecen **4**, 161–162 (1956).

[57a]   McCoy, N. H.: A note on finite unions of ideals and subgroups. Proc. Amer. Math. Soc. **8**, 633–637 (1957).

[57b]   McCoy, N. H.: Annihilators in polynomial rings. Amer. Math. Monthly **64**, 28–29 (1957).

[64]   McCoy, N. H.: Theory of Rings. McMillan, New York 1964.
          McCoy, N. H. (see Brown).

[69]   Mewborn, A. C., Winton, C. N.: Orders in self-injective semiperfect rings. J. Algebra **13**, 5–9 (1969).

[66]   Michler, G. O.: On maximal nilpotent subrings of right Noetherian rings. Glasgow Math. J. **8**, 89–101 (1966).

[69a]   Michler, G. O.: On quasi-local noetherian rings. Proc. Amer. Math. Soc. **20**, 222–224 (1969).

[69b]   Michler, G. O.: Structure of semiperfect hereditary Noetherian rings. J. Algebra **13**, 327–344 (1969).

[69c]   Michler, G. O.: Idempotent ideals in perfect rings. Canad. J. Math. **21**, 301–309 (1969).

[69d]   Michler, G. O.: Asano orders. Proc. Lond. Math. Soc. **19**, 421–443 (1969).
          Michler, G. O. (see Lambek).

[73]   Michler, G. O., Villamayor, O. E.: On rings whose simple modules are injective. J. Algebra **25**, 185–201 (1973).

[65]   Mitchell, B.: Theory of Categories. Academic Press, New York 1965.

[68a]   Mitchell, B.: On the dimension of objects and categories I. Monoids. J. Algebra **9**, 314–340 (1968).

[68b]   Mitchell, B.: On the dimension of objects and categories II. Finite ordered sets. J. Algebra **9**, 341–368 (1968).

[72]   Mitchell, B.: Rings with several objects. Advances in Math. **8**, 1–161 (1972).

[65]   Miyashita, Y.: On quasi-injective modules. A generalization of the theory of completely reducible modules. J. Fac. Sci. Hokkaido U. **18**, 158–187 (1965).

[69]   Mohamed, S. H.: A study of q-rings. Ph. D. Thesis. U. of Delhi 1969.

[70a]   Mohamed, S. H.: q-rings with chain conditions. J. Lond. Math. Soc. (2) **2**, 455–460 (1970).

[70b]   Mohamed, S. H.: Semilocal q-rings. Proc. Nat. Inst. Sci. India.
          Mohamed, S. H. (see Jain).

[72]   Monk, G. S.: A characterization of exchange rings. Proc. Amer. Math. Soc. **35**, 349–353 (1972).

[56]   Morita, K.: On group rings over a modular field which possess radicals expressible as principal ideals. Sci. Rpts. Tokyo Kyoiku Daigaku **4**, 155–172 (1956).

[58]   Morita, K.: Duality for modules and its applications to the theory of rings with minimum condition. Sci Rpts. Tokyo Kyoiku Daigaku **6**, 83–142 (1958).

[62]   Morita, K.: Category-isomorphisms and endomorphism rings of modules. Trans. Amer. Math. Soc. **103**, 451–469 (1962).

[66]   Morita, K.: On S-rings in the sense of F. Kasch. Nagoya Math. J. **27**, 688–695 (1966).

[67]   Morita, K.: The endomorphism ring theorem for Frobenius extensions. Math. Z. **102**, 385–404 (1967).

[69]   Morita, K.: Duality in QF-3 rings. Math. Z. **108**, 385–404 (1967).

[71b]   Morita, K.: Flat modules, injective modules, and quotient rings. Math. Z. **120**, 25–40 (1971).

[70, 71a] Morita, K.: Localizations in categories of modules I, III. Math. Z. **114**, 121–144 (1970); **119**, 313–320 (1971).

[57]   Morita, K., Kawada, Y., Tachikawa, H.: On injective modules. Math. Z. **68**, 217–218 (1957).

[56]   Morita, K., Tachikawa, H.: Character modules, submodules of a free module, and quasi-Frobenius rings. Math. Z. **65**, 414–428 (1956).

[56]   Mostow, G. D.: Fully reducible subgroups of algebraic groups. Amer. J. Math. **78**, 200–221 (1956).

[64, 68] Müller, B. J.: Quasi-Frobenius-Erweiterungen I, II. Math. Z. **85**, 345–368 (1964); **88**, 380–409 (1968).

[70a]   Müller, B. J.: On semiperfect rings. Ill. J. Math. **14**, 464–467 (1970).

[70b]   Müller, B. J.: Linear compactness and Morita duality. J. Algebra **16**, 60–66 (1970).

[71]   Müller, B. J.: Duality theory for linearly topologized modules. Math. Z. **119**, 63–74 (1971).

[74]   Müller, B. J.: The quotient category of a Morita context. J. Algebra **28**, 389–407 (1974).

[62]   Murase, I.: On the extension of Wedderburn's structure theorem. Scientific Papers of the College of General Education. U. of Tokyo **12**, 1–16 (1962).

[62, 63, 64] Murase, I.: On the structure of generalized uniserial rings I. Scientific Papers of the College of General Education. U. of Tokyo **13**, 1–22 (1963); II *ibid.* **13**, 131–158 (1963); III *ibid.* **14**, 12–25 (1964).

[65]   Murase, I.: Generalized uniserial group rings I. Scientific Papers of the College of General Education. U. of Tokyo **15**, 15–28 (1965); II *ibid.* **15**, 111–128 (1965).

[53]   Nagao, H., Nakayama, T.: On the structure of (M_0)- and (M_u)-modules. Math. Z. **59**, 164–170 (1953).

[50]   Nagata, M.: On the theory of semilocal rings. Proc. Japan Acad. **26**, 131–140 (1950).

[51a]   Nagata, M.: Note some studies on semilocal rings. Nagoya Math. J. **2**, 23–30 (1951).

[51b]   Nagata, M.: Note on subdirect sums of rings. Nagoya Math. J. **2**, 49–53 (1951).

[52]   Nagata, M.: Nilpotency of nil-algebras. J. Math. Soc. Japan **4**, 296–301 (1952).

[60]   Nagata, M.: On the fourteenth problem of Hilbert. Proc. Internat. Congress of Math., Cambridge U. Press 1958.

[62]   Nagata, M.: Local Rings. Interscience, New York 1962.

[39, 41] Nakayama, T.: On Frobeniusean algebras I, II. Ann. of Math. **40**, 611–633 (1939); **42**, 1–21 (1941).

[40a]   Nakayama, T.: Note on uniserial and generalized uniserial rings. Proc. Imp. Acad. Tokyo **16**, 285–289 (1940).

[40b]   Nakayama, T.: Algebras with antiisomorphic left and right ideal lattices. Proc. Imp. Acad. Tokyo **17**, 53–56 (1940).

[42]   Nakayama, T.: On Frobeniusean algebras III. Japan J. Math. **18**, 49–65 (1942).

[50a]   Nakayama, T.: Supplementary remarks on Frobeniusean algebras II. Osaka Math. J. **2**, 7–12 (1950).

[50b]   Nakayama, T.: On two topics in the structural theory of rings (Galois theory and Frobenius algebras). Proc. ICM Vol. II, pp. 49–54 (1950).

        Nakayama, T. (see Eilenberg).

        Nakayama, T. (see Ikeda).

        Nakayama, T. (see Jans; also Nagao).

[74]   Narkiewicz, W.: Algebraic Numbers. Monografie Matematyczne, Warsaw 1974.

[71]   Năstăsescu, C.: Quelques remarques sur la dimension homologique des anneaux, Eléments réguliers. J. Algebra **19**, 470–485 (1971).

[70]   Năstăsescu, C., Popescu, N.: On the localization ring of a ring. J. Algebra **15**, 41–56 (1970).

[46]   Nesbitt, C. J., Thrall, R. M.: Some ring theorems with applications to modular representations. Ann. of Math. **47**, 551–567 (1946).

        Nesbitt, C. J. (see Brauer).

[68]   Nobeling, G.: Verallgemeinerung eines Satzes von Herrn E. Specker. Invent. Math. **6**, 41–55 (1968).

[21]   Noether, E.: Idealtheorie in Ringbereichen. Math. Ann. **83**, 24–66 (1921).

[29]   Noether, E.: Hyperkomplexe Größen und Darstellungstheorie. Math. Z. **30**, 641–692 (1929).

[60]   Northcott, D. G.: Homological Algebra. Cambridge U. Press. Cambridge, England 1960.

[62]   Northcott, D. G.: The centre of a hereditary local ring. Proc. Glasgow Math. Assoc. **5**, 101–102 (1962).

[70]   Oberst, U.: Duality theory for Grothendieck categories and linearly compact rings. J. Algebra **15**, 473–542 (1970).

[70]   Oles, F. J.: Characterizations of bounded hereditary Noetherian prime rings. Ph. D. Thesis, Cornell U., Ithaca, 1970.

[73]   O'Meara, K. C.: Primeness of right orders in full linear rings. J. Algebra **26**, 172–184 (1973).

[76a]   O'Meara, K. C.: Right orders in full linear rings. Trans. Amer. Math. Soc. (1976).

[76b]   O'Meara, K. C.: Intrinsic extensions of prime rings. Pac. J. Math. (1976).

[69]   Onodera, T.: Über Kogeneratoren. Arch. Math. **19**, 402–410 (1968).

[71]   Onodera, T.: Eine Bemerkung über Kogeneratoren. Proc. Japan Acad. **47**, 140–141 (1971).

[72]   Onodera, T.: Linearly compact modules and cogenerators. J. Fac. Sci. Hokkaido U. Ser. I **22**, 116–125 (1972).

[73a]   Onodera, T.: Linearly compact modules and cogenerators II. Hokkaido Math. J. **2**, 243–251 (1973).

[73b] Onodera, T.: Koendlich erzeugte Moduln und Kogeneratoren. Hokkaido Math. J. **2**, 69–83 (1973).
[31]   Ore, O.: Linear equations in non-commutative fields. Ann. of Math. **32**, 463–477 (1931).
[33a] Ore, O.: Theory of non-commutative polynomials. Ann. of Math. **34**, 480–508 (1933).
[33b] Ore, O.: On a special class of polynomials. Trans. Amer. Math. Soc. **35**, 559–584 (1933).
[35, 36] Ore, O.: On the foundations of abstract algebra I, II. Ann. of Math. **36**, 406–437 (1935); *ibid.* **37**, 265–292 (1936).
[36]   Ore, O.: Direct decompositions. Duke Math. J. **2**, 581–596 (1936).
[67]   Ornstein, A. J.: Rings with restricted minimum condition. Rutgers Ph. D. Thesis 1967.
[68]   Ornstein, A. J.: Rings with restricted minimum condition. Proc. Amer. Math. Soc. **19**, 1145–1150 (1968).
[64a] Osofsky, B. L.: On ring properties of injective hulls. Canad. Math. Bull. **7**, 405–413 (1964).
[64b] Osofsky, B. L.: Rings all of whose finitely generated modules are injective. Pac. J. Math. **14**, 646–650 (1964).
[65]   Osofsky, B. L.: A counter-example to a lemma of Skornjakov. Pac. J. Math. **15**, 985–987 (1965).
[66a] Osofsky, B. L.: Cyclic injective modules of full linear rings. Proc. Amer. Math. Soc. **17** 247–253 (1966).
[66b] Osofsky, B. L.: A generalization of quasi-Frobenius rings. J. Algebra **4**, 373–387 (1966); Erratum **9**, 120 (1968).
[68a] Osofsky, B. L.: Endomorphism rings of quasi-injective modules. Canad. J. Math. **20**, 895–903 (1968).
[68b] Osofsky, B. L.: Noncommutative rings whose cyclic modules have cyclic injective hulls. Pac. J. Math. **25**, 331–340 (1968).
[68c] Osofsky, B. L.: Homological dimension and the continuum hypothesis. Trans. Amer. Math. Soc. **132**, 217–230 (1968).
[68d] Osofsky, B. L.: Upper bounds on homological dimensions. Nagoya Math. J. **32**, 315–322 (1968).
[68e] Osofsky, B. L.: Erratum. J. Algebra **9**, 120 (1968) (see Osofsky [66b]).
[69]   Osofsky, B. L.: A commutative local ring with finite divisors. Trans. Amer. Math. Soc. **141**, 377–385 (1969).
[70]   Osofsky, B. L.: Homological dimension and cardinality. Trans. Amer. Math. Soc. **151**, 641–649 (1970).
[71a] Osofsky, B. L.: Homological Dimensions of Modules. viii + 89 pp. Amer. Math. Soc., Providence 1971.
[71b] Osofsky, B. L.: Loewy length of perfect rings. Proc. Amer. Math. Soc. **28**, 352–354 (1971).
[71c] Osofsky, B. L.: On twisted polynomial rings. J. Algebra **18**, 597–607 (1971).
[74]   Osofsky, B. L.: The subscript of $\aleph_n$, projective dimension, and the vanishing of $\underleftarrow{\lim}^{(n)}$. Bull. Amer. Math. Soc. **80**, 8–26 (1974).
[59]   Papp, Z.: On algebraically closed modules. Pub. Math. Debrecen 311–327 (1958).
[66]   Pareigis, B.: Radikale und kleine Moduln. Sitzungsber. Bayer. Akad. Wiss. München, math.-naturwiss. Kl. 1965. München 185–199 (1966).
[71]   Pareigis, B.: When Hopf algebras are Frobenius algebras. J. Algebra **18**, 588–596 (1971).
[62]   Passman, D. S.: Nil ideals in group rings. Mich. Math. J. **9**, 375–384 (1962).
[71]   Passman, D. S.: Infinite group rings. Marcel Dekker, New York 1971.
[72]   Passman, D. S.: On the ring of quotients of a group ring. Proc. Amer. Math. Soc. **33**, 221–225 (1972).
[73]   Passman, D. S.: Primitive group rings. Pac. J. Math. **47**, 499–506 (1973).
       Passman, D. S. (see Hampton).
       Pierce, R., Beaumont, R. (see Beaumont and Pierce).
[39]   Pontryagin, L.: Topological Groups. Princeton U. Press, Princeton 1939.
[60a] Posner, E. C.: Prime rings satisfying a polynomial identity. Proc. Amer. Math. Soc. **11**, 180–183 (1960).
[60b] Posner, E. C.: Differentiably simple rings. Proc. Amer. Math. Soc. **11**, 337–343 (1960).
       Preston, G. B. (see Clifford).
       Price, D. T. (see Kruse).
[63]   Procesi, C.: On a theorem of Goldie concerning the structure of prime rings with maximal condition. Acad. Naz. Lincei Rend. **34**, 372–377 (1963) (in Italian).
[64]   Procesi, C.: Sugli anelli principali ed un teorema di Goldie. Acad. Naz. Lincei Rend. **36**, 804–807 (1964).

[65]   Procesi, C.: On a theorem of Faith and Utumi. Rend. Math. e. Appl. **24**, 346–347 (1965) (in Italian).

[66]   Procesi, C.: The Burnside problem. J. Algebra **4**, 421–425 (1966).

[67a] Procesi, D.: Non commutative Jacobson rings. Scuola Norm Sup. Pisa **21**, 381–390 (1967).

[67b] Procesi, D.: Non commutative affine rings. Atti Acad. Naz. Lincei VII, 239–255 (1967).

        Procesi, C., Amitsur, S. A. (see Amitsur and Procesi).

[65]   Procesi, C., Small, L. W.: On a theorem of Goldie. J. Algebra **2**, 80–84 (1965).

[68]   Procesi, C., Small, L. W.: Endomorphism rings of modules over PI-algebras. Math. Z. **106**, 178–180 (1968).

        Procesi, C. (see Amitsur).

[23]   Prüfer, H.: Untersuchungen über die Zerlegbarkeit der abzählbaren primären abelschen Gruppen. Math. Z. **17**, 35–61 (1923).

[25]   Prüfer, H.: Theorie der Abelschen Gruppen II. Ideale Gruppen. Math. Z. **22**, 222–249 (1925).

[73]   Rangaswamy, K. M.: Abelian groups with self-injective endomorphism ring. Proc. Second Internat. Conference, Theory of Groups. Springer, Berlin-Heidelberg-New York 1973, pp. 595–604.

        Rangaswamy, K. M. (see Fuchs).

[72]   Rao, M. L. R.: Azumaya, semisimple, and ideal algebras. Bull. Amer. Math. Soc. **78** (1972).

[73]   Rao, M. L. R.: Semisimple algebras and a cancellation law. Preprint, E.T.H. Zürich 1973.

[74]   Raphael, R.: Rings which are generated by their units. J. Algebra **28**, 199–205 (1974).

[61]   Reiner, I.: The Krull-Schmidt theorem for integral representations. Bull. Amer. Math. Soc. **67**, 365–367 (1961).

[62]   Reiner, I.: Failure of the Krull-Schmidt theorem for integral representations. Mich. Math. J. **9**, 225–231 (1962).

[70]   Reiner, I.: A survey of integral representation theory. Bull. Amer. Math. Soc. **76**, 159–227 (1970).

        Reiner, I. (see Curtis).

        Reiten, I. (see Auslander; also Fuller).

[11]   Remak, R.: Über die Zerlegung der endlichen Gruppen in direkte unzerlegbare Faktoren. J. reine u. angew. Math. (Crelle's Journal) **139**, 293–308 (1911).

[70]   Renault, G.: Sur les anneaux de groupes. Preprint. Faculté des Sciences, 86-Poitiers, Vienne 1970.

        Renault, G. (see Cailleau).

[66]   Rentschler, R.: Eine Bemerkung zu Ringen mit Minimalbedingung für Hauptideale. Arch. Math. **17**, 298–301 (1966).

        Rentschler, R. (see Borho).

[65]   Richman, F.: Generalized quotient rings. Proc. Amer. Math. Soc. **16**, 794–799 (1965).

[72]   Richman, F., Walker, E. A.: Modules over PID's that are injective over their endomorphism rings, Ring Theory. Academic Press 1972.

[65]   Rieffel, M. A.: A general Wedderburn theorem. Proc. Nat. Acad. Sci. USA **54**, 1513 (1965).

[62]   Riley, J. A.: Axiomatic primary and tertiary decomposition theory. Trans. Amer. Math. Soc. **105**, 177–201 (1962).

[65]   Riley, J. A.: Reflexive ideals in maximal orders. J. Algebra **2**, 451–465 (1965).

[62]   Rinehart, G. S.: Note on the global dimension of a certain ring. Proc. Amer. Math. Soc. **13**, 341–346 (1962).

        Ringdal, F. (see Amdal).

[74]   Ringel, C. M.: Commutative QF-1 rings. Proc. Amer. Math. Soc. **42**, 365–368 (1974).

[74]   Ringel, C. M., Tachikawa, H.: QF-3 rings J. reine u. angew. Math. **272**, 49–72 (1974).

        Ringel, C. M. (see Dlab).

[67a] Robson, J. C.: Artinian quotient rings. Proc. Lond. Math. Soc. **17**, 600–616 (1967).

[67b] Robson, J. C.: Pri-rings, and Ipri-rings. Quarterly J. Math. **18**, 125–145 (1967).

[68]   Robson, J. C.: Non-commutative Dedekind rings. J. Algebra **9**, 249–265 (1968).

[71]   Robson, J. C.: A note on Dedekind rings. Bull. Lond. Math. Soc. **3**, 42–46 (1971).

[72a] Robson, J. C.: Idealizers and hereditary Noetherian prime rings. J. Algebra **22**, 45–81 (1972).

[72b] Robson, J. C.: Idealizer Rings, Ring Theory, 309–317, Academic Press, New York 1972.

[74]   Robson, J. C.: Decompositions of Noetherian rings. Comm. Algebra **4**, 345–349 (1974).

        Robson, J. C. (see Eisenbud).

        Robson, J. C. (see Griffith; also Gordon).

[72] Roggenkamp, K. W.: An extension of the Noether-Deuring theorem. Proc. Amer. Math. Soc. **31**, 423–426 (1972).

[68] Roiter, A. V.: Unboundness of the dimensions of indecomposable representations of algebras having infinitely many indecomposable representations. Isv. Akad. Nauk. SSSR Ser. Mat. **32**, 1275–1282 (1968).

[52] Rosenberg, A.: Subrings of simple rings with minimal ideals. Trans. Amer. Math. Soc. **73**, 115–138 (1952).

[54] Rosenberg, A.: Finite-dimensional simple subalgebras of the ring of all continuous linear transformations. Math. Z. **61**, 150–159 (1954).

[58] Rosenberg, A.: The structure of the infinite general linear group. Ann. of Math. **68**, 278–294 (1958).

[61] Rosenberg. A.: Blocks and centres of group algebras. Math. Z. **76**, 209–216 (1961).
       Rosenberg, A. (see Auslander; also Chase).
       Rosenberg, A. (see Eilenberg).

[59] Rosenberg, A., Zelinsky, D.: On the finiteness of the injective hull. Math. Z. **70**, 372–380 (1959).

[61] Rosenberg, A., Zelinsky, D.: Annihilators. Portugalia Math. **20**, 53–65 (1961).

[73] Rosenblade, J. E.: Group rings of polycyclic groups. J. Pure and Applied Algebra **3**, 307–328 (1973).

[72] Roux, B.: Sur les anneaux de Köthe. Preprint, Dept. Math., Montpelier U., Montpelier, 1972.

[74] Rowen, L. H.: Maximal quotient rings of semiprime PI-algebras. Trans. Amer. Math. Soc. **196**, 127–135 (1974).
       Rutter, E. A. (see Colby).
       Sai, Y. (see Harada).

[60] Samuel, P.: Un exemple d'anneau factoriel. Bol. Soc. Mat. Sao Paulo **15**, 1–4 (1960).

[65] Sanderson, D. F.: A generalization of divisibility and injectivity in modules. Canad. Math. Bull. **8**, 505–513 (1965).

[64] Sandomierski, F.: Relative injectivity and projectivity. Ph. D. Thesis. Penna, State U., U. Park 1964.

[67] Sandomierski, F. L.: Semisimple maximal quotient rings. Trans. Amer. Math. Soc. **128**, 112–120 (1967).

[68] Sandomierski, F. L.: Nonsingular rings. Proc. Amer. Math. Soc. **19**, 225–230 (1968).

[70] Sandomierski, F. L.: Some examples of right self-injective rings which are not left self-injective. Proc. Amer. Math. Soc. **26**, 244–245 (1970).

[72a] Sandomierski, F. L.: Modules over the endomorphism rings of a finitely generated projective module. Proc. Amer. Math. Soc. **31**, 27–31 (1971).

[72b] Sandomierski, F. L.: Linearly compact modules and local Morita duality. Ring Theory. Academic Press, New York 1972.
       Saciada, E. (see Cohn).
       Schopf, A. (see Eckmann).

[28] Schmidt, O.: Über unendliche Gruppen mit endlicher Kette. Math. Z. **29**, 34–41 (1928).

[28] Schreier, O.: Über den Jordan-Hölderschen Satz. Abh. Math. Sem. U. Hamburg **6**, 300–302 (1928).

[55] Serre, J. P.: Sur la dimension homologique des anneaux et des modules Noethériens. Proc. Internat. Sympos. Algebraic Number Theory, Tokyo 1955.

[68] Serre, J. P.: Corps Locaux (Second Edition), A. S. I. N° 1296. Hermann, Paris 1968.

[58] Seshadri, C.: Triviality of vector bundles over the affine space K^2. Proc. Nat. Acad. Sci. USA **44**, 456–458 (1958).

[71] Sharpe, D. W., Vámos, P.: Injective Modules. Cambridge U. Press, Cambridge 1971.

[51] Shepherdson, J. C.: Inverse and zero divisors in matrix rings. Proc. Lond. Math. Soc. **61**, 71–85 (1951).

[71a] Shock, R. C.: Injectivity, annihilators, and orders. J. Algebra **19**, 96–103 (1971).

[71b] Shock, R. C.: Nil subrings in finiteness conditions. Amer. Math. Monthly **78**, 741–748 (1971).

[71c] Shock, R. C.: Essentially nilpotent rings. Israel J. Math. **9**, 180–185 (1971).

[72a] Shock, R. C.: Orders in self-injective cogenerator rings. Proc. Amer. Math. Soc. **35**, 393–398 (1972).

[72b] Shock, R. C.: Polynomial rings over finite dimensional rings. Pac. J. Math. **42**, 251–258 (1972).

[72c] Shock, R. C.: A note on the prime radical. J. Math. Soc. of Japan **24**, 374–376 (1972).

[71]   Shores, T. S.: Decompositions of finitely generated modules. Proc. Amer. Math. Soc. **30**, 445–450 (1971).

[74]   Shores, T. S.: Loewy series of modules. J. reine u. angew. Math. **265**, 183–200 (1974).

[73]   Shores, T. S., Wiegand, R.: Decompositions of modules and matrices. Bull. Amer. Math. Soc. **79**, 1277–1280 (1973).

[74]   Shores, T. S., Wiegand, R.: Rings whose finitely generated modules are direct sums of cyclics. J. Algebra **32**, 57–72 (1974).

[72]   Shock, R. C.: The ring of endomorphisms of a finite dimensional module. Israel J. Math. **11**, 309–314 (1972).

[74]   Shock, R. C.: Dual generalizations of the Artinian and Noetherian conditions. Pac. J. Math. **54**, 227–235 (1974).

[67]   Silver, L.: Noncommutative localizations and applications. J. Algebra **7**, 44–76 (1967).

[74]   Singh, S.: Quasi-injective and quasi-projective modules hereditary Noetherian prime rings. Canad. J. Math. **26**, 1173–1185 (1974).

[75]   Singh, S.: Modules over hereditary Noetherian prime rings. Canad. J. Math. **27**, 867–883 (1975).

[67]   Singh, S., Jain, S. K.: On pseudo-injective modules and self-pseudo-injective rings. J. Math. Sci. India **2**, 23–31 (1967).

[70]   Singh, S., Wasan, K.: Pseudo-injective modules over commutative rings. J. Indian Math. Soc. **34**, 61–66 (1970).

Singh, S. (see Jain).

[69]   Slover, R.: The radical of row finite matrices. J. Algebra **12**, 345–359 (1969).

[65]   Small, L. W.: An example in Noetherian rings. Proc. Nat. Sci. USA **54**, 1035–1036 (1965).

[66a]   Small, L. W.: Hereditary rings. Proc. Nat. Acad. Sci. USA **55**, 25–27 (1966).

[66b]   Small, L. W.: Orders in Artinian rings. J. Algebra **4**, 13–41 (1966); corrections and addendum *ibid.* **4**, 505–507 (1966).

[66c]   Small, L. W.: One some questions in Noetherian rings. Bull. Amer. Math. Soc. **72**, 853–857 (1966).

[67]   Small, L. W.: Semihereditary rings. Bull. Amer. Math. Soc. **73**, 656–658 (1967).

[68a]   Small, L. W.: Orders in Artinian rings II. J. Algebra **9**, 266–273 (1968).

[68b]   Small, L. W.: A change of rings theorem. Proc. Amer. Math. Soc. **19**, 661–666 (1968).

[69]   Small, L. W.: The embedding problem for Noetherian rings. Bull. Amer. Math. Soc. **75**, 147–148 (1969).

[72]   Small, L. W.: Localization in PI-rings. J. Algebra **18**, 269–270 (1972).

[73]   Small, L. W.: Prime ideals in Noetherian PI-rings. Bull. Amer. Math. Soc. **79**, 421–422 (1973).

Small, L. W. (see Goldie).

Small, L. W. (see Herstein).

[71]   Smith, M.: Group algebras. J. Algebra **18**, 477–499 (1971).

[50, 51, 52]   Snapper, E.: Completely primary rings I–IV. Ann. of Math. **52, 53, 55** 25–42, 207–234, 46–64 (1950, 1951, 1952).

[75]   Snider, R. L.: Algebras whose simple modules are finite dimensional over the commuting rings. Comm. Algebra **2**, 15–25 (1975).

Snider, R. L. (see Fisher).

Snider, R. L. (see Formanek).

[50]   Specker, E.: Additive Gruppen von Folgen ganzer Zahlen. Portugaliae Math. **9**, 131–140 (1950).

[60]   Srinivasan, B.: On the indecomposable representations of certain class of groups. Proc. Lond. Math. Soc. **10**, 497–513 (1960).

[48]   Steinitz, E.: Algebraische Theorie der Körper. Chelsea, New York 1948.

[70]   Stenström, B.: Coherent rings, and FP-injective modules. J. Lond. Math. Soc. **2**, 323–329 (1970).

[71]   Stenström, B.: Rings and modules of quotients. Lecture Notes in Mathematics, vol. 237. Springer, Berlin-Heidelberg-New York 1971.

[36]   Stone, M. H.: The theory of representations for Boolean algebra. Trans. Amer. Math. Soc. **40**, 37–111 (1936).

[69a]   Storrer, H. H.: A characterization of Prüfer domains. Canad. Bull. Math. **12**, 809–812 (1969).

[69b]   Storrer, H. H.: A note on quasi-Frobenius rings and ring epimorphisms. Canad. Math. Bull. **12**, 287–292 (1969).

[71a]   Storrer, H. H.: Rational extensions of modules. Pac. J. Math. **38**, 785–794 (1971).

[71b]   Storrer, H. H.: Rings of quotients of perfect rings. Math. Z. **122**, 151–165 (1971).

[73]   Storrer, H. H.: Epimorphic extensions of non-commutative rings. Comment. Math. Helv. **48**, 72–86 (1973).

[71]   Stringall, R. W.: The categories of *p*-rings are equivalent. Proc. Amer. Math. Soc. **29**, 229–235 (1971).

[66]   Strooker, J. R.: Lifting projectives. Nagoya Math. J. **27**, 747–751 (1966).

[63]   Suprunenko, D.: Soluble and Nilpotent Linear Groups. Amer. Math. Soc., Providence 1963 (English translation of the Russian Edition, Minsk 1958).

[59]   Swan, R.: Projective modules over finite groups. Bull. Amer. Math. Soc. **65**, 365–367 (1959).

[60]   Swan, R.: Induced representations and projective modules. Ann. of Math. **71**, 552–578 (1960).

[62]   Swan, R.: Projective modules over rings and maximal orders. Ann. of Math. **75**, 55–61 (1962).

[63]   Swan, R.: The Grothendieck ring of a finite group. Topology **2**, 85–110 (1963).

[68]   Swan, R. G.: Algebraic *K*-Theory. Lecture Notes in Mathematics, vol. 76. Springer, Berlin-Heidelberg-New York 1968.
         Swokowski, E. W. (see Feller).

[60, 61, 63] Szász, F.: Über Ringe mit Minimalbedingung für Hauptrechtsideale I. Publ. Math. **7**, 54–64 (1960); II. Acta Math. Acad. Sci. Hung. **12**, 417–440 (1961); III *ibid*. **14**, 447–461 (1963).

[61a]   Szász, F.: Bemerkungen zu assoziativen Hauptidealringen. Proc. Koninkl. Nederl. Akad. Wet. **64**, 577–583 (1961).

[61b]   Szász, F.: Die abelschen Gruppen, deren volle Endomorphismenringe die Minimalbedingung für Hauptrechtideale erfüllen. Monatsh. Math. 150–153 (1961).

[58]   Tachikawa, H.: Duality theorem of character modules for rings with minimum condition. Math. Z. **68**, 479–487 (1958).

[59]   Tachikawa, H.: On rings for which every indecomposable right module has a unique maximal submodule. Math. Z. **71**, 200–222 (1959).

[60]   Tachikawa, H.: A note on algebras of unbounded representation type. Proc. Japan Acad. **36** (1960).

[61]   Tachikawa, H.: On algebras of which every indecomposable representation has an irreducible one, as the top or bottom Loewy constituent. Math. Z. **75**, 215–227 (1961).

[62]   Tachikawa, H.: A characterization of QF-3 algebras. Proc. Amer. Math. Soc. **13**, 101–103 (1962).

[69a]   Tachikawa, H.: On splitting of module categories. Math. Z. **111**, 145–150 (1969).

[69b]   Tachikawa, H.: A generalization of quasi-Frobenius rings. Proc. Amer. Math. Soc. **20**, 471–476 (1969).

[70a]   Tachikawa, H.: Double centralizers and dominant dimensions. Math. Z. **116**, 79–88 (1970).

[70b]   Tachikawa, H.: On left QF-3 rings. Pac. J. Math. **32**, 255–268 (1970).

[71]   Tachikawa, H.: Localization and Artinian quotient rings. Math. Z. **119**, 239–253 (1971).

[73]   Tachikawa, H.: Quasi-Frobenius Rings and Generalizations of QF-3 and QF-1 Rings. Lecture Notes in Mathematics. Springer, Berlin-Heidelberg-New York 1973.
         Tachikawa, H. (see Morita; also Ringel).

[57]   Taft, E. J.: Invariant Wedderburn factors. Ill. J. Math. **1**, 565–573 (1957).

[64]   Taft, E. J.: Orthogonal conjugacies in associative and Lie algebras. Trans. Amer. Math. Soc. **113**, 18–29 (1964).

[68]   Taft, E. J.: Cohomology of groups of algebra automorphisms. J. Algebra **10**, 400–410 (1968).

[63]   Talintyre, T. D.: Quotient rings of rings with maximal condition for right ideals. J. Lond. Math. Soc. **38**, 439–450 (1963).

[66]   Talintyre, T. D.: Quotient rings with minimum condition on right ideals. J. Lond. Math. Soc. **41**, 141–144 (1966).

[37]   Teichmüller, O.: Diskret bewertete perfekte Körper mit unvollkommenem Restklassenkörper. J. reine u. angew. Math. **176**, 140–152 (1937).

[70]   Teply, M. L.: Homological dimension and splitting torsion theories. Pac. J. Math. **34**, 193–205 (1970).

[71]   Teply, M. L.: Torsionfree projective modules. Proc. Amer. Math. Soc. **27**, 29–34 (1971).
         Teply, M. L. (see Fuelberth).

[48]   Thrall, R. M.: Some generalizations of quasi-Frobenius algebras. Trans. Amer. Math. Soc. **64**, 173–183 (1948).
         Thrall, R. M. (see Artin; also Nesbitt).

[72]   Tits, J.: Free subgroups in linear groups. J. Algebra **20**, 250–270 (1972).

[70]   Tol'skaja, T. S.: When are cyclic modules essentially embedded in free modules. Mat. Issled. **5**, 187–192 (1970).

[56]   Utumi, Y.: On quotient rings. Osaka Math. J. **8**, 1–18 (1956).

[60a] Utumi, Y.: On continuous regular rings and semi-simple self-injective rings. Canad. J. Math. **12**, 597–605 (1960).

[60b] Utumi, Y.: A remark on quasi-Frobenius rings. Proc. Japan. Acad. **36**, 15–17 (1960).

[61] Utumi, Y.: On continuous regular rings. Canad. Math. Bull. **4**, 63–69 (1961).

[63a] Utumi, Y.: A theorem of Levitzki. Math. Assoc. of Amer. Monthly **70**, 286 (1963).

[63b] Utumi, Y.: On rings of which any one-sided quotient rings are two-sided. Proc. Amer. Math. Soc. **14**, 141–147 (1963).

[63c] Utumi, Y.: A note on rings of which any one-sided quotient rings are two-sided. Proc. Japan Acad. **39**, 287–288 (1963).

[63d] Utumi, Y.: Prime J-rings with uniform one-sided ideals. Amer. J. Math. **85**, 583–596 (1963).

[65] Utumi, Y.: On continuous rings and self-injective rings. Trans. Amer. Math. Soc. **118**, 158–173 (1965).

[66] Utumi, Y.: On the continuity and self-injectivity of a complete regular ring. Canad. J. Math. **18**, 404–412 (1966).

[67] Utumi, Y.: Self-injective rings. J. Algebra **6**, 56–64 (1967).

Utumi, Y. (see Faith).

[63] Uzkov, A.I.: On the decomposition of modules over a commutative ring into a direct sum of cyclic submodules. Math. Sb. **62**, (104), 469–475 (1963) (in Russian).

Valette, J. (see Goursaud).

[68] Vámos, P.: The dual of the notion of finitely generated. J. Lond. Math. Soc. **43**, 643–646 (1968).

[71] Vámos, P.: Direct decompositions of modules. Algebra Seminar Notes. Dept. of Math., U. of Sheffield, 1971.

[75] Vámos, P.: Classical rings. J. Algebra **34**, 114–129 (1975).

[76] Vámos, P.: Rings with duality. Preprint U. of Sheffield 1976.

Vámos, P. (see Sharpe).

[68] Vasconcelos, W.V.: Reflexive modules over Gorenstein rings. Proc. Amer. Math. Soc. **19**, 1349–1355 (1968).

[69] Vasconcelos, W.V.: On finitely generated flat modules. Trans. Amer. Math. Soc. **138**, 505–512 (1969).

[70a] Vasconcelos, W.V.: Flat modules over commutative Noetherian rings. Trans. Amer. Math. Soc. **152**, 137–143 (1970).

[70b] Vasconcelos, W.V.: Simple flat extension. J. Algebra **16**, 106–107 (1970).

[72] Vasconcelos, W.V.: The local rings of global dimension two. Proc. Amer. Math. Soc. **35**, 381–386 (1972).

[73a] Vasconcelos, W.V.: Coherence of one polynomial ring. Proc. Amer. Math. Soc. **41**, 449–456 (1973).

[73b] Vasconcelos, W.V.: Finiteness in projective ideals. J. Algebra **25**, 269–278 (1973).

[58] Villamayor, O.E.: On the semisimplicity of group algebras. Proc. Amer. Math. Soc. **9**, 621–627 (1958).

Villamayor, O.E. (see Michler).

[73] Viola-Prioli, J.: On absolutely torsion-free rings and kernel functors. Rutgers U. Ph.D. Thesis, Rutgers, The State U., New Brunswick, N.J.

[36a] von Neumann, J.: On regular rings. Proc. Nat. Acad. Sci. (USA) **22**, 707–713 (1936).

[36b] von Neumann, J.: Examples of continuous geometries. Proc. Nat. Acad. Sci. (USA) **22**, 101–108 (1936).

[60] von Neumann, J.: Continuous Geometry. Princeton Mathematical Series No. 25. Princeton U., Princeton, N.J. 1960.

[70] Wagner, G.B.: Radicals related to V-rings. Technical Report. Dept. of Math., U. of Maryland, College Park 1970.

[66] Walker, C.L.: Relative homological algebra and abelian groups. Ill. J. Math. **10**, 186–209 (1966).

Walker, E.A. (see Faith; also Richmond).

[69a] Warfield, R.B., Jr.: Purity and algebraic compactness for modules. Pac. J. Math. **28**, 699–719 (1969).

[69b] Warfield, R.B., Jr.: Decompositions of injective modules. Pac. J. Math. **31**, 263–276 (1969).

[69c] Warfield, R.B., Jr.: A Krull-Schmidt theorem for infinite sums of modules. Proc. Amer. Math. Soc. **22**, 460–465 (1969).

[70] Warfield, R.B., Jr.: Decomposability of finitely presented modules. Proc. Amer. Math. Soc. **25**, 167–172 (1970).

[72a]   Warfield, R. B., Jr.: Rings whose modules have nice decompositions. Math. Z. **125**, 187–192 (1972).

[72b]   Warfield, R. B., Jr.: Exchange rings and decompositions of modules. Math. Ann. **199**, 31–36 (1972).

[75]   Warfield, R. B., Jr.: Serial rings and finitely presented modules. J. Algebra **37**, 187–222 (1975).
        Wasan, K. (see Singh).
        Watanabe, Y. (see Endo).

[70]   Webber, D. B.: Ideals and modules of simple Noetherian hereditary rings. J. Algebra **16**, 239–242 (1970).

[08]   Wedderburn, J. H. M.: On hypercomplex numbers. Proc. Lond. Math. Soc. (2) **6**, 77–117 (1908).

[09]   Wedderburn, J. H. M.: On the direct product in the theory of finite groups. Ann. of Math. **10**, 173–176 (1909).

[14]   Wedderburn, J. H. M.: A type of primitive algebra. Trans. Amer. Math. Soc. **15**, 162–166 (1914).

[37]   Wedderburn, J. H. M.: A note on algebras. Ann. of Math. **38**, 854–856 (1937).

[74]   Weil, A.: Two lectures on number theory, past and present. L'enseignement Math. XX, 87–110 (1974).

[39]   Weyl, H.: The Classical Groups. Princeton U. Press, Princeton, (revised) 1946.
        Whaples, G. (see Artin).
        Wiegand, R. (see Shores).

[65]   Willard, E.: Properties of generators. Math. Ann. **158**, 352–364 (1965).

[68]   Williams, R. E.: A note on weak Bezout rings. Proc. Amer. Math. Soc. **19**, 951–952 (1968).

[59]   Wong, E. T., Johnson, R. E.: Self-injective rings. Canad. Math. Bull. **2**, 167–173 (1959).
        Wong, E. T. (see Johnson).

[71]   Woods, S. M.: On perfect group rings. Proc. Amer. Math. Soc. **27**, 49–52 (1971).

[66]   Wu, L.: A characterization of self-injective rings. Ill. J. Math. **10**, 61–65 (1966).
        Wu, L. (see Jans).

[73]   Yamagata, K.: A note on a problem of Matlis. Proc. Japan Acad. **49**, 145–147 (1973).

[56]   Yoshii, T.: On algebras of bounded representation type. Osaka Math. J. **8**, 51–105 (1956).

[68]   Zaks, A.: Semiprimary rings of generalized triangular type. J. Algebra **9**, 54–78 (1968).

[69]   Zaks, A.: Injective dimension of semiprimary rings. J. Algebra **13**, 73–86 (1969).

[71a]   Zaks, A.: Dedekind subrings of $k[x_1, \ldots, x_n]$ are rings of polynomials. Israel J. Math. **9**, 285–289 (1971).

[71b]   Zaks, A.: Some rings are hereditary. Israel J. Math. **10**, 442–450 (1971).

[72]   Zaks, A.: Restricted left principal ideal rings. Israel J. Math. **11**, 190–215 (1972).

[74]   Zaks, A.: Hereditary Noetherian rings. J. Algebra **30**, 513–526 (1974).

[75]   Zalesskii, A. E., Neroslavskii, O. M.: On simple Noetherian rings (in Russian). Izv. Akad. Nauk. USSR, 38–42 (1975).

[38]   Zassenhaus, H.: Beweis eines Satzes über diskrete Gruppen. Abh. Math. Sem. Hansische U. **12**, 289–312 (1938).

[67]   Zassenhaus, H.: Orders as endomorphism rings of modules of the same rank. J. Lond. Math. Soc. **42**, 180–182 (1967).

[53]   Zelinsky, D.: Linearly compact modules and rings. Amer. J. Math. **75**, 79–90 (1953).

[54]   Zelinsky, D.: Raising idempotents. Duke Math. J. **21**, 315–322 (1954).
        Zelinsky, D. (see Eilenberg; also Rosenberg).

[67]   Zelmanowitz, J. M.: Endomorphism rings of torsion-less modules. J. Algebra **5**, 325–341 (1967).

[71]   Zelmanowitz, J.: Injective hulls of torsion free modules. Canad. J. Math. **23**, 1094–1101 (1971).

[72]   Zelmanowitz, J.: Regular modules. Trans. Amer. Math. Soc. **163**, 341–355 (1972).

[73]   Zelmanowitz, J.: Semiprime modules with maximum conditions. J. Algebra **25**, 554–574 (1973).

[04]   Zermelo, E.: Beweis, daß jede Menge wohlgeordnet werden kann. Math. Ann. **59**, 514–516 (1904).

[08a]   Zermelo, E.: Neuer Beweis für die Möglichkeit einer Wohlordnung. Math. Ann. **65**, 107–128 (1908).

[08b]   Zermelo, E.: Untersuchungen über die Grundlagen der Mengenlehre I. Math. Ann. **65**, 261–281 (1908).

[73]   Zöschinger, H.: Moduln, die in jeder Erweiterung ein Komplement haben. Algebra-Berichte-Seminar Kasch und Pareigis. Math. Inst. München **15**, 1–22 (1973).

Register of Names

This register does not include references in the Introduction, pp. 1–5.

Index

This index contains entries only for words defined in this volume. Consult Volume I for concepts not found here.

Die Grundlehren der mathematischen Wissenschaften in Einzeldarstellungen mit besonderer Berücksichtigung der Anwendungsgebiete

Eine Auswahl